ROCKS IN OUR HEADS

Stories of Exploring for Mineral
Deposits in Exotic Lands

Published in 2020 by Connor Court Publishing Pty Ltd

Copyright © Andrew Drummond as editor of the collection, Australia 2020

Connor Court Publishing Pty Ltd

PO Box 7257

Redland Bay QLD 4165

sales@connorcourt.com

www.connorcourtpublishing.com.au

Phone 0497 900 685

ISBN: 9781922449405

Front Cover Design: Maria Giordano

Cover photo: Taken by Andrew Drummond

Printed in Australia

ROCKS IN OUR HEADS

Stories of Exploring for Mineral
Deposits in Exotic Lands

Collected & Edited
By Andrew Drummond

Connor Court Publishing

CONTENTS

INTRODUCTION

Taking a Eurocentric view as a starting point, intrepid adventurers and navigators radiated out from about the beginning of the second millennium AD. The Crusaders sailed to the Middle East, Marco Polo walked to East Asia and beyond. The Portuguese Golden Age of exploration resulted in the discovery of the Atlantic, then the Indian Ocean, margins of Africa and then they pushed on to India, the Philippines, and the Far East.

The Spaniards, and a lesser extent the Portuguese again, discovered the Americas and initially settled the central and southern parts of it. The English, French and Spanish colonised North America. The Dutch then colonised the East Indies and found Western Australia, and the Englishman Cook mapped that continent's eastern margin, found New Zealand and significant parts of Polynesia. And they met and interacted with the original inhabitants of those 'new' lands.

During the 19th and earliest parts of the 20th Century many expeditions gradually filled in the hitherto unknown interiors of the terrestrial part of the globe. Some of the exploration was politically driven, such as Rhodes' effort to establish continuous British control of Africa from south to north, from Cape Town to Cairo. And more resulted from the race by mainly European countries to colonise the remaining parts of the globe and to claim the Poles. And some was by curious, brave, or desperate men pursuing a variety of scientific, evangelical, or fame or wealth-seeking aims.

There are no remaining geographical reasons for a modern day Vasco da Gama, Columbus, Cook, Tasman, Lewis and Clark, Speke or Burton, Sturt or Eyre or Stuart, Scott or Mawson to risk life and treasure in trying to find new corners of the land surface of the Earth and hidden mysteries therein. We can now all sit at a desk and call up Google Earth instead. But there are still people who are adventurous, curious, educated, willing to take a risk, and be a little different. Among them are mineral explorers, and this book contains a collection of stories by them.

There have never been as many wealthy people on the earth as there are today – people with spare money to spend to improve their homes and health, to gain an education, to enjoy leisure time and to travel almost anywhere. The upper and middle classes have become huge consumers of just about anything mankind can think of to produce, and the lower classes and the poor still consume to the best of their financial ability – whether it be on food, shelter, transport or consumer items such as televisions, mobile phones and so on.

Now while the relative demand for different minerals can vary, for instance think of the greatly increased demand in recent years for metals such as cobalt, nickel and lithium for the manufacture of rechargeable batteries, absolute consumption of almost everything tends to increase. If one assumes that the average rate of consumption increase of any mined commodity rises at about the same rate of increase as the does the Earth's population, then that at is around 3% per year. Due to the wonders of compound interest, the resultant rather salutary statistic is that for any mined product, the world will have to find, develop and mine as much of it in the next 23 years as it has since the dawn of history. If we round that to 25 years and look at the year 2025 as a starting point; then take that, for any particular mined commodity or metal, the world's population has consumed one unit to that year, then it will consume a second unit by 2050. And two more units by 2075, and four more units by 2100!

While obviously that leads to pressures for substitution and recycling, the fact remains that many very large deposits of a whole range of minerals will need to be found and developed during the remainder of this 21st Century.

Now the large outcropping ore bodies of the First World countries, think of the Australian examples of Broken Hill, Mount Isa, Kalgoorlie, and the Pilbara, have generally already been located and have been exploited to varying degrees. The world's largest diamond mine, by weight of production, is Argyle in the Kimberley and that was found 40 years ago. The giant Olympic Dam copper, gold and uranium deposit in South Australia was found by very clever exploration under 300 m of solid rock cover at about the same time, and so on. In these advanced countries, sophisticated techniques are now used by well-educated professional exploration individuals and companies to explore deeper and under concealing cover to replace the orebodies we are mining out.

Geologists are the spearhead of the effort required to find these new and vital deposits, and Second and Third World countries are the regions in which to now look for relatively easily discoverable new ore bodies.

For a variety of reasons, Australian born, or adopted, geologists are relatively well equipped and often have the inclination and drive to do so. The reasons include the paucity of remnant large ore bodies outcropping at surface, and hence relatively easy to find, within this country. Its geologists are fairly substantial in number and are generally well-trained and experienced. The industry has developed highly sophisticated and innovative exploration techniques, such as the BLEG low-level gold detection assaying method which has resulted in the discovery of very many large gold deposits around the world. We also speak English, generally the universal language, and tend to be comparatively honest and ethical, easy-going, and politically neutral. The ones considering working overseas generally are likeable, curious, and have a can-do attitude. And they, or their employers, can bring or attract development capital to parts of the world where that is scarce.

The work can be hazardous, arduous and can play havoc with family and social obligations. It is often frustrating on a day to day and whole of project basis with, statistically, most efforts resulting in failure. Australian geologists and the companies they work for generally pull beyond their weight, and the world is a much better place for it when they make the discoveries and they are eventually brought into production.

This book contains a collection of 32 short stories by 26 geologists and a mining engineer who have given some, and often much, of their professional careers in the endeavour to find and help develop economic mineral deposits in lands remote from Australia. The stories are not technically oriented, and the standard of the prose is variable. The general theme is the adventures, frustrations, inevitable dangers, and different cultural challenges which have been experienced. Sometimes we overcame these challenges, and sometimes vice versa.

It may lead the general reader to an appreciation of what is required at the personal level to find the minerals which are the basis of the consumer goods on which we all rely these days to a huge extent. Collectively, they may answer the questions I asked myself in Siberia in 1993, as described by my story in the book: *"What the hell am I doing here, and why?"*

The genesis of the book was occasioned by a reunion of about half of an original 50 or so geologists in Perth in 2015: they had been employed by Australian Consolidated Minerals Ltd. That company had been taken over in the early 90s and the geologists had subsequently dispersed around Australia and to the four corners of the world. As we shared some stories over libations, I put it to some that our collective stories could be the basis of an interesting book, written for a non-technical readership, and focussed upon overseas experiences, as many had been in rather exotic locations or jurisdictions. It met a generally enthusiastic response, and the initial efforts to get their stories on to paper began at the end of 2019. We reached out to other industry colleagues to expand the range of stories.

The collection has been grouped by geographic region, then by time sequence. Readers will be exposed to wild animals and wild people; shady, generous, and lucky individuals; dodgy jurisdictions, revolutions, and war; dangerous, frustrating, and challenging situations; harsh to idyllic work surroundings; and more. We hope you will enjoy doing so and, through it, gain an appreciation of what we do, where, how and why as we go about earning a living and trying to find the mineral deposits the world needs.

SUB-SAHARAN AFRICA

A CHOPPER IN POLI

ROB DUNCAN
CAMEROON 1975 - 1977

Rob Duncan grew up in rural Victoria and graduated as a geologist from Melbourne University in 1970. He began his career exploring for uranium around Camooweal and the Nullarbor Plains.

In a 40-year working life, he has been involved in exploration, development and mining. His career-long geological obsession was the pursuit of the unusual intrusive rock called carbonatite which is often associated with rare earths mineralization. Often a lonely and unloved quest, continually interrupted by financial droughts and corporate machinations, his greatest satisfaction has been the intermittent exploration of the Mt Weld carbonatite in Western Australia from first drilling to commencement of mining in 2007. The mine is the world's highest grade rare earths deposit and the most significant producer outside China.

He in now retired in Alpine Victoria, a far cry from the flat arid regions of WA and the Kazakhstan steppes, the Rajasthan desert of India, the extremes of the permafrost of Siberia and the jungles of the Congo River in southern Cameroon. He believes he lives in Paradise.

Cameroon. What magic that name conjured. Memories of Gerald Durrell and the Bafut Beagles, a real live steaming volcano, jungles with pygmies and gorillas, everything a boy-adventurer could want and, in 1975, still at the vanguard of exciting post-colonial African independence. Of course, I really knew little of fact about the country outside a Lonely Planet introduction where Cameroon was lumped in with the rest of Central Africa, but when Utah Development Company threw an invitation around the office to lead a uranium exploration project in northern Cameroon, I instantly put my hand up.

I had come out of Melbourne Uni with a degree in geology four years previously and loved my work in exploration geology, around Camooweal, western Queensland; the Eucla Basin in WA and a year in the Northern Territory in uranium resource exploration; but this was a chance to see the world!

Interpretation of French mapping by the BRGM, its government geological survey department, appeared to locate a small secondary uranium occurrence in Archaean crystalline basement rocks just below a major regional unconformity with overlying sedimentary strata. At a stretch, it could fit the 'Proterozoic Sub-unconformity' model for high grade uranium deposits such as those found around Kakadu in the Northern Territory in Australia and in the Athabasca Basin of Canada.

It took a couple of years to set the project in place, sort political and corporate arrangements, run an extensive airborne radiometric survey with a French geophysical company and, as an afterthought, send me on a two-week intensive French language course at Monash University in Melbourne. Returning to my university city, I'm afraid that doing the rounds with old friends and drinking mates didn't progress my French much past my Year 10 at high school.

Allison and I had been two years in Brisbane, living in our first home and enjoying a champagne life with the spoils of two full salaries, a far cry from our previous life on and around the Nullarbor in Western Australia. She was working full-time as a solicitor, while I made various geological forays into the Northern Territory and Cape York, but mostly I hung around head office in Eagle Street in a suit and tie, compiling exploration reports for Utah and doing any administrative tasks that came my way. It was a good job, as was any geological job following the demise of the Nickel Boom, but I was keen to get my teeth into a solid field-based exploration project. Perhaps a bit selfishly as far as Allison was concerned, I abandoned the good life in Head Office to work in the wilds of blackest Africa.

At the time, uranium was the primary alternative to high-priced oil as an energy source and the world could not get enough. I was also enamoured with the concept of reducing pollution from fossil fuels, especially particulate matter and nitrous chemicals from coal-burning plants, though the effects of carbon dioxide on climate change were not known to me at the time.

I was technically prepared for the work ahead, conversant with the science of natural radiation, uranium deposit geology and even had a small find in the Northern Territory to my credit. I had experience with working in tropical conditions and remote locations with minimal infrastructure, and my farm background gave me a head start in 'fixing things with fence wire' in a low-tech environment, but nothing could have prepared me for the African physical and cultural environment in which I worked for the next twelve months.

Our project was based in the small remote village of Poli, some 150km south of the city of Garoua, capital of North Province of Cameroon. Poli was well past its French colonial heyday, the few substantial buildings in the main street staggering with sagging corrugated iron roofs and cracked dirty render, and a post office open sometimes for telegram transfer and occasional mail delivery. There was no telephone service. The town generator was defunct, the water supply limited to a couple of wells, and anything that worked was only for the convenience of the deeply corrupt sous-prefect and his household. There were a handful of kiosk shacks that sold cigarettes, soap and kerosene for lighting, mostly contraband smuggled from Nigeria to avoid the substantial Cameroonian VAT. There was an open bar made of woven branches that sold Trente-Trois, or 33, brand beer cooled under wet sacks in the shade. Footpaths wandered off in all directions to adobe mud family compounds under shady mango trees where children and chickens played around in the dust.

Our house was round, 13m in diameter, and made of white painted cement-rendered concrete blocks with a high conical roof supported by a fan of water pipes rising to a central apex. Thick thatch sandwiched between sheets of chicken wire made a thoroughly weather-proof roof. There were two small bedrooms to one side of a cavernous space, a small bathroom and steel shuttered windows without glass. On the south side, two big steel doors opened outwards to a stunning view over the little stream directly below, and on across green fields and savannah bush to magnificent Hosseré Vokré mountain. It rose 2000m high into the tropical sky where clouds gathered daily around it during the two annual wet seasons. On many days the clouds thickened and darkened, tongues of lightning flickered, thunder rumbled, and a discrete storm would tumble down one of the valleys from the peak to the plain below, sometimes straight down the northern valley to Poli where we would watch from the open doors.

The data acquired by the airborne radiometric survey had been interpreted to select about 150 uranium-sourced anomalies, potentially indicative of economic deposits. It was to be my job over the following months to locate, visit and assess the source and significance of each of these anomalies. In those pre-GPS days this was facilitated by a downward-pointing camera in the plane's belly which produced a black and white strip film of a swathe approximately 100m wide of the ground it had passed over, clearly showing narrow intercepts of streams, pistes, clumps of trees and the occasional thatch-roofed cluster of huts in isolated villages. These clues would enable me to search aerial photography of the region and locate the anomaly where the two images coincided. I would then plan the best ground access route.

On our early reconnaissance visit to Poli, we had driven down tracks and roads radiating out from Poli towards the places we would be investigating. But the vehicle tracks rarely went more than a few hundred metres before terminating at steep creek banks where the former colonial bridges were never rebuilt after floods. After negotiating the creeks, the old tracks then continued as walking paths. It was obvious that nearly all the work would have to be done on foot or, better, by trail bike with a radiation scintillometer strapped to my chest and pack of navigation and sampling gear on my back. Trail bikes were unknown in Cameroon at the time, until I found a consignment destined for the American Peace Corp for a project subsequently abandoned when politics intervened. I bought three of them, two for riding and one for spare parts.

The trail bikes worked a treat, French geologist Michel Marchat and I riding the pistes on the plains from village to village, or scrambling up steep slopes to access all but the most remote mountain sites. We worked steadily through the dry season and the "little wet", and at first thought the biggest problems of the "big wet" season were the daily fording of rising rivers, slippery walking paths and soggy savannah soils bogging the motorbikes. But then the tropical grass grew, at first a gentle swishing beneath the bikes, and then making a difficult ride through waist-high cane grass with hidden ant hills and holes causing ever-increasing falls and spills. Finally we were riding completely blind through two-metre high cane grass, navigating by stopping at the base of small trees, leaning the bike against the trunks and standing on the saddle to get a view over the grass. The blind riding was particularly hazardous near streams where the cane grass and tussocks grew particularly thickly. In one instance I pulled up right on the edge of a near-vertical drop to a large flowing pool at the outside bend of a stream. The bank crumbled and bike and all slid three metres down into the stream. As I hit the water, I glimpsed Michel behind me, wide eyed, legs splayed at the top of the bank trying not to follow me. When he eventually found a better route across a gravel bar and joined me, I still hadn't found my bike, presumably sunk in the bottom of the pool, but repeated diving hadn't located it. I eventually found it twenty or thirty metres downstream where it had semi floated and snagged on a gravel bank just below the water surface.

As much as I planned our traverses to avoid them, stream crossings were always a challenge, sometimes one of us having to swim across with one end of a long rope which was tethered to a tree on the far bank and the near end tied to the handlebars of the first motorbike. It would then be eased into the stream after turning the fuel off and sealing the tank, and towed across the stream underwater with the rope. We would swim back and tie onto the second bike to repeat the process.

The next stage was to restart the bikes. The spark plug would be removed, the bike stood on its rear wheel and the engine and exhaust drained by hand pumping the kick-start lever. The magneto cover had to be removed, water drained and the magneto dried before opening the fuel tap and kick-starting the motor. This could take an hour or more, but as we refined techniques we managed to get both bikes going in twenty minutes.

Finally, as site visit productivity fell to near zero, it was clear that the only way to get the project back on schedule was to employ helicopter support. This would be difficult and expensive, but justified if we could get back to assessing five or more sites per day. The problem was, apart from military helicopters, in 1975, there were virtually none in West Africa. Head Office made enquiries from Australia and suggested that, on my next field break in London, I negotiate with a helicopter company that had a small Bell 47G currently working in Sierra Leone. That worked well for me as I could reunite with Allison after nearly six months' separation and escort her back to Cameroon and life together again.

While with the managing director of the helicopter company in his office, I made a phone call back to Australia to confirm costs, timing etc for the helicopter to be utilized, but additionally, my boss Chris Gregory was persistent in quizzing me about my impending hospitalisation in London for minor surgery. I was a bit shy of being specific, but finally, in front of that director, I blurted out that I was going under anaesthetic to get work done on a "busted arsehole", a result of the unrelenting daily beating up on a motorbike. Chris went very quiet while I explained what was to happen and how I had found a specialist surgeon who was confident he could stitch me up through the London School of Hygiene and Tropical Medicine. The helicopter MD was not the least embarrassed, and explained he had been through a similar procedure himself with great results. His comments were a wonderful confidence boost.

The plan was that the little helicopter would be dissembled in Sierra Leone and air freighted to Cameroon to arrive in about a month's time and the company would keep me informed of progress through telegrams to Poli. The company would supply a pilot, mechanic and essential maintenance supplies, but I would be responsible for the high-octane avgas the chopper required.

The day after repairs to my posterior, Allison and I took a taxi all over London buying replacement gear and parts for the field work, and managing to take a quick glimpse at the London historical sites before taking off from Heathrow on an evening flight to Douala. I was wearing a motor bike tyre around my neck because that particular size tyre was not made by French Michelin and therefore not available in Cameroon. I was pretty uncomfortable from my operation, and at 3am in the morning, Allison was trying not to be too shocked by her first time in an African airport terminal. The noise, the heat and humidity, the crowd and the tropical smell of rotting vegetation were overwhelming and the Douane, the customs officials, were giving me a hard time about the tyre. I had a receipt and was happy to pay duty, but they couldn't find the appropriate tax rate, saying something like I would have to leave it behind. I knew it would never be seen by me again. I grabbed the customs price catalogue, spotted the raw rubber import duty cost and pointed out that at 20 francs per kilogram, 50 francs was plenty, and as an after-thought, "accidently" lost 1000 francs in the paperwork and managed to leave with my tyre.

Our next airleg was to the capital city of Yaounde. The flight was long-delayed, and when we were finally boarded, we went nowhere and had to disembark again. We were finally transferred to a 4-engined Douglas DC4, looking like a relic of the WWII Berlin Airlift and fitted out with tubular steel and canvas troop seating. We finally lumbered into the air and flew low and slow beneath the tropical cloud, keeping visual contact with the Yaounde bitumen road just below. Allison, in the window seat, soon noticed blobs or thick black oil, from the inner starboard motor, rippling across the wing and out into the slipstream. We pointed it out to the hostess who ran up to the cockpit. When she returned, she told us that the crew were aware of the problem, but there should be enough oil to reach Yaounde and, anyway, the plane could fly perfectly well on three motors. All very comforting.

Flying in Cameroon was rarely straightforward, Camair being the only operator running a relatively modern fleet of Boeing or Airbus jets commonly supported by temporarily hired European jets and aircrew. There were persistent problems with fuel supply and quality. Casually chatting with a Shell jet-fuel engineer on the tarmac at Yaounde one day, we watched a swarm of fuel trucks emptying and refilling the jet's wing tanks. A batch of contaminated fuel necessitated a complete draining and refilling. Then I remembered that, on a flight a couple of weeks earlier, we had flown from Garoua in the north at what felt like a couple of thousand metres, through smoke from grass fires and dust from the fallout from the annual Arabian dust storms. The engineer suggested the denser air at such a low flight level helped overcome the fuel power deficiency, but the dust was a problem for engine longevity.

In 1977, seats were not allocated for domestic flights, and since there were invariably more tickets distributed than seats available, there was an almighty break of bustling bodies from the terminal doors when they were opened and a stampede to the aircraft stairs. I used to carefully coach my crew on the boarding procedure. First, stand in a flying wedge behind me, about a metre back from the double, inward opening doors. No matter the crowd in front pressed up against the glass. Then when the flight attendant turned up on the outside, and two burly gendarmes forced the doors open, the crowd would be forced back behind the doors leaving me and my crew able to saunter through in the vanguard and rapidly march to the aircraft stairs. Then our crew had to split up and dive into window seats, strap in, and casually open a newspaper. This virtually ensured none of us would get kicked off to make room for any unseated VIPs or a "gift-giving" businessman.

Still sweating from the tarmac rush, I watched over the top of my Cameroon Times, with increasing mean amusement, a dignified-looking businessman in a suit, and with a Samsonite briefcase, stroll up the aisle looking for a vacant seat. Then, reaching the cockpit, he turned around and standing with his briefcase in front of him, surveyed the full house for fully ten minutes, looking for a vacant seat. Finally, as the take-off procedure commenced he was gently, but firmly, escorted by two flight attendants to the stairs at the rear of the aircraft and ejected agitated and sweating. He didn't know the drill!

But I have digressed from Allison's introduction to Cameroon. Ultimately, we reached our destination at the thatched-roof Poli house, Allison settled in to life in the village and the month went by and no sign of the helicopter. I finally received a telegram from London; "Will relocate machine overland, ETA one week". I could hardly believe it! More than 3,500km across six countries with dubious international cooperation credentials, at least one unfolding military coup, and the whole lot an impossible mess of West African bureaucracy. I just about gave up hope on the helicopter program but, remembering my promise to have fuel available, made enquiries and found avgas for private aircraft was unobtainable in northern Cameroun. Finally, with the help of Royal

Dutch Shell, we bought sheet steel in Nigeria, had sixty 200 litre drums manufactured in Lagos, filled with certified avgas and set on the road to be trucked 3000km to northern Cameroon. I was not confident of timely delivery, but since there was no news of the helicopter, there was no immediate problem.

But one evening driving back to Poli on the one lane dirt road, I spotted out above the tall roadside cane-grass, a red reflector floating low in the sky. As we drove we could see it was the rear tail light of a big Mercedes truck lying on its side after having run off the muddy track. To my dismay, scattered about in the bush and grass, were about 60 red fuel drums thrown from the truck during the roll over. With sinking heart, I knew it was Utah's helicopter fuel, and an English-speaking Nigerian truck driver confirmed it.

It was a miracle that no-one riding atop the load, as was the universal African custom, was killed or even badly injured and that there had not been a massive explosion and fire. The pushy driver kept thrusting a delivery docket and pen at me to sign. I pointed out that the load was still in transit to Poli and, from the smell of petrol in the air, at least one of the drums was split and still leaking and there was no chance of the cart note being signed until the whole consignment was delivered to our compound in Poli. It was up to the driver and transport company to sort the roadside mess.

Over the next few days, Michel and I continued our work in the field, but when there was no sign of anything being done at the wreck site, I finally offered to right the truck with the Landcruiser winch, but first the driver had to organize a lot of digging of the mud under the truck to assist the roll-back. A day later things were progressing, the mud was cleared away, I anchored the Landcruiser into the field on the opposite side of the road and rigged cables to slowly roll the truck back onto its wheels, and then back onto the track. I left the crew to load the drums over the next couple of days, but still the fuel was not delivered. The batteries had leaked acid and gone flat. Being a 24 volt system, the Landcruiser could not help, but the enterprising driver had somehow located an army construction camp in the bush 50 kilometres away. If I took him there, he would be able to borrow a set of batteries.

It was a Sunday at the military base and, after animated discussion, a half dozen off-duty soldiers jumped on board with batteries taken from a grader and bulldozer and some lengths of thick copper cable for jumper-leads. We drove back to the truck and, with typical African mechanical ingenuity, distributed battery acid from the army batteries to the dry truck batteries, and amid lots of 24-volt sparks, got the truck started. It was a miracle of strength and cooperation as the soldiers and Nigerians loaded the sixty drums onto the high truck tray and headed up the road to Poli.

As the truck pulled up at our compound, everyone dispersed into the village to celebrate. I don't know how long the soldiers had been confined to their remote camp but, as evening fell, it was difficult job to round them all up from bars and non-conjugal beds to get them back to camp before the commander returned in the morning.

But I had the fuel on site, and there was only a helicopter missing.

We were on a supplies run in Garoua a couple of weeks later and eating lunch at the Hotel de la Benoué, when the manager came rushing in with news that there was a helicopter at the airport. And, sure enough, a little Bell 47G was having its bubble washed down by a pilot in white shirt and gold epaulettes while an aged English mechanic fussed about the engine, a vertically mounted six cylinder piston motor behind the plastic bubble.

Over the next few hours back at the hotel, Chris Balak, the Austrian ex-army helicopter pilot, and Dennis, a career helicopter mechanic doing his last field posting before retirement, related some of their adventures flying the west African coast from Freetown to Douala. Most of the time they had skimmed along the beaches, noting crocodiles and sharks, while flying high and out to sea around military sites. They carried extra fuel in jerry cans on the skids but, at one stage due to an inexplicable extra fifty kilometres not shown on their maps, they had to land outside a roadside service station and make up a shandy of low octane motor fuel and their remaining avgas to get them to the next airport.

In Sierra Leone, Chris had been hired by the police to fly around the diamond fields to surprise illicit diamond hunters and coordinate ground troops closing in for the arrests. However, at one point he came under small arms fire while on the ground dropping off the over-weight commander of the operation, and took off under full throttle to get clear. The machine laboured inexplicably until the commander's face suddenly appeared outside the cockpit, his arms stretched out clawing at the bubble for a handhold and looking very white-eyed terrified. He had tried to shelter from the gunfire behind the chopper and got caught with the skid between his legs as it took off. Not hurt, but very frightened, it was tricky putting down at a safe distance without breaking the Commanders legs or attracting further gunfire.

That and many other stories of his Austrian army attack helicopter days, gave me confidence in his flying skills, especially in the hot, thin air around Hoseré Vokré. In the event, the air was normally too thin to fly the high plateau of the mountain and we were restricted to the surrounding foothills and plains. Dennis also was sanguine about the chopper's mechanical condition, although he cut the second magneto test from the pre-flight check routine. Chris tried it once and the engine ran like a chaff-cutter. "Leave it on magie two", shouted Dennis, and thereafter I always made sure the magneto switch was on No 2. Dennis also quickly found water in the fuel in his daily early morning mechanical check routine. It appeared that the unscrupulous Nigerian truckie had topped up, with water, two leaking fuel drums at the roll-over site. I didn't make it obvious, but I kept a close watch over Dennis' shoulder when he did his daily fuel test.

I had great confidence in Dennis and Chris, and gained even more in Chris's flying ability when, at the expense of a bad fright, one night approaching Poli he killed the engine and auto-rotated down to a safe landing on the football oval, pulled on the collective, rose a metre, turned the machine around and finally settled as the rotor momentum weakened below flight threshold.

The helicopter made a great sensation in the village, there having being only one other ever seen in the district before. That one had flown around the remote villages of the region, landing at each, and some proselytizing Americans read from the Bible, played some recorded music, climbed back into the machine and flew off to never return again. The dominantly animist Duoayo communities were puzzled and the remote mountain animists too terrified to come out of the bush. The Muslim Fulani in Poli were bemused by the American antics but not at all concerned they might represent any sort of threat to Allah.

I did the rounds of the village establishment, first the Suprefect, then the police commandant and the traditional chief, explaining what we proposed to do and offering an aerial view of their domain. Everyone professed support for the program of visiting the radiometric anomalies, but only the commandant ultimately took up my offer of a flight. There was a long litany of excuses from the chief and Suprefect, but the commandant had done military service and was the only one not deeply fearful of flying. The fear was otherwise universal amongst the villagers, and even 15-year-old Arbu, my best field assistant was loath to come with me early one morning when flying conditions were perfect. At the distant landing site, we carried out scintillometer gridding, collected samples and prepared to leave for Poli. But nothing would induce Arbu to climb aboard, preferring to run nearly fifty kilometres back to Poli. Overnight he did just that, fronting up at breakfast ready to walk to the next site.

Our productivity increased as planned, although not everything was smooth or easy. I would fly out early in the morning when the air was cool and dense, normally just with Chris, to keep the load minimized. We would locate our next target radiometric anomaly from a plotted point on a black and white aerial photograph. With a portable scintillometer held between my feet, we would circle around to locate the most intense part of the anomaly and select a landing point as close as possible. This was sometimes impossible in steep hill country or dense scrub, and I would have to do a long walk.

Even on the loess-covered plains, as the dry season came on, the cane grass between the scrubby trees grew taller, two metres or more, making choice of landing sites very difficult. The usual procedure was to make a 'nest' by blowing the canegrass flat with rotor downdraft, then lift off, turn around, and blast another nest adjacent to the first over which he would set the tail rotor. Usually this was safe and straightforward, but I would always jump out and check under the motor where grass might be touching the red-hot engine exhaust manifold. If there were signs of smoke or flames I would immediately step back and signal Chris to take off urgently, which he would do in a maelstrom of flying debris and burning embers. I would frantically stamp out any burning remnants before he returned and landed.

If there was no possibility of landing safely, I would don my field gear, hang the scintillometer around my neck, open the door and carefully clamber down to hang swinging off the skid by my hands, while Chris hovered as low as possible. I would let go and plunge into deep grass and invisible rough ground. On feeling my weight drop off, Chris would roar off and land somewhere safe, like a gravel island in a nearby river, and come back for me in an hour.

Sometimes it took an inordinate time to locate a safe return landing pad, which meant me clearing the cane grass, termite mounds and small trees for the chopper to land and to let me climb aboard. So we hatched a plan to pick me up from the tall grass without actually landing. I would climb to the top of the tallest termite nest away from any trees, knock the top off with my geology hammer and stand on the little platform with my hands in the air to signify I was ready to board. Chris would hover by as slowly as possible. I would grab a skid with both hands and give a brief weight test. If Chris didn't pull on the collective, I would let go and drop to the ground to repeat the process. If the machine roared as Chris pulled on the collective, I would hang on and as the machine rose strongly out over the trees, swing a leg up onto the skid and get a good grip. The next part was the hardest, stretching to reach the doorhandle while half kneeling on the skid, opening the door, clambering into the bubble and collapsing onto the seat.

There were times flying back to Poli at the end of the day, when I could take my eyes off maps and aerial photographs to look at the exotic wildlife. Chopping along low over the savannah, I saw elephants, herds of buffalo, solitary antelope cheval, gazelles, dik-diks and lots of others. Michel told me Chris once put the chopper down to grass-top level in front of a lion on the prowl which, with its tawny mane rippling back in the rotor wash, made angry fanged roars and clawed swipes at the bubble. Michel said the plastic bubble felt awfully thin.

Our exploration area lay between two large game reserves and, being lightly populated, was a corridor for wild animals moving around the mountain. The chopper gave unparalleled views of ungainly hippos on a riverbank diving into deep water holes and then gracefully powering along the bottom like streamlined submarines with a cloud of muddy wake billowing behind. We saw an old bull elephant pounding through the savannah, panicked by the chopper, plough full tilt into a thicket of vines festooned all over a tall dead tree; the tree exploded, branches and vines flying through the air; and the elephant went on pounding along without missing a step.

Crocodiles sunned themselves on gravel banks of the rivers and herds of black buffalo would run from the noise overhead. We saw all sorts of antelope, little dik-diks to horse-sized antelope-cheval, others identifiable by their distinctive horns; back-curved, straight-pointed or spiral. An occasional herd of giraffe would gallop to tree thickets and disappear and all sorts of monkeys were seen leaping around in large isolated trees. I'm a bit sorry we didn't take time to properly identify many of the animals, but we had a job to do and, besides which, the chopper obviously instilled great panic in wildlife that generally remained placid when encountered on the ground.

The Big Walk Out

By far the most memorable incident was the big walk I had to lead from a crippled chopper left in the bush.

Chris had run his flying hours to the maximum and had to be replaced by an Indian crop-dusting pilot flown in from England. Vijay Arvie was a competent pilot, but lacking confidence after a life-threatening crash some years previous. His injuries still inhibited his agility and his gross overweight state made chopper work in the thin, hot air more challenging than ever. I tried to schedule our flights to carry only a single passenger, and a light-weight at that, ideally the lightly built Cameroonian geologist, Jean-Blaise Nyobie, on secondment from the Cameroonian scientific research organization, or myself, at the time tipping the scales at 60kg.

The plan for this particular day was to fly to the village of Guijiba which lay on the main highway between Garoua and Ngoundéré where Jean-Blaise was living. I would wait there while the chopper flew Jean-

Blaise to his field mapping area and returned for me to carry on our usual radiometric site assessments. However mechanic Dennis was not completely happy with the chopper and wanted to come, at least for the flight to Guijiba. Dennis was a light 70kg and I was happy for the three of us to fly together. It was a regular ferry flight, but navigating west from Poli, I spotted the target site for investigation later that morning. Arvie immediately spiralled down, and landed in short green grass a few hundred metres from the target area. Breaking our flight plan was not a good idea, but for the sake of a half hour delay I could complete work on the site right now, or come back later if necessary.

Leaving Arvie and Dennis comfortably resting in some shade, I took off on foot towards the site. To get there, I had to walk up a dry gully with steep rocky banks and heavy scrub overhanging. Suddenly a tribe of baboons ambushed me, dropping out of the trees, screaming and hooting and making threatening rushes at me. I tried taking a big stick and yelling just as loudly, threw rocks and sticks at the mob, but that just set them right off. Amongst renewed baboon obscenities they started hurling sticks and stones from all around and from the trees overhead. I backed down the gully, sat down in a shady spot and waited for things to quieten down. The tribe eventually moved off with a few parting screams and I was able to resume my traverse.

A good hour had passed before I returned, hot and sweaty, to the chopper, stowed my backpack and gear and settled into the cabin with Arvie and Dennis. The starter motor gave one weak groan, the motor backfired once and silence reigned. The battery was near dead flat. My idea of tying a rope onto the end of the rotor blades and running rings around the chopper to bump-start it like an old car was negated by the autorotation mechanism, but the idea of heating the battery to tease out a little more power met with some interest. But Dennis said the high energy-density battery needed to be cooled to get extra power. I suggested wrapping it in a wet towel to produce evaporative cooling, but before I could start down the hill to get the water from a creek, Arvie emptied our emergency water bottle all over the towel-wrapped battery.

I settled down under a shady bush and reviewed my air photographs. I knew exactly where we were and could see that by following the winding

ridge top, a few hundred metres above us I could, by myself, hike out
to the highway about seven or eight kilometres to the west. I estimated
that I could do it in about two hours, arriving at the highway by midday, flag down one of the numerous trucks going south past Gijiba,
where I could probably rent a jeep and driver to return overland to the
helicopter, and pick up Dennis and Arvie. But neither would have a bar
of the idea, being absolutely petrified at the thought of being stranded in
the African bush with wild animals, and a slowly approaching grass fire
in the distance. I could argue that there was no danger from either, the
animals not daring to approach the chopper, and the area around it being
too short and green to burn; and that a 68 eight year old mechanic and
an overweight, half-crippled pilot would slow me down considerably.
But they were adamant they were not staying behind.

I led the way up the hill towards the ridge, an easy walk through grass
only knee high, but within fifteen minutes the chopper crew were sitting
down to have a rest and light a cigarette. I realized we would be spending
a lot more than a couple of hours walking to the highway.

I made it clear that this was not a walk in the park. We would have to
do one-hour walking stints and they would need every bit of breath they
had. Smoking was off the agenda, and so was a drink of water.

Dennis was reasonably fit for an old man, and Arvie knuckled down
for the next couple of hours. From the highest point on the ridge we
could see the highway in the distance, but the ridge top with its short grass
and relatively easy walking, snaked about a lot, and sometimes we were
walking at a tangent to our goal before the ridge turned back to a more
direct route. At one point we were even having to walk in the completely
opposite direction to the highway. From where we were we could see a
long sweeping curve on the highway in the valley below, much closer to
us than the ridge top route was taking us, and with some trepidation,
we headed down to the valley. I knew the grass would be high and the
going potentially difficult, but there was a fair chance there would be a
village path leading out to the highway and there was a long straight spur
leading down to the valley and, under unrelenting encouragement from
the tired crew, we headed down to the valley.

It didn't take long to regret the decision. At the bottom of the slope we plunged into head-high cane grass, so thick it was barely possible to push through. For Arvie it was impossible. The only way to make progress was for me to stamp a track, sometimes pushing backwards, for 5 or 10 metres and call the guys up and do the same again. Sometimes elephant trails of trampled grass made for easier going when they more or less coincided with our direction to the highway, but generally they were randomly wandering grazing tracks we would have to abandon to maintain our route. Vertical-sided, dry water courses were tricky to cross, usually easy enough to drop down into, but difficult to climb from on the opposite side. The only way to get Arvie up was to slide my shoulder between his fat thighs and bodily heave him up to sprawl into the grass on top.

We were making less than five hundred metres each hour but, as evening came on, we were greatly encouraged by the sound of trucks on the highway. But I well knew they would be stopping soon to camp on the roadside as wildlife and ghosts made night-time travel hazardous.

By now as I was pushing and stamping the trail ahead while Dennis helped the staggering Arvie. They would start calling out if I was out of sight and I would have to go back and help them along with a shoulder for the few metres they could manage, but then the ground started to rise steeply and I realized we had reached the toe of the high embankment on the highway curve we were aiming for. I abandoned the crew and clambered upwards, but still deep in the cane grass, heard and saw a vehicle roaring by in a swathe of bright lights, headed south. It was a Toyota Landcruiser with the unusual feature of driving lights mounted on the roof-rack; it had to be our Landcruiser and I realized that Michel, back in Poli, had become worried and come looking for us. Unfortunately, we had missed him by less than twenty seconds, the time it would have taken for me to stumble up the last few metres of embankment and jump on to the road.

I helped the others up to the road, Dennis pulling the semi-comatose Arvie by the arms, and me heaving from behind and we collapsed onto the warm bitumen to contemplate our next move. The night was dark and moonless with the glow of the grass fire silhouetting hills and trees. There was a lot of mysterious rustling in the roadside grass and bush beasts and birds were protesting our presence with shrieks and squeals and Dennis nervously asked how would we get a ride, would anyone stop? Without thinking, I answered: 'Europeans are rare out here, a helicopter crew and geologist unprecedented; we just have to look white and they'll stop". Arvie, who was particularly dark-skinned for an Indian, grumbled: "That rules me out!"

A truck going north stopped, but no cash offer I could make would get them to turn around, but eventually another heading south towards Guijiba not only stopped, but hoiked the cab passengers into the cattle crate on the back and welcomed us to a twenty kilometre ride in comfort.

At Guijiba we dropped down from the truck cab into the crowded town square to be greeted by a very relieved Michel and Jean-Blaise who had thought of all sorts of crash scenarios but, most importantly, there was Allison who gave me the biggest hug of my life. The villagers brought out the water we desperately needed and bottles of cool 33 beer. A spontaneous party started and old Dennis got his second wind and was soon dancing around the bonfire to the African beat with half a dozen of the village ladies. Poor Arvie was a cot case, in fact it was two days before he could walk again, and then only with difficulty and, ultimately, he had to be repatriated to London. Two beers and I was not much better, sleeping on Allison's shoulder for the midnight journey back to Poli.

Two days later, I had fully recovered and drove to Garoua where I knew Chris would be doing R&R in one of the airconditioned bars, and sure enough there he was at the Benoué front bar with a couple of European tourists boasting loudly about his helicopter exploits from the Austrian Alps to Africa. With obviously more than a few rum-and-cokes in him, and me a long way from where he thought I should be, it took a while before he burst out, "Where's the chopper? Where's Arvie?" I answered coolly as I could: "The chopper's in the bush. Arvie's in bed". He nearly fell off his stool, not just from the rum, but assuming the chopper was crashed and Arvie was injured.

Dennis said that to recover the chopper, assuming it hadn't been burnt by the grassfire, we would need only two heavy duty 12-volt truck batteries and two sets of jumper leads and a pilot. I was already familiar with borrowing batteries. Only this time I borrowed them legitimately without half the Army Construction Division tagging along.

We drove back to Poli and equipped the Landcruiser for a cross-country excursion to the chopper. I plotted a route on air-photographs, avoiding the deepest creeks and ravines, a route considerably longer than our walk-out one, but putting the vehicle within a couple of kilometres of the chopper. So with Dennis and Chris and Michel, we returned to Guijiba village, hired a bunch of porters and headed out through the scrub along the ridge tops. Two of the porters rode on the roof-rack directing me through areas of grass higher than the roo-bar, as we crawled at less than walking pace to minimise damage from hidden hazards and, more than once, had to winch the vehicle off a rock or out of a gully. The last part was easy: parked at the edge of a steep ravine, the porters jumped down from the roof, slung the batteries on poles shouldered between them and headed on towards the chopper. As I had guessed, the grass fire had petered out at the edge of the green and short grass area and, within an hour, we had the chopper running and, shortly after, flying Chris and Dennis back to Poli. It was easier on foot this time retracing our path back to the vehicle and following the wheel tracks back to Guijiba, and ultimately back to Poli.

We had only a few more days left in the helicopter contract and I concentrated on visiting the sites most difficult to access with the motorbikes. Constant burning of grass by the villagers preparing new fields made access easier in many areas and I tidied up my list of obligations to people to whom I had promised joy flights. I suspect bank manager Camburi from Garoua, and his wealthy trucking company friend, mainly wanted to locate game for later hunting, but the commandant of the Poli garrison accepted my offer for more legitimate reasons. There were many isolated areas on the mountain which fell within his jurisdiction, but he had never been able to visit. We endeavoured to get into those up to the highest elevation the helicopter could manage. We could not usually attain the highest plateau where the air was usually too thin to support flying. But it was a reasonably cool morning and suddenly Chris said he had caught a favourable updraft and, with its help, coaxed the little machine up onto the plateau where we could fly on ground air cushion at grass-top height between widely scattered trees.

Over a long time previously I had observed through binoculars, a small cliff-face on the plateau which might have been a section through layered sediments overlying older basement rocks. If that proved true, the sedimentary outcrop might have been important to verify our model for uranium mineralization deposition. However, brief glimpses of the terrain from the helicopter and radiometric readings indicated a series of basalt flows constituted the layer cake outcrop and was therefore irrelevant to potential uranium mineralisation.

But the next problem was to descend safely from the plateau. We couldn't just fly over the edge and plunge through thin air: we had to ride the ground effect cushion down a spur off the plateau to an elevation where the air density was sufficient to support flying. It was a slow and careful descent, dodging around trees and at times having to follow a narrow razor-back ridge down to the foothills.

The plateau visit marked the end of the helicopter program, for me more than 300 hours flying on an exciting phase of our exploration project. This left only work we could complete on the ground. Following a somewhat subdued celebratory meal at the Poli house, Chris and Dennis flew off to Yaoundé to dissemble the chopper, load it to airfreight and return to England. Michel and I continued to walk and ride the remaining program of site visits: it was slow hard work, but at least I could do it without the thought that during those many hours choppering around Poli, this day might be my last.

THE WAR DIARY
OF A
DE BEERS GEOLOGIST

MANFRED MARX
ANGOLA 1975

Manfred Marx is a retired diamond geologist who was born in South Africa but lives with his wife, Kate, in Perth, Western Australia.

After graduating from the University of Cape Town in 1965, he joined the De Beers exploration team in Botswana. He was fortunate to lead the field team that discovered the Orapa kimberlite pipe in April 1967 that was destined to become the world's largest gem diamond mine.

In 1970, he transferred to Angola as a senior member of the very large De Beers exploration team. On the commencement of the Angolan civil war in 1975 he was transferred to the De Beers operations in Australia.

Later in life he established his own consulting company which took him to many countries in the world. He still retains a close interest in developments within the mining industry and enjoys giving lectures on diamond exploration in schools and social clubs.

War Breaks Out

I had heard a faint and garbled message on the Cuilo camp SSB radio on a late afternoon in June 1975. The Mussunuige base camp had been invaded by soldiers. Several times I asked Mike Shearing, the senior geologist, to repeat the message but heard nothing except the radio static crackling ominously back at me. Lindsay Doig, the ever-optimistic Scottish geologist, said that it might have been a practical joke and that I should at least stay for dinner. I was not reassured because the situation in Angola had been steadily deteriorating over the last year and it was always going to be just a matter of time before we would be drawn into this civil war. I asked Lindsay to stand by his radio until he heard from me again.

I turned the wheel of my Land Rover and headed back along the bush track to the Mussunuige camp. It had grown dark by the time I neared the western edge of the valley. I switched off the headlights and walked until I could see the lights of the camp on the other side of the valley. It was a clear night with only the usual comforting African bush noises mingling with the welcome sound of the camp generator. All seemed reassuringly normal in the camp. I felt annoyed that I had not taken Lindsay's advice to overnight in the Cuilo camp and share a bottle of scotch with him. I hurried back to the vehicle. Maybe the stress of living with gathering war clouds over Angola was taking its toll and it was affecting my judgment. More than ever now I was looking forward to the Christmas holidays in Cape Town. The vehicle raced down the steep track and across the creaky wooden bridge that Fernando had so lovingly built out of felled tree logs. I could smell the smoke from the campfires and was looking forward to the evening meal of bacalhau and rice with a Cuca beer or two. I parked the Rover next to the office and, as I switched off the engine, I looked out of the side window into the barrel of an AK-47 assault rifle held by a black soldier in a camouflaged uniform.

How had it come to this?

The Background

Angola was colonized in 1575, as a trading post and as a source of slaves who were mainly shipped to Brazil. The first stirrings of rebellion against the Portuguese colonists dated back to 1961 when a group of coffee plantation owners were massacred by their black workers in the northern province of Uige. Lisbon rushed troops into Angola and the rebellion was quickly and brutally crushed. However, the desire for freedom was not extinguished. At first the exiled liberation movements (called terrorists by the Portuguese) were united under a schoolteacher and poet called Agostinho Neto, who was based in the neighbouring Congo and supported by the Soviet Union. It was during the Cold War that factions arose within this MPLA - dominated group based, not only on their desire to liberate Angola, but also to side with the superpower that promised the most in aid and military assistance.

The leader of UNITA was a bearded charismatic giant of a man called Jonas Savimbi who was supported by the USA and "white" South Africa. The FNLA was led by an enigmatic man called Holden Roberto who was fond of wearing dark glasses and Mao jackets. He was said to be the brother-in-law of President Mobuto Sese Seko of Zaire (formally called Congo). He received his backing from the communist Chinese initially, but when they judged that FNLA was too reluctant to engage in combat they withdrew it, and subsequent support came from the CIA.

The Portuguese decided to quit Angola soon after the 1974 coup de etat which overthrew the Caetano government in Lisbon and installed the flamboyant General Antonio de Spinola. He had earlier written a book in which he stated that the unaffordable colonial war was also unwinnable. In a desperate effort to avoid a civil war in Angola, Lisbon convened a government of national unity for the latter. Within a year the agreement had predictably broken down and the scene was set for a rapid withdrawal of the Portuguese troops and the unleashing of the dogs of war. On 11 November 1975, the MPLA-led government in Luanda declared independence with Neto as Angola's first president.

There was a political situation that could not last.

The first Angolan diamonds were found in 1912, by Belgian explorers in the north-eastern Lunda Province, by following the diamond trails up-stream across the Congo border along the main rivers draining from the Angolan plateau into the Congo River basin. In 1917 the Portuguese government in Lisbon established DIAMANG, the company that was granted a monopoly over all diamond exploration and mining. From its headquarters in Dundu, near the Congo border, DIAMANG geologists discovered numerous rich alluvial diamond deposits along the terraces of the Chicapa, Luachimo, Cuango and other rivers draining the remote Lunda Province. Whilst they realized that diamonds were originally hosted by kimberlites, many of which they had recognized within the diamond-rich river valleys, their focus and expertise lay in the mining of alluvial diamonds, not pipes.

By the 1960s the winds of change were howling across Africa as one colonial country after the next gained independence. That transition had not always progressed smoothly, as was tragically witnessed in the neighbouring Belgian Congo. By 1970, only the Portuguese colonies of Angola, Guinea-Bissau and Mozambique, as well as South Africa and Rhodesia, were grimly hanging on to "white" rule. In this politically charged environment it belatedly occurred to the DIAMANG authorities that their diamond monopoly could be threatened in a post-colonial period. Also, it was evident that most of Angola had never been explored for kimberlites.

What should DIAMANG do next?

In the world of diamond exploration the name De Beers loomed large in those days. So it was to this company that DIAMANG turned in its hour of need and established the joint venture company called CONDIAMA, jointly financed but managed by De Beers. The team was led by Roy Edwards, an enthusiastic and very bright 35 year-old geologist whose British army training would stand him in good stead in the Angolan campaign. I met Roy in 1969, when he was on a visit to the Orapa diamond mine in Botswana. I was fortunate to have been a member of the discovery team led by Dr Gavin Lamont in 1967, and was then subsequently involved in the economic evaluation of the kimberlite pipe. Much to my surprise, Roy chose me as his second in command.

I was 27 years old, and buoyed by my Orapa discovery, I was about to embark on an exciting new project in West Africa with the promise of many more diamond discoveries. Could life get any better than this?

But the De Beers geologists were caught in an unpredictable situation for which they were totally unprepared. Little did I know what life had in store for me and it wasn't to be all plain sailing.

Hostilities

The Land Rover was surrounded by a group of heavily armed MPLA soldiers. I gingerly stepped out of the vehicle to be confronted by a small, plump, sweating man, who introduced himself as "the commissar". He wore civilian clothes, a Lenin-style cap and the ubiquitous reflective sunshades, even in the dark. In Africa there is nothing more intimidating than an armed man with hostile intent and reflective Ray Bans. This was a situation for which there is no rehearsal and where the wrong word or a sudden move can have unhappy consequences. I blurted out. "Senhor commissar, would you and your comrades like to step into the dining room for a beer and a bite to eat?"

Sensing my fear, Guy Lathbury, our English mining engineer, had an opportunity to whisper to me. "Manfred, look tall and show no fear as they are as afraid of us as we are of them." I thought with a wry smile: "With my height of 6 foot 4 inches, at least the first task was achievable!"

For some much-needed comic relief, I introduced myself as Senhor Marx, the nephew of Karl. That well-worn joke raised a smile from the little commissar and eased the tension somewhat, before he got to the serious point of his uninvited visit.

"Comrade Marx, some of your workers have informed us that you have bombed the garimpeiros (artisanal diggers) from the air using helicopters. Furthermore, that you have a mortar hidden in the room where you sort your diamonds"

Guy growled at me. "He is referring to the X Ray tube in the SORTEX machine."

Before I had an opportunity to reply, Fernando, one of the more excitable Portuguese field assistants interjected. "That is a lie, the explosives we have on site are used for mining purposes only and the so-called mortar is part of our diamond recovery machine!"

The commissar whipped out his pistol and held it to Fernando's head and barked. "One more word from you and it will be your last!"

The commissar then abruptly got up and said. "My comrades and I have had a long day so we will leave now for the nearest village but be assured that we will return tomorrow to continue our investigation."

Evacuation

Never before or since have I experienced such a mixture of both elation and fear. The die had been cast, we were leaving Mussunuige on the following day. How would the MPLA soldiers react if they returned before our departure? Our rapid exit could well be interpreted as an admission of guilt. If we stayed we could be held hostage or worse. That night I slept fitfully, willing the dawn to break so that the evacuation could commence. Lindsay had driven over from his camp with a bottle of Portuguese S'Bell whisky and was now snoring peacefully in the room next door seemingly unfazed by the adventures that awaited us at daybreak. "Tough fellows, these Scots", I thought enviously, as I tossed and turned in my bed battling both the uncertainty and fear.

We did not have to waste any time on breakfast as the troops had devoured the provisions, as well as the last bottles of Cuca beer and a good bottle of Dao Gran Vasco wine. The orderly evacuation plan that I had envisaged never eventuated, as some of the staff had fled during the night and the rest were frantically packing their belongings onto the vehicles. These were now in short supply as the commissar had "borrowed" two Land Rovers on the night before.

Any doubt that may have existed about the wisdom of our hasty departure was erased when later that morning Antonio, the field assistant from the drillers' camp, staggered into the base camp. He was in a bad way, only barely conscious and bleeding heavily from a head wound. His misfortune was to meet the militia on the road after their departure from Mussunuige the night before. His crime was being in possession of a UNITA membership card in an MPLA-controlled area. This type of rough justice became common practice by all of the liberation movements. Having no card was no defense against this thuggery.

I had managed to contact Roy to advise him of developments. He promptly flew up on a chartered plane and landed at noon on the emergency airstrip we had prepared some time before, never believing that we would one day actually be needing it. The decision was to fly Guy and some of the Portuguese wives back to Nova Lisboa, while the rest of us would use the old army bush track down south into UNITA-held territory.

Roy and I had loaded the last maps and exploration records onto the Land Rover as the dust of the rapidly departing convoy settled over the now deserted Mussunuige camp. The abandoned local black workers watched in silence with inscrutable faces as we stepped into the vehicle, waved goodbye and sped down the track knowing that that this was to be a one-way trip south. Were they happy to see the last of us? Was this the freedom that had eluded them for 400 years? There was no time for recriminations or analysis. Our sole aim was to put as much distance between us and the returning MPLA troops.

The first obstacle to overcome was the crossing of the main road from Luanda to Henrique de Carvalho (now called Saurimo) near the town of Cacolo. The UNITA troops had recently been driven out of that town and the only road west was now blocked by the MPLA soldiers. Fortunately we knew of a 200km bush track the Portuguese army had cut through to the trans-Angola railway that linked the coastal port of Lobito to the Congo and Zambian copper mines. From there we needed to travel a further 400km along established roads to our headquarters in Nova Lisboa (now called Huambo). Fortunately, I had spent some time earlier that week on a reconnaissance trip down that rough army bush track.

It was a miracle that the militia in Cacolo failed to spot the dust cloud raised by the fleeing convoy. Maybe they did not care. This was a time when confusion ruled supreme and the lines between friend and foe were often not clearly defined. That is the nature of a civil war.

The trip south was mercifully uneventful except for the heavy trucks that periodically bogged down in the sand. The adrenalin-charged men made quick work of freeing the vehicles and sent them on their way with whoops of joy and sheer exuberance. By midnight we felt that the danger had passed. Happy but exhausted, I dozed off on a blanket beside the Land Rover, periodically awakened by the stragglers, whose engines had been overheating, while they passed on their way to safety. By daybreak we had accounted for all of the vehicles, which even included a grader and a front-end loader. Later that day we were met at the agreed rendezvous town by a relieved Guy Lathbury, who welcomed us with an ample picnic meal washed down with ice-cold Cucas.

We were now safely back in friendly territory. The question was, what now?

Aftermath

The period between our escape from Mussunuige and our final exit from Angola two months later was one of confusion, uncertainty and often a denial to accept the horrible reality of the rapidly deteriorating situation. As the country slid into the civil war that was to last 27 years the public mood swung from optimism to desperation and back. There are too many stories to tell in this chapter but of interest are two incidents that illustrate the "war mentality" that we had adopted. It is strange how the mind adapts to perceive its surroundings in a positive way by screening out the unpleasant images.

I was playing a game of table tennis in the back yard of the Ozellos, Italian friends of mine, when a bullet struck the wall behind me. I casually picked up the bullet and cried out: "Ouch, it's still hot!" I then threw it away … and carried on playing. All this happened while the skirmishes between the rival soldiers were raging in the black favellas down the road.

Late one night, while a group of us were sleeping in Roy's house waiting for the evacuation plane from Namibia, a bomb exploded in the neighbouring UNITA delegation's quarters. At breakfast the next morning we discussed "the situation" until someone remarked: "By the way, did any of you hear that explosion next door last night?". In our "war-weary" minds it was no more newsworthy than a thunderstorm.

Finally, in August I and the two Italian geologists, Max Rotolo and Paulo Battino, flew out of Nova Lisboa in the twin-engined Cessna plane that had previously taken me to all the corners of this country. Our friend and pilot, Captain "Baldy" Carceiju, was in tears when we said our final farewells on the tarmac of the South African army base at Rundu on the Namibian border. Roy Edwards remained in Angola for a few more days before even his irrepressible optimism was shattered when he realized that the game was up and our dream of discovering an Angolan diamond mine was never to be realized.

ONLY EMPLOYABLE IN WARTIME

BRIAN LEVET RHODESIA (ZIMBABWE) 1976 - 1979

Brian was born in Kenya in 1953, educated at Allhallows School in Lyme Regis in the UK and completed a degree in geology through the University of London in 1974. His interest in geology began by collecting specimens while on safari throughout Kenya in the 1950s and 60s and was cemented in England, where the school he attended not only offered geology as a high school subject, but was located on a remarkable coastal sequence of Jurassic rocks, made famous by the famous fossil discoveries of Mary Anning in the early 1800s.

Throughout his 45 year career, Brian has maintained this passion for geology, exploration, discovery and wildlife. His exploration philosophy is summed up by a quote from the Hungarian Nobel Prize winner, Albert Szent-Gyorgoi, (the discoverer of Vitamin C), who famously said that "Discovery is seeing what everyone else has seen but thinking what no one else has thought."

He emigrated to Australia in 1983 during a mining and exploration down turn, he was unemployed and seemingly unemployable. A friend at the University of Western Australia commented that he was perhaps only employable in war time. However, a three week contract with Newmont later that year turned into a 27 year career with the company, working all over the world with direct involvement in a number of team-driven discoveries.

Two years out of university with a degree in geology, and absolutely no practical geological experience was pushing the envelope when it came to finding a job as an exploration geologist. Proficiency at cleaning British Rail trains and being an expert at packing KitKat at Rowntree's chocolate factory in York created little impact during career interviews.

The tedium of working twelve hour shifts in a factory in England was a means to an end, it was geared to saving enough money to buy, renovate and drive an 11 year old Land Rover on a year's journey across Africa with a group of friends. The physical and bureaucratic hurdles of travelling through developing Africa, crossing the Sahara, being detained in Nigeria during a major coup and keeping the Land Rover mechanically sound over extremely rough roads for a year were great challenges. In retrospect, this proved to be a remarkably useful foundation for a geologist who was passionate about exploration.

Returning to Kenya to look for work as a geologist also proved futile, as it was a country with little or no mining history. As a Kenyan citizen, the easiest option would have been to join the safari business, an occupation for which I had both passion and experience. Wisdom however dictated otherwise and I continued to drive south to look for employment on the goldfields of South Africa. Arriving in Rhodesia, as it was called in 1976, proved fortuitous. There were opportunities in the mining industry mainly created by the number of technical people leaving the country due the to an escalating guerrilla war.

Rhodesia was a truly remarkable country and, as fortune favoured the bold, I decided to stay and look for work. My first interview was with Rio Tinto; their offer was compelling and I needed work, security and money, so I accepted. The company chose to send me to an emerald mine called Sandawana. I was flown by a light aircraft to the isolated mine located in the Belingwe (now Mberengwa) district in the central south part of the country. After a very short handover period, I found myself a Resident Geologist for one of the largest resource companies in the world working on one of the more important emerald discoveries in Africa, an unusual commodity that was economically strategic for both company and country.

The emeralds had been discovered in the Mweza Range in 1956, when two South African prospectors, looking for beryl and lithium occurrences, found emeralds associated with unusual suite of rocks known as pegmatites.

A bit of introductory geology first: the Mweza Range was made up of predominantly volcanic rocks and sediments forming what is known as an Archean Greenstone Belt, which is approximately 2.7 billion years old. To the south of this range, extending down towards South Africa, is a belt of rocks known as the Limpopo Mobile Belt. This belt is a "cooked up" series of gneissic rocks that separate the stable blocks of the Rhodesian Craton to the north from the Kaapvaal Craton to the south. The belt has been a zone of crustal weakness and shearing over a long geological period.

Later in the geological history of the area, associated with remobilisation of the Limpopo Mobile Belt, the pegmatites had been injected into the anticlinal crests of a series of folded mafic and ultramafic lavas that now form the southern edge of the Mweza range. The emeralds form on the upper (hanging wall) contact of the pegmatite bodies in a rock known as tremolite schist. This tremolite schist is very distinctive and is apple green in colour, whereas the base of the pegmatite or the footwall is micaceous and rarely contained quality emeralds.

Emeralds are bright green beryl crystals and throughout history have been one of the most prized of precious gems. The distinctive green colour has been derived by incorporating traces of chromium and vanadium into the molecular structure of beryl during crystallization. The Sandawana emeralds are unique because when cut even the smallest emeralds maintain a vivid green colour, which makes them highly sought after.

The first poor quality emeralds were discovered in the Mweza Hills at a prospect named Vulcan. After a rainstorm in 1957, a local worker found a deep green crystal protruding from a muddy footpath. This area would later be known as the Zeus mine. Initially emeralds were mined by washing the emeralds out of the black soil, later they were mined by a small open pit and finally by underground mining.

The pegmatite body at Sandawana was about a kilometre in length and plunges underground from surface to a depth of about 200 metres. Thirty kilometres of drives, raises and stopes had been mined by the time I had arrived and the mine resembled a three dimensional maze. The Resident Geologist's job was to guide the mining activities to find the emerald occurrences underground by mapping and drilling from these drives. The drives were located in the centre of the white pegmatite with a drill array located every 5 metres along the drive, strategically placed to explore the hanging wall contact of the pegmatite. The initial drilling was completed using a percussion long hole drill to determine exactly where these contact zones were, and then with a diamond drill, to take core to determine whether it was a geologically favourable location. To complement the underground exploration, two surface diamond rigs were used to determine the location of the pegmatite from the surface in an effort to plan future underground mining. The drill rigs and geological team were entirely the responsibility of the Resident Geologist. He was expected to order drill rods, drill bits and spare parts in order to keep the drill rigs operational. There were no contractors, so the geologist had to very quickly learn everything there was to know about drilling.

Using regional geological mapping, geophysics and the two surface diamond rigs, the team was also tasked with exploration along the edge of the densely tree-covered Mweza Range over a strike distance of approximately 12 kilometres, where a number of other small emerald occurrences were known.

In considering the old adage "If only I knew then what I know now", I am sure I could have made a great contribution to exploration at Sanadwana. It was, however, an experience that proved extraordinarily valuable that I still remember some forty years on. The isolation of the mine and the escalating bush war made it virtually impossible for geological mentorship. The other mentorship all came from the miners and surveyors at the mine, and the rest was done by intuition. The senior staff village was located on the top of the range with extraordinary views to the north over the Belingwe Tribal Trust Land and, to the south, over the huge 1.2 million acres Liebig's cattle ranch, touted to be largest single fenced cattle ranch in the world. The ranch was divided into a number of manageable sections each with its own section manager. It was a wildlife paradise and an extraordinary place to explore by jeep and motorbikes at weekends.

There were about twelve mine residences located on the top of the hill, which also included a club and single quarters. A large village, housing the workforce, was located to the south close to the mine area, with many of the workers being expatriates from Malawi.

On being issued my geological equipment, which included the mandatory geological pick and a hand lens, I was also presented with a single barrelled shotgun. Although guerrilla activity in the district was sporadic at this time, the shotgun was to provide some form of protection in case of an ambush while prospecting in the district. The only occasion I actually had to use the shotgun was to shoot a very large black mamba, that moved like lightning across my veranda during breakfast and into a nearby tree. The snake was about as thick as one's wrist and at least three meters in length; even then I managed to shoot the wrong end of the snake, which required a second shot at what was, by then, a very irate mamba. This obviously caused some consternation in the village, with the families in the village thinking they were under attack.

The first major guerrilla attack did occur soon after my arrival. Periodically, a visiting vicar would drive the 70 kilometres from West Nicholson to the mine to conduct a Sunday service. On this occasion, he was ambushed quite close to the mine. His car was riddled with bullets by a group of guerrillas with AK47s but, other than shooting off the oil filter, he was very lucky and still able to drive away from the ambush to safety, before proceeding on foot to the mine.

My shotgun was soon supplemented with a pistol, which was certainly more useful than the single break shotgun. The pistol was to become part of my geological equipment for the next three years. The only time I had reason to use it was to shoot a rabid dog. As I approached the security fence one day, which by now had been erected around the village, a very thin dog, foaming at the mouth, chased me. I climbed rapidly up on to the Land Rover bonnet only to see the dog enter the fenced area via a drain to where all the families lived. There was no choice and I finally got to use the pistol. Having very carefully buried the dead dog, someone suggested I that I should dig it up and send it to the vet for testing. Sure enough, three days later, it was identified as rabid.

The war in the district escalated rapidly from that point on. My fairly useless shot gun was replaced by a semi-automatic rifle and I carried at least four spare ammunition magazines wherever I went.

I had bought a 1958 short wheel-based Land Rover with my first pay packets and I would often spend weekends staying with the section managers on the Liebig's ranch. Looking back on it, it seemed crazy to drive the 25 to 100 kilometres to their homestead, usually on my own, in such dangerous circumstances. My Land Rover resembled a jeep used by the desert rats in North Africa during the Second World War, with a rifle mounted on the side of the door. Such was the stupidity of youth!

My first contact with the guerrillas came one night, when they attacked the mine. We were woken by shooting and the sound of explosions. My friend went to one side of the single quarters and I went to the other. In fact, the group were not actually attacking the village but the mine facilities using rocket propelled grenades and filling the sky with tracer bullets as they did so. Not wanting to miss out on the action, my friend managed to decimate the manager's wife's cactus garden and, in the darkness of my veranda, I managed to fall over my own spent cartridge cases as they rolled over the concrete floor. The mine was left remarkably intact, the village store had been ransacked and burned down and the roof of the generator and compressor house had been blown off. The rockets had clearly missed both the compressor and mine generator, I think largely because the guerrillas feared that they might explode in their faces.

Tragedy and reality struck really struck home when my friend, the ranch manager of Section 7 at Liebig's, was ambushed and killed at a gate near his homestead. Over the next two years, incidents increased from monthly to weekly. These included landmines and ambushes, aimed largely against the neighbouring ranchers as they were a very much easier target than we were.

The modus operandi at the mine changed completely, as we were now clearly living in the war zone. All mine personnel were exempted from military call up; we now had the permanent duty of protecting the mine and its workers because the operation generated critical foreign currency for the country. Initially the police and the army had reacted quickly to any incident but, gradually, protection of the infrastructure was left to the mine personnel, and the reaction time by the police became days not hours.

A command bunker was built on the top of the hill, which monitored the mine and all the security fences. A belt-fed Browning machine gun was set up near the bunker and the men took it in shifts to monitor the operation at night, and the women did so by day. A 30 mm mortar pit was set up at the single quarters. All the mine supplies and personnel were escorted to and from the local town by a duty team in an armoured vehicle. The vehicles were clad with high tensile steel plates with bullet proof glass windscreens. A heavy screen was attached to the steel on the sides of the vehicles to tumble armour piercing bullets. A strange assortment of armoured personnel carriers, which were land mine protected and bullet-proofed, were used to ferry personnel to and from the mine.

Our vehicles were fitted with what was known as a spider, one on each side of the vehicle. They were tubes with loaded shotgun cartridges, that could be released from inside the cab, falling one at a time, spraying out shotgun pellets. The tubes were covered using a condom to keep out the rain and the dirt. Some of the ranch vehicles were fitted with AK 47 rifles mounted in the engine compartment and on either side in the rear of the vehicle, and which could be triggered by a switch and a solenoid during an ambush. Neither of these were particularly effective but they acted as some kind of a deterrent to the guerrillas when they were mounting an ambush close to the road.

Initially the mine was left alone: I believe this was because we were not a soft target; the relations with the local people were very good, and the guerrillas levied their own tax on the mine in the form of emeralds which were stolen to supplement their cause.

Exploration continued throughout this period and you might well ask why. I must have been the last person conducting greenfield exploration in Rhodesia in a war zone in 1979. I think it was a combination of the boiling frog syndrome and the exuberance of youth for exploration and discovery. The tenacity and bravery of my fellow peers and their families, and the great respect I had for them, also had a lot to do with it.

We completed a geophysical gravity survey over the volcanic rocks of the Mweza range. The idea was to see if we could locate the lower density target pegmatites amongst a sea of denser volcanic rocks. To do this, I used the armour-plated Land Rover with six guards armed with semi-automatic rifles, as we carried the gravimeter up and down the hills of the Mweza range. We varied the time that we went out on the survey and always varied the route, making it much more difficult for the guerrillas to plan ambushes.

The main mine access road was always our Achilles heel and it was on this road that I was ambushed while standing in the back of one of our armour-plated Land Rovers. There were five bullets holes in the vehicle, but all bullets missed me and I lived to tell this tale.

Soon after the ambush, I was transferred to the Rio Tinto Headquarters in Gatooma (Kodoma). I took a year off to backpack through South America in 1980, and returned to Zimbabwe to work at an iron ore mine close to Sandawana. It was extraordinary that I then worked with the guerrilla fighters, instead of against them. It was like someone had turned the tap and the animosity off, and it taught me the absolute futility of war. I did hold great hope for the new Zimbabwe.

ANGOLAN ESCAPE

ZLAD SAS
ANGOLA 1994 - 1995

Zlad Sas graduated with an Honours degree in geology in 1974 from the University of Western Australia. He has over 40 years experience in diamond exploration having worked extensively on projects in Australia, Africa, India, and Brazil.

For 13 years, he was the Managing Director of an ASX listed diamond explorer, Australian Kimberley Diamonds NL. During the majority of this period, AKD was involved in various diamond projects, both in Australia and overseas, some of which were in joint venture with Rio Tinto and De Beers.

Zlad is currently Managing Director of unlisted diamond explorer, GeoCrystal Limited, which has discovered a large kimberlite field in the Gibson Desert of Western Australia.

It was 1994, a year from when I floated my latest venture, Australian Kimberley Diamonds ("AKD") on the Australian Stock Exchange. I was 42, a geologist with a young family, firmly established in a quiet suburb of Perth, Western Australia.

After the listing on the ASX, I felt euphoric and had visions of building AKD into a substantial diamond company. Already the company had established a portfolio of advanced diamond projects in Australia, some of which were in joint venture with CRA (now Rio Tinto), but I was looking for more and my ambition led me to Angola.

At the time, Angola was in the grip of a civil war between the communist "MPLA" army, headquartered in the capital Luanda and backed by the Soviets and Cuba; and the opposing Unita rebels, occupying most of the country and backed by the South Africans and the United States. The war first began immediately after Angola became independent from Portugal in 1975 and was often referred to as a surrogate Cold War between the super powers. There was much at stake and they were each trying to establish a political/military presence on the west coast of Africa through the backing of their militia. The prize was the rich resources of Angola, and diamonds and oil headed the list.

Tragically, by the time the MPLA had declared victory in 2002, more than 500,000 people had been killed in the war in Angola. Little did I know back in 1994 that I could easily have been one of those deadly statistics.

Looking back now I guess I was a risk taker, and at the time I calculated that the risks were minimal if I stayed in the capital Luanda and had my pre-arranged meetings with the State diamond company, Endiama, and with others as organised by my Australian/Portuguese minder.

I arrived in Luanda on a hot, muggy day in April 1994 on a TAP flight from Lisbon, Portugal, and the bouncy plane touched down on a runway flanked by intimidating anti-aircraft guns. At other times, I arrived in Angola on flights from Harare, Zimbabwe, on an Air Angolan F28 plane that seemed to just hold together while navigating treacherous weather and scaring the African passengers who were desperately holding onto their oversized cabin luggage.

Luanda airport was managed by the military and all passengers were instructed to disembark and line up on the runway with their passports. After an army officer collected the passports, we were led into the terminal for immigration clearance by the military who had AK-47's slung over their shoulders, and some of whom were intoxicated. They seemed uninterested in the protocols and returned my stamped passport without much fuss. It dawned on me straight away that entry into Angola was relatively easy, if you were not an enemy of the State, but getting out might be tricky.

My minder, who had arrived earlier in Angola, met me outside the airport terminal with his driver and took us to a small Portuguese-owned hotel in Luanda. After booking in, we immediately phoned and confirmed our appointment for the next day with the head of Endiama. Our objective was to obtain diamond concessions in the Lunda Norte province of north-eastern Angola, especially within the highly prized alluvial diamond areas of the Cuango Basin.

The following day we had our meeting with him and the discussions were positive but required a follow-up meeting in a day or two.

The next evening, after spending most of the day poring over maps supplied by Endiama, my minder convinced me to have a meal at a downtown restaurant in Luanda renowned for its Portuguese food. I had heard reports that some areas in Luanda were being terrorized by armed gangs at night, some reportedly made up by rogue MPLA army members and even suggestions that they were ex-police. But I thought we were safe to go to the restaurant as long as we stayed in the car and took the main roads to our destination.

The food at the restaurant was certainly good and after a few glasses of Portuguese red wine, we decided to head back to my hotel around 10.00pm.

Our driver was called and met us outside the restaurant. As we were about to get into the car, two African militiamen appeared out of the shadows across the street and approached us shouting in Portuguese and waving their AK-47's. My minder quickly intercepted them and pushed me back towards our car and interpreted that I needed to hand over my backpack and empty my pockets - no hesitations, but just quick. In my backpack was my passport, airline tickets, my Nikon camera, wallet and US$10,000 cash! Everything I needed to get out of this country during a crisis.

I hesitated for a moment, but having a muzzle of an AK-47 pointed close to my head was enough to make me react and hand over everything that I had. As the two armed militiamen started to flee down the street and into the darkness, our driver reached into the glove box and pulled out a revolver, aimed it at the bandits and squeezed the trigger, but nothing happened - a misfire maybe. Just as well, as had he started shooting, we would have been no match for the AK-47s and shot dead by the bandits in revenge.

The shock set in but I gathered my senses and we decided to get out of downtown Luanda and head back to the safety of my hotel. Along the way, my minder and driver suggested that we go to the police and make a report - was it going to be manned by the same armed bandits that robbed us? Paranoia was starting to bite but I thought it was worth the risk and we entered the poorly lit dark tomb of the central Luanda police station at 11.00pm and met the officer in charge.

He led us past a maze of rooms and into the grisly depths of the police station passing cells in which detainees were being interrogated and from which we could hear terrifying sounds.

The officer found a room with a desk and one overhead light and we sat down flanked by other police and answered his questions about the robbery. "Sign this", he demanded in Portuguese as he thrust my statement into my trembling hands - it was intimidating stuff and I felt alone and vulnerable. I just wanted to get out of this place and head back to the hotel to regroup. After a while, we left the police station shaken but in one piece.

Back in the hotel, I had a few drinks and went to bed trying to figure out an escape plan. The next morning I woke with some clarity and had plan A - to head to the British Embassy in Luanda as there was no Australian Embassy or Consulate in the city. Plan B was to wait for the Australian authorities to get me out, which would probably take weeks.

I made my way to that British Embassy and stood outside the iron gates and kept calling out to the guards who couldn't understand what the hell I was saying.

Eventually the assistant to the Consulate General came out to see what all the commotion was about and asked who I was. I gave him my story and he ushered me into the Embassy but shook his head saying that there was essentially nothing he could do. However, as a last resort, he would discuss it with his superior and see if there was anything in the official protocol rules that could help. After a couple of hours waiting in the interview room, both he and the Consulate General walked in carrying a large embossed folio that contained a diplomatic book of rules and regulations. To my surprise, they said they had a solution.

Rather than wait for weeks in Luanda whilst the Australian Embassy in Zimbabwe organised a new passport for me, there was another way. The UK Government could issue me with a British passport stamping my citizenship in the passport document as "Commonwealth Citizen" under a rarely used diplomatic regulation. This could be done in consultation with the Australian Government, which could be arranged by phone to the Australian Embassy in Zimbabwe.

I was flabbergasted - this was my saviour, travelling back to Australia via Europe on a British passport! It did not take me long to say yes. "Where do I sign?" I said with enthusiasm. The assistant to the Consulate General gave me a cup of tea and a scone and indicated that the passport forms would be ready for me to sign the next day. However, for the new passport to be issued, I would need passport photos and US$75 cash to cover the application fee. I said that I did not have any money and, although he was sympathetic to my plight, he could not bend the rules.

I left the Embassy despondent and headed into the city. Maybe there were someone in town who could loan me some money, but who? I walked the streets of central Luanda and eventually passed an impressive bank that had many customers inside, one of who was talking loudly in English in a strong accent.

I waited outside the bank until he came out and boldly tapped him on the shoulder and introduced myself. He was an Indian businessman who was importing cars into Angola. After listening intently to my plight, he could see I was desperate and he suggested that I meet him the next day at his office, which I did.

I arrived on time for my appointment and, after hearing my story again, he reached into the drawer of his desk and pulled out a wad of notes and gave me US$500 cash. I insisted that I write an IOU for US$500 on a scrap of paper and then I thanked him several times and left with the promise that one day I would return and pay him back with interest. He just laughed and shook his head with disbelief but I guess he surmised that I was genuine in my intentions.

The next day I had my passport photos completed and went immediately to the British Embassy to pay the application fee and collect my British passport. Incredible - the name "Zlatomir Aurel Sas" from Czech descendants on a British passport and my nationality stamped as "Commonwealth Citizen", valid for 6 months! I had gone from Australian citizen to a Pom in a matter of days.

The next major hurdle was to have an official of the Angolan Immigration Department backdate my entry stamp into Luanda so that I could show valid entry into Angola and allow me to exit the country without drama. This was left to my Australian/Portuguese minder who armed himself with $US100 cash and his own devices. At the end of the day, I had a stamped entry on my British passport, no questions asked.

I collected my new airline tickets the next day which were re-issued by TAP and British Airways and organized by my office in Perth. By this stage I was ready to go - it had taken just three days to get everything organized and my flight left Luanda the next day for Lisbon, Portugal.

I checked out of the hotel that morning and headed to the Luanda airport to catch my flight. On entering the terminal, I had a terrible feeling that getting through immigration and customs was going to take a lot of nerve and composure.

The whole building was controlled by the military and my first challenge was the intoxicated immigration officer with his AK-47 in one corner of his booth. I was told by people in Luanda that to exit smoothly at Luanda airport, some cash was needed to "oil the wheels", so I slipped a US$100 note into my passport and held my breath. The glazed eyes of the immigration officer looked into mine and hesitated, but then he stamped my passport and handed it back to me with a satisfied grin. My heart was racing and I grabbed my passport and headed to the customs section, US$100 poorer. The same fate awaited me at customs so by the time I got to the "other side", I was again US$100 poorer and it left me with only US$100 cash to get me home.

After a final examination of my documents and boarding pass by the Angolan military, I was allowed to board the TAP plane destined for Lisbon, the first stage of my journey home back to Australia. I sat in my window seat staring at the chaotic scene outside and wondered whether I would ever return to this hostile country. As the plane taxied onto the runway, I closed my eyes and relaxed, and even before the plane took off, I was asleep, the first sound one in many days since my unfortunate incident.

I landed in Lisbon waving my new UK passport at immigration officials and I was ushered through the terminal with ease. Next stop was London with British Airways and it was so easy transiting Heathrow with my passport that I thought about staying for a night to see the sights, but the money was tight so I departed for the long haul back to home in Perth.

Arriving there a day later, I handed my UK passport to the immigration officer at the terminal and proudly stated I was an Australia citizen who had just been granted "Commonwealth Citizenship" for six months by the UK Government through their Embassy in Angola. Well, it did not go down well with the Australian immigration officials who had never seen such a document before and thought it was a fake. I was ushered away to an interview room and interrogated for a few hours on how I had obtained this document. Several hours later, after confirmation from the UK Embassy in Angola and the Australian Embassy in Zimbabwe, it was verified and I was allowed to enter Australia. My UK passport was stamped by immigration officers and I left the airport terminal, relieved and exhausted.

FOOTNOTE:

I was told later by some trusted Angolans that the bandits who robbed me were caught by the authorities while still in possession of my stolen items - tragically they were taken into the bush and executed, I was told. However, the cash, Australian passport and camera were never returned to me but in the end that did not matter - I never wanted them back after what had happened.

I returned to Luanda the following year to meet again with Endiama and I was also able to repay the car importer the money he had loaned me. He greeted me with a big smile because he didn't think I would ever return. This time around in Angola I was very cautious and, with the war still on foot, I started to look elsewhere. Sadly, the legacy of the civil war was a country in ruins and infested by land mines that saturate and litter the countryside, and which still remain a hazard today.

I thought I had learnt my lesson in Angola but I went to neighboring Zaire soon after and met up with a general who had diamond concessions on the border with Angola. The Zairians were allies of the South Africans and the CIA at the time and were also fighting the Angolans. I met the general's armed escort early in the morning at my hotel in the capital Kinshasa and they ushered me into their convoy of black limos and machine gun-toting trucks and took me on a three hour road trip following the Congo River. Along the way I silently said my prayers and chastised myself for allowing myself to again get into such a high-risk situation. It had a happy ending though and I ended up meeting the general at his fish farm on the banks of the Congo River and drinking beer and having a BBQ with his farm workers and troops.

I left later that day vowing never to get involved in high-risk situations like that again. In hindsight, it was also morally wrong to conduct business in African countries like Zaire at the time as it was controlled by a corrupt tyrant, President Mobutu, who embezzled US$5 billion from his country and lived an extravagant and opulent lifestyle whilst his people starved and lived in poverty.

Today, I still explore for diamonds but only in Australia.

SNAPSHOTS OF EXPLORING
FOR ZIM DIAMONDS

MARTIN SPENCE
ZIMBABWE 1990 - 2000

I arrived in Australia in 1972 with the vague idea of training in TV production as by then the Americans had put a man on the moon but South Africa was yet to get TV. I studied Film & Television in Sydney but never worked at it nor went back to South Africa to work or live.

Venturing around Australia in the latter 1970's I discovered Perth and have made this my home ever since. A degree in geology from the University of WA gave me a wonderfully exciting life of hard work and constant learning. Working in gold, nickel, other base metals, diamonds, uranium, platinoids, rare earths, lithium and the many various types of occurrences thereof.

I have worked throughout Australia, most countries in Southern Africa, parts of Eastern Europe and am currently working back in Australia.

This short story is about some of the adventures my wife and I had in Zimbabwe while exploring for diamonds in the 1990's. It was a time of exhilaration, blood, sweat and tears spent exploring for minerals in a third world country.

It has been and continues to be a wonderful life.

Never sure why I was approached to run Delta Gold's diamond exploration in Zimbabwe, probably was a case of "grab that big, fat Yaapie, he'll know what to do in Africa and so shorten the learning curve" Hmmmm....

It was a wonderful opportunity to return to Africa after leaving there in 1972 plus we were the first diamond explorers in Zimbabwe since independence in 1984, so the available land to explore was unlimited.

I arrived at Harare International in early 1990 loaded with spare tubes and tyres for my 1960's Peugeot trayback exploration vehicle.

The first exploration area was in the Midlands where our camp had been set up in the shadow of a small portion of the Great Dyke with convenient rounded boulders that we used as seats.

So off I went, all curious to explore where I hadn't been for many years. Naturally all the forgotten bugs and diseases latched on as soon as I arrived without any form of care and prevention. Great big sores on my legs portended the start of tick bite fever, eyes closed and refused to unglue in the morning, headaches, the shakes and an arthritic-type back pain kept me awake sitting in a chair from midnight to dawn.

My men said they noticed I had discomfort at night and they had booked a treatment at the local N'anga or witchdoctor at three that afternoon…what do you say? Their concern was genuine and I grew to wonder at their kindness and forgiveness of us whitey's stupidity and lack of manners. So promptly at 1500 I was marched to the witchdoctor's hut, stood outside and gave the traditional good afternoon greetings and requests to enter his property. He seemed fairly normal, and was stooped with grey hair so one presumed he had been around the block in witchdoctoring. He gave me a jam tin with some liquid in it to drink. My guys saw my hesitation and proved it was safe by taking a sip each: it seemed to consist of aloe juice which I had once painted on my fingernails to stop chewing them because it was so foul. Anyway the next portion of the treatment was being whipped across my lower back with shredded sisal; not sore, uncomfortable, but ok.

A few days later a chief turned up to tell me the round boulders we were camped on were ancestors' graves, "Oh no no no no NO!!! Now we are in trouble", I thought but, no, he just wanted to perform a simple ceremony to ask the spirits' help in our endeavours. Could I get some snuff and some alcohol? I felt so bad I would have bought Cuban cigars and Grange but the local outlet didn't run to much so, armed with some snuff and a plastic bottle of gin, his N'anga performed a simple ceremony; bit of chanting and hand clapping. Poured some snuff and hooch on the ground, that was it, never to be seen again.

This ceremony was to occur again in various areas we worked, even in Mocambique. Though the Mocambiquans make a huge fiesta of this and the amount of chooks and goats slaughtered can put a serious dent in the budget.

Our arrival at a site would excite much curiosity with people looking for work immediately surrounding us. We only ever got rid of one guy as he was wanted for murder. The remainder were a delight to work with and any problems were generally due to a language misunderstanding. Shona bears no relation to any other language and as such it was a real bugger to learn but perservere I did until I got to the stage one of my geologists said I was speaking Shonglish.... compliment? Maybe not.

I employed a Scottish guy for a while who was a keen hunter, fisherman type and wanted to get his kids away from the influences of modern Glasgow. He asked me if he could employ his own guys and explain the job to them. Of course, says I, so off he goes; "orrite min oi wan ewe an ewe an oghh wun mooore. Fello mi ta mi trok an...." Somehow they all got the job done but my men were quite perplexed how we spoke this language and struggled with Shona.

Our main exploration tool was river sampling. Drive to a point as near as we could get using the excellent government maps and air photos in that pre-GPS time, then tramp along the river to where rocks would cause the flood waters to tumble and slow, depositing the specific heavier mineral grains that would tell us there might be a diamond pipe up-stream. We would take about 30kg back to camp, screen it to various size fractions and concentrate these using a jig. Then reduce the size further by using a heavy media liquid to separate the lighter gains from the heavier ones, and the latter would be examined under microscope to see if they were the specific indicator minerals we were looking for. Having no power everything was done by hand, even the sample drying on an open fire in tin soup bowls, which was generally undertaken by the oldest madala we employed.

Our whole process was water dependent. If we couldn't get access to a river, there was always an ox cart for hire and these stoic beasts would plod back and forth with barrels of water from a nearby dam.

That's the technical bits nearly over.

On one of these walks up a river, my group of six were walking ahead of me around a bend when I heard almighty screams and panic: running to the bend I saw some of the guys hanging from trees while others were lying on the ground all laughing hysterically. They had walked onto a female leopard and two cubs, dropped everything and headed for the hills. One guy was so panicked he climbed a tree that fast a branch penetrated his head and he fell out the tree in the path of Mrs Leopard who thankfully had her mouth full of kittens. More yelping and scrambling, with blood pouring down his face, had the others in hysterics. From then on they would just look at him and start howling with laughter.

So on we went exploring vast swags of Zims: we had developed a great team of permanent men who ran the jigs, wet lab, exploration and the camp. The initial results were not positive, maybe sampling in the wrong places or not controlling our own microscope end of the process. To counter the latter we got Lynda Frewer and the dear departed Bubbles Garlick to come over from Diatech, Perth, to train fourteen local church lasses in the art of microscope indicator mineral identification, which provided us with control over the entire process.

As the overseas explorer numbers in the country increased, the government became paranoid about exporting samples for assay. Convinced these bags of samples contained undeclared gold, the hold-ups started. These sort of occurrences were not unusual throughout Africa as the deep seated suspicion of us included wondering why are these overseas companies suddenly spending big money in their countries on exploration, with no apparent return. It didn't make a lot of economic sense. Maybe they were right.

Our camps had grown from two guys with a meat and beer fridge to luxury tent-style camping with sheets, table cloth, hot showers, ironed clothes washed by a local lady who would beat the bejesus out of them on a rock at the local creek. A varied diet, which included green stuff; and a sit-down toilet with a view, were all due to the arrival of the lady I would shortly marry. Sandy was brought up in the West Australia Wheatbelt near a signpost that says Baandee, located on the Great Eastern Highway west of the town of Merredin. Anyway, this rather lovely lady settled into this life like a true explorer managing the data flow and the camp while attempting to get me house trained.

Our wildlife adventures included being charged by elephants. There are two things about elephant charges that you should know:

1. if its ears are laid back and the trunk is down, you are going to die;
2. if the ears are splayed out and the trunk is up then it is a mock charge and you can basically carry on doing what you were doing until a mountain of screeching greyness surrounded by a sandstorm stops one metre in front of you, then retreats and decides that you probably thought that was a bit whoosie and comes back at you another three times. Some time in the distant future when time re-starts and she decides to trot off to join her gang you realize with dismay that as soon as your heart starts again and you regain the use of your limbs, you are going to have to change your undies.

Snakes generally scrammed on our approach apart from the odd mamba that stood up for its territorial rights, and puff-adders which were just too damn lazy to move. I was driving down a bush track doing about 70kmh with a local policeman as passenger and saw a mamba on the side of the road. I moved over to give it space and, as I passed, it struck the passenger window just where his head was: boy did that clear the sinuses. I since learnt that female mambas in egg are known to get a bit ratty.

One hippo took a bite out of a truck, which was a surprise, and a buffalo calf we pulled out of a mud hole promptly charged us in gratitude. Then there was the tug o' war we had with a croc over a goat: we won and had goat for supper.

During the '92 drought we camped at a large dam that we shared with every type of animal including the Big Five. They roamed through our camp without concern and at night the local lion pride would wander in after lights-out, having been tempted by the smell of our BBQ, and make themselves at home. They once managed to open the freezer, but didn't take anything, which was curious. Then they killed a buffalo next to our shower, which made ablutions interesting. We always felt quite safe and, at this camp, had an armed Parks Ranger to look after our well-being. Though when I checked the condition of his rifle, I discovered he didn't have any ammunition.

One night Sandy was trying to wake me up as an elephant was delicately walking over our tent's guide ropes with its rumble, rumble, tummy sounds but, having had a couple of bottles of sleeping tablets with supper, her "darling, darling" whispers didn't penetrate. What did though was the ellie letting go right outside our tent, which brought a shriek of laughter from Sandy and a satisfied "urk" from Mr Ellie.

To be honest every day was an adventure, recounted around the campfire nightly as Kennedy Mtetwe invariably beat us at Scrabble.

In 1995 Delta decided to withdraw from diamond exploration so Sandy and I formed Cratonic NL with the view to raising our own capital to continue exploration. We were starting to make discoveries and used Rob Ramsay in Perth as our consultant on the indicator mineral geochemistry to assist in identifying the diamond potential of these discoveries. We formed a joint venture with a Canadian company to look at their data: this outlined a highly prospective kimberlite in western Zimbabwe that they had somehow missed. Off we rushed leaving three of our own other recent discoveries untested to this day. This pipe produced a lot of small diamonds from surface sampling. Delta chose to continue funding us after all so we could concentrate on the job at hand without the distraction of corporate fundraising. The future looked pretty rosy until the farm invasions struck. However we had a pipe to test; we hired some underground miners who sunk four shafts in each quarter of the pipe to take a bulk sample. Dragged a five tonne per hour crusher onto site and hired a Dense Media Separation plant from South Africa to concentrate the diamonds from the crushed bulk sample. Our site was 200km southwest of Bulawayo and partly roaded, the rest we had to fix dragging weighted-down old tyres behind our Hiluxes.

Finally, we got the plant set up and the rains came, and we went on a 24hour shift. When the river flooded most nights, we had to drag the equipment to higher ground, get our tent flaps rolled up and personal gear in trees. Thankfully it was summer and quite warm.

It became exhausting all right but we only had one shot at this, so worked until finally our 500tonne stockpile of material was tested or had been washed away.

We had very good relationships with De Beers and Rio Tinto which were genuine in their offers of help, of course with a view to an eventual agreement that suited both parties, if we found an economic body, but there wasn't any pressure. Just fellow explorers getting on together, curious about each others' progress and always willing to share. Up till this stage we were explorers, kimberlite pipe development was a different ballgame so their advice was appreciated.

Our lab girls went into a really competitive mode after identifying a number of diamonds from our Ngulube kimberlite pipe. It was a professional looking lab, white sheets on the tables for any mineral spillage, modern microscopes, good lighting and the girls all in white laboratory coats. Professor Trashliev, our Bulgarian academic who, with his wife, Jana, ran the lab, were fantastic people with a rigid discipline and a very good training. She was a mineralogist who, I subsequently discovered, was big in black market forex dealings.

One day Constance, from the lab, asked if she could talk to Sandy, "One moment please Constance," became half an hour before Sandy finally went down to the lab. The door was jammed so Sandy gave it a shove to see Constance on her back and giving birth while shouting "I am not pregnant!". From somewhere a woman appeared who seemed to know what to do and scissors, towels and hot water were forthcoming. "Push the medical alarm" shouted Sandy, so both medical and security alarms were sounded resulting in armed guards, an ambulance and ZimTV arriving at the same time. Constance had a little boy.

Trouble had been brewing for a while throughout Zimbabwe, food riots due to shortages and cost increases. Fuel shortages led to spending frustrating hours chasing rumours of availability. We were OK in the bush, problems became more apparent in the major centres. Sandy was running our data and administration in Harare, her safety was paramount, but she used her farmgirl common sense and steered clear of trouble. She went back to Perth on her quarterly catch-up with her grown up children and to check on our house renovations that we had being managing via fax. She experienced a vicious home invasion and quickly returned to the relative safety of Harare.

The expat girls had an intelligence network so they knew which shopping centres to avoid as they were going to have a spontaneous riot and subsequent tear gassing and probable shooting.

There were outlying farm butchers who we got to know and could get meat and some veggies, as well as French, South African and West Australian wines at reasonable prices from some of the better Harare suburbs: we paid in forex. All a bit surreptitiously, but everyone seemed to know about them. Our contract smuggler would bring in technical gear or delicacies from South Africa when required, with a surprisingly small margin, so business must have been good.

The apparent siding by white farmers with the opposition party, MDC, and the jeering at Mugabe by his war veterans really lit the fuse and, in 2000, the farm invasions took on a very brutal phase. Military roadblocks sprung up on most major roads, the blockers not too sure what to do with the blockees so that turned into a pigs' breakfast. One of my mates tied some chicken feathers soaked in red paint to a tree where a fairly obnoxious roadblock was established: that chased them away. The currency started to collapse and shortages really became quite pronounced.

The turmoil was sponsored by a rabid bunch of mongrels whose history included killing some 20,000 N'debele after independence in retribution for being independent of thought, being Russian affiliated, from a different tribe and that is what you did in those days to solve these differences. These same perpetrators are running the place today.

Our workers were petrified, public transport broke down and there were large crowds at each hamlet struggling to get home, wherever that was. We drove our men to their homes, along with a truckload of other poor benighted souls whose common fears were the brutal bashings, kidnappings and rapes that were occurring with increasing intensity. We were generally unmolested, maybe given a bit of whitey bastard stuff but most of the cops seemed quite embarrassed by the situation. When you did hit a smart arse, keep your mouth shut until he ran out of oomph and let us through. The one time I was cuffed while they searched my vehicle for weapons I said to the MP corporal that when I got free I was going to invade his farm, he was still waving and laughing as I drove away.

Our foreign capital was appropriated and houses started to go up for sale as people decided to leave. Most of our expat friends left, the Aussie embassy was devising evacuation plans but told us that we were probably more aware of what was going on in the bush than they were. They had bigger problems than us and really we could look after ourselves. We had our own plans if it got too bad. Out via Zambia to the north, the South African border was always frantic but under pressure would be a fiasco, or across Kariba by a mate's boat if it really got that bad, which it never did for us.

We hung on hoping the problems would prove to be short-term and we had a couple of farmers who had lost their property staying in our guest cottage. They had been chased off with not much more than the clothes they wore and their lives: their productive farms, employees, millions of dollars of investment, and years of hard work gone in an instant of howling fury.

Did our 26 acres qualify as a farm? Who knew, it was really at the wishes of the mob that were led at that stage by the delightfully named Hitler Hunzvi.

An exclusive small group was to benefit and the majority of Zimbabweans just thought it was more Mugabe madness but what caused us the most anxiety was the uncertainty; not knowing what was happening or was to come, what the rules were. Were there any rules? When I did neighborhood night patrols, who was friend or foe, were there any friends, what was protective action? Just made it up as you went along really.

Years later in 2005 I drove through former white farming centres; mostly all derelict and burnt out. With no assistance to grow any crops, the new farmers were reduced to selling firewood from once graceful trees or the window and door frames from demolished homesteads.

Our Delta-Cratonic funding ceased by mutual agreement and, recognizing the futility of trying to get investors interested in Zimbabwe, Sandy and I returned to Western Australia in 2000.

I was to return a few years later to get involved in the Marange diamond fiasco. But that's another African story.

Yes, the witchdoctor's treatment did work.

THE LIPSTICK PIRANHA

WOLF MARX
ZIMBABWE 1997 AND 1999

Wolf Marx graduated from the University of Cape Town in 1969 and worked for De Beers in Botswana, Angola and Zimbabwe until 1973, when he emigrated to Australia. He continued working for De Beers in Australia until 1980, when he joined Freeport of Australia. In 1983 he discovered the Bow River alluvial diamond deposits, which were mined by Freeport. In 1987 he left that company and joined a private consultancy group and was CEO of several junior mining companies until he retired in 2008.

Now that I am retired, older and, hopefully, wiser, I often think about the crazy situations I found myself in when I worked as a geologist. When there was well documented corruption at the highest levels of society in a country, was it not obvious that the lower levels of that society would also be corrupt? It is every geologist's dream to be the one who discovers a bonanza mineral deposit and be praised and richly rewarded for his intelligence and perseverance. This lust for fame and fortune clearly blinds us to the stark realities of life. It clearly blinded me as I embarked on two ventures in Zimbabwe when that country was under the rule of Robert Mugabe.

In 1997 I was introduced to a gold deposit to the north of Harare, which looked interesting, by a local geologist. It had been mined by "the Ancients" long before colonisation. Evidence of this could be seen by the little shafts and tunnels that were obviously dug a long time ago. More recently, undocumented shallow diggings had been conducted on numerous outcropping quartz reefs. Clearly no systematic modern exploration had been conducted even though the geology looked promising for substantial gold mineralisation.

I was encouraged enough to apply for a mining lease over the deposit and our company was soon granted exclusive rights to explore the property. We employed a team of local workers and started sampling and trenching to determine the size and nature of the mineralisation.

Everything seemed to be progressing well so I decided to fly to Zimbabwe and visit our team on site to see how things were going. The resident geologist and I drove to the site and along the way passed the fertile farmland that was once the breadbasket of Africa. We discussed the land "invasions" that were orchestrated by Mugabe and his cronies and never once thought that those illegal appropriations of legal title could also be applied to our mining lease.

On arrival none of our team was to be seen. Instead there was a group of about twenty men, women and children digging away on our mining lease and filling up the back of a dilapidated truck with buckets of dirt. Children were scratching with little shovels at the quartz reefs exposed in the crumbling and unsafe tunnels left by "the Ancients".

I managed to identify the leader of the group who turned out to be a large muscular gentleman wearing intimidating mirrored sunglasses. I told him that we were the Australian owners of a mining lease over the property and that his presence there was illegal and that he and his group should leave immediately. He stepped a little closer to me and said: "No, you are Australian and we are Zimbabweans. We own the land so you should leave immediately."

I thought it was probably not an appropriate time to discuss the finer points of mining law, as the group was now standing behind the leader with shovels, pick-axes and machetes at the ready. We left immediately as recommended.

Back in Harare I went to the Mines Department and asked to speak to the officer in charge of management of mining leases. I was asked to what it was in reference and was told the officer was busy but could meet me at my hotel after work. At six in the evening he met me in the bar of the hotel and I explained the day's events to him and asked what he could do about it. He took another sip of the whiskey I had ordered for him and took out a letter from his coat pocket and handed it to me without comment. The letter was supposedly from a doctor telling him that his daughter was gravely ill and that it would cost US$20,000 to heal her.

I gave the letter back to the officer and said it was very sad to read about his daughter's condition but that it did not relate to our discussion about the mining lease. He then said that he would help me if I could give him the money to pay the doctor. I told him that I would think about it and talk to him the next day. I then went to my room, packed my bags and got a flight out of Zimbabwe that evening.

I thought that would be last time I would venture into that country, but the lust for fame and fortune blinded me once again. In 1999 a diamond mine on the southern border of Zimbabwe was placed into receivership. It had been mined by an Australian company, but after some time the diamond grades started to fall without any obvious geological reason, and the mine was placed into receivership. I heard that the receiver had called for expressions of interest to restart the mine and pay the creditors so I started due diligence on the deposit.

It seemed to me that the diamond grade fall was possibly a result of theft rather than geological variation. There was also a stockpile of about a million tonnes of ore that had been mined but not yet processed, so that seemed to give a bonus for whoever restarted the mine as it would require very little capital to begin recovering diamonds again. I therefore contacted the receiver and asked for a meeting and flew to Harare.

The receiver was a formidable woman who still called herself "Rhodesian" in spite of the country having been independent for decades. She also told me that she was the toughest receiver in the country and that she was called "The Lipstick Piranha". She gave me the details of the mine and creditors and, together with some investors in Melbourne, we decided to propose a settlement to the creditors and take over the running of the mine.

The creditors' meeting was held in Harare in July 1999 and there were only two other proponents besides our group. One was a South African prospector who did not seem to have any backing. The other was a husband and wife team from Bulawayo. Their expertise and financial backing were based on a hardware store in Bulawayo. I thought we had a good chance of winning the creditors over to our side. Little did I know that the husband and wife were best friends of the Lipstick Piranha.

After the creditors, stacked with friends of the receiver, voted for the Bulawayo couple I again left Zimbabwe empty-handed.

Six months later, in February 2000, Cyclone Eline slammed into Mozambique and kept going over southern Zimbabwe and into Botswana and even Namibia. In doing so it killed many people and countless animals and extensively destroyed crops. It also flooded huge parts of territories it passed over, including the diamond mine. The open pit at the mine was filled with water, the stockpile of ore was washed onto the surrounding paddocks and the infrastructure of the mine was destroyed.

I phoned the Lipstick Piranha and profoundly thanked her for helping me dodge the proverbial economic bullet.

Sometime later the mine did get back into production and Robert's wife, Grace Mugabe, decided it was time to take over the mine "in the National interest". I again phoned the Lipstick Piranha to thank her for helping me again to dodge the bullet both figuratively and, perhaps, literally.

SAFARI IN KENYA, OF SORTS

PRUE LEEMING
KENYA 2003-2004

Prue is 65 years old and graduated as a geologist from University of Melbourne in 1975. She joined Newmont and worked Australia-wide, in Fiji, New Zealand and Spain mostly in gold, tin and base metal exploration. After completing an MSc at Rhodes University, South Africa in 1984, she joined Freeport, followed by Homestake and worked in open pit gold mines and exploration. Drifting to the "dark side" in 1993, she became a consultant with World Geoscience working with airborne geophysics in the Eastern Goldfields, India, Indonesia and Oman and, in 1999, took a posting to Botswana. After a two-year contract ended, she set up a consultancy, working on project generation and various assignments in Botswana, Zambia, Namibia, Kenya and Nigeria. Based in Perth, projects in coal, iron ore and manganese were staked for Aquila in Mozambique, South Africa and Botswana. From 2010 she consulted to ASX-listed and private companies on project review and procurement for coal, copper and gold in Western Australia, Queensland, Colombia, Ecuador, Peru, Chile and Brazil.

"Your first seeing of a country is a very valuable one. Probably more valuable to you than anyone else, is the hell of it." E. Hemingway In Green Hills of Africa.

Introduction

Consulting within the airborne geophysical company World Geoscience ("WGC") since 1991, the opportunity arose for a transfer to Botswana in late 1999. I was to assist the team in marketing airborne EM (electromagnetic geophysical surveys), preparing tenders for government surveys and undertaking interpretation projects. The company had recently won a tender for 1,000,000-line km in Namibia, one of the largest airborne geophysical surveys of its type ever undertaken by the company.

After accepting a two-year contract, I travelled from Perth to

Johannesburg, only to find my on-going flight with Air Botswana to Gaborone was cancelled. Air Botswana made alternative arrangements with South African Airways so eventually I boarded a flight. As we circled the airport at Gaborone prior to landing, a major anomaly appeared below. Wreckage of three ATR-42 turbo propeller planes lay sprawled and incinerated in front of the terminal. I was grateful for the safe landing by SAA! Bob Blizzard met me at the airport with a wry smile. After my please explain, he said "Oh, some grumpy Air Botswana pilot managed to steal a plane, get it airborne and then crash landed into the rest of Air Botswana's fleet! Welcome to Gabs!"

My term with WGC was short-lived as, two months later, Fugro Airborne took over the company besides numerous other airborne companies, including Geodass in Johannesburg. Work continued as before – the tender was in force, crew and aircraft mobilised, data acquisition proceeded, new tenders were written, and marketing airborne surveys continued. My contract ended, and after eight months working in the old Geodass Woodmead office, I returned to Gaborone, set up a consultancy and returned to mineral exploration.

Approaching my fifth year living and consulting in Africa, I was advised by a WGC colleague Paul Larkin "Always make sure you have enough money to return home. If your situation becomes dire, book an airfare immediately!" My principle was to not to draw upon savings from Australia, but I was close to booking a ticket to Perth at least twice. Somehow a job always cropped up, either through colleagues in the Botswana Geological Survey or from companies for which I had previously worked. I opened my email in late November 2003 and was puzzled to see a message from Freeport colleague, John Nettle. Typically, he was direct. "Bruce wants me to go to Kenya to supervise a diamond drilling program on a project called Migori. I told him I am retired! You're in a much better position based in Africa than me (Perth). Let me know if you're interested and I'll get Bruce to make contact." Kenya was always on my wish list to visit – devouring books by Blixen, Huxley, Hemingway and Mathieson to mention a few only sharpened my interest. It was easy to respond in the affirmative and let John enjoy his retirement.

Preliminary Visit To Migori Project

Bruce Walsham, the Managing Director of Kansai Corporation soon emailed and set up a proposed meeting in Nairobi for a week later. Gaborone is surprisingly accessible to East Africa with an hour-long flight to Johannesburg to meet connections with regular Kenyan Airways flights to Nairobi. Kenyan Airways is well regarded having a long-standing joint venture with KLM (since 1995). A delayed Air Botswana flight however, left me with no time to meet the Kansai accountant whom I was intended to meet in Johannesburg prior to flying to Nairobi.

The last to board the Kenyan Airways four-hour flight, I settled in my window seat and ordered a gin and tonic. A short, dark, nuggety man approached me from the aisle and introduced himself as Basil Crawford, the company accountant. I motioned the seat next to me and we ordered lunch. Basil spent years with the South African Defence Forces (SADF) as a commando and had seen action on most fronts including Angola, the Caprivi Strip and Namibia. On retirement from the defence forces he became an accountant and worked in the resources industry. Hired most recently by Bruce with Diamond Works, he described how he accompanied shipments of large sums of US dollar cash by air, to fund a diamond operation in north east Angola. He then told me that in 1998, the mine at the Yetwene kimberlite pipe in Lunda Norte was ambushed by 50 soldiers, some belonging to UNITA. The rebels were likely backed by another party and, although not publicly acknowledged, Basil suggested it was Endiama, an Angolan company backed by De Beers. Eight personnel were killed, and five others kidnapped (also believed killed), including a British geologist. "Diamond Works spent $1M looking for the kidnapped people but never found a trace", Basil added. The civil war in Angola lasted 27 years, from 1976 to 2002, with the root of its activities focused on diamond and iron ore operations. By 2002, foreign companies held 80% of all licences held for both mining and exploration and most were militarised. Diamond Works morphed into another company called Energem, but never recovered from the incident.

My tales of working with Bruce were tame coming from the Western Australian Archaean. My involvement with Freeport of Australia was with gold exploration in the Murchison (Cuddingwarra) and at Karonie Gold Mine, with a reputation around the Eastern Goldfields for extremely hard rock especially around drilling companies. "I hope Kenya is not like Angola" I remarked to Basil. "I understand Nairobi is not safe – in fact, I believe it is as dangerous, if not worse, than Johannesburg. Well, you've lived in Pretoria for many years and I too have lived in Johannesburg for many months and we're still here. Guess we just have to be careful". He grinned. "Actually, I have not been to Kenya before as Kansai have just taken over this property – I am looking forward to it."

We duly arrived at Jomo Kenyatta International Airport, which was bustling with activity, Africans, European tourists and a few businesspeople of mixed backgrounds, but despite the chaos we found our luggage carousel. Wheeling our bags through customs and immigration, we were approached by a tall, thin, laconic African named Lucas. He resembled one of those exaggerated, thin carved statues you can buy in African markets. Bruce must have described us well for Lucas to recognise us! We boarded a Toyota LandCruiser 4WD ex safari vehicle and Lucas drove us out of the airport complex. "So, are you our driver, Lucas?", Basil inquired. "Yes, but I am also a mechanic", Lucas replied as he weaved between the chaotic traffic. "You must have to be brave to drive in this Nairobi traffic, Lucas?" "No sir", smiled Lucas in the mirror. "You have to be *courageous!*" We smiled back – obviously, we were in excellent hands.

Lucas dropped us at the Intercontinental Hotel in downtown Nairobi where we checked in. We met Bruce Walsham and James Pincher, (not his real name) both British, later in the poolside tropical garden and bar. A sales convention was winding up in the hotel, showcasing Kenyan coffee. Bruce Walsham has a long history in the mining industry with Freeport McMoran, and eventually as Managing Director of Freeport of Australia until 1989. He then managed other groups including Diamond Works until in 2003 he became CEO of Kansai and today remains President. James's background was less known, but he left Diamond Works' diamond operation in Angola, a week before it was invaded. He was appointed Logistics Manager for the project. Easy conversation and a few beers later we dined early at the hotel.

Next morning, after breakfast our first stop was the company town house found up hill and north of the main city centre in a security complex of eight similar properties. Nairobi (Maasai for "cool water") is a typical ex-British colonial city located in hill country at an altitude of 1,795m (5,889ft). Originally built to suit colonial settlers escaping heat and humidity, Nairobi's location on a mountainside is hopeless for the practicality of a growing capital city. Mountain tops, steep valleys and slopes pose an engineering nightmare with roads and highways zig zagging around the city and causing traffic jams everywhere. Minibuses and government vehicles were fitted with governors to maintain speeds of 60kmph and so reduce the accident rate.

The Migori project was formerly explored by Auvista, an ASX-listed company which undertook geological mapping, extensive geochemical programs and drilling between 1995 and 1998. Licences covered 65km of strike over the ESE-trending Archaean Nyanza Greenstone Belt from Lake Victoria (west) to the Great Rift Valley (east). The greenstone belt hosts the Macalder massive sulphide deposit (VMS-type), which was mined for zinc, copper and gold from the 1950's to 1960's and several small-scale gold prospects at Mikei, which were also mined between 1990 and 1992. The most discerning feature of the Kenyan Archaean greenstone belts was their lack of modern gold exploration compared to their counterparts in Tanzania. The Nyanza belt lies only 40km north of the North Mara Greenstone Belt, which hosts a rich gold mine across the border in northern Tanzania.

Several reports on the Migori Project were produced. A large, hastily rolled, pile of maps and sections materialised from the corner of the office and several generations of sections were identified but fortuitously one set was meticulously interpreted and coloured. While the section assays and drill traces were computer-generated, the geologist added welcome details and, after some cross-checking with drill logs, the continuity of mineralisation was believable.

We walked to the local Westgate Shopping Centre crossing a bridge crammed with pedestrians, over a polluted stream below and thence across a major road. The same shopping centre was targeted by gunmen from the terrorist Islamic organization al-Shabaab in September 2013. The attack resulted in 71 deaths and 200 wounded people, remaining one of the worse mass shootings in recent history. The attack was claimed as retribution for deployment of the Kenyan military in Somalia, home of the extremist al-Shabaab. Crowds swarming both in and out of the mall made them easy targets while on our first shopping visit ten years earlier we felt no threat at all. We purchased some basic stationery, printing paper, graph paper and a new A3 printer/plotter for maps and sections for the Migori office. Out of curiosity, I bought a large picture book detailing the people and tribes of Kenya. Forty-two peoples are recognised of which about thirteen are well known (see below). Between them they belong to three different ethnic groups recognised as either Bantu, Nilotic or Cushitic-speaking.

1.	Kikuyu people	8.	Turkana people
2.	Kalenjin people	9.	Nilotic people
3.	Luo people	10.	Samburu people
4.	Luhya people	11.	Kipsigis people
5.	Kamba people	12.	Kuria people
6.	Maasai people	13.	Kisii people
7.	Mijikenda people		

Marathon runners are almost purely Kalenjin in origin, a small, minority tribe in Kenya with a population of about 5 million. Maasai, Samburu and Turkana are cattle-raising tribes known for their fearless tolerance of lion and other predators. The Migori project was located in Luo country, which overlaps both Uganda and Tanzania, and makes a mockery of the original colonial and now national borders. We returned on the same bustling footpath, relieved that we averted stalled traffic during the rush hour. A lively barbeque followed after packing up the roll of maps and sections, reports, and digital data ready for the trip to site next morning.

An early departure to Wilson Airport with Bruce and James was scheduled to meet a charter flight to Migori. Our English pilot was young, with a slight swagger, but seemed professional enough as he escorted us through to the Cessna 210. Flying in Kenya is not for the faint hearted, especially in the rift valley and mountains where severe tropical storms can develop from nowhere. The plane took off amid a few bumps but soon we were airborne over a rich green and ochre landscape. With altitude and cloud, differentiating landmarks became less easy. We landed on a surprisingly rough, red dirt airstrip and taxied to a Toyota LandCruiser personnel carrier. Lucas, having driven up the previous afternoon and evening, stood by ready to drive us to the house.

The town of Migori follows a main road which travels northward to Kisii and onto Kisumu, a regional centre found on the shores of Lake Victoria. Lush green vegetation on red lateritic soil showed evidence of subsistence farming and market gardening, a consequence of widespread deforestation. Migori itself is not dissimilar to many African towns – stalls lining the main road, shops unpainted and either barred with steel mesh or sliding security doors, wares spilling onto the street from doorways, a few two or three-story concrete or rendered buildings with peeling paint, slow-moving pedestrians, sleeping dogs and a tangle of electrical wires overhead. The most striking feature of Migori's main street were the coffin sellers – shop after shop crafting coffins from local timber. HIV AIDS took a heavy toll in this town while cerebral malaria is also endemic.

The company house was located out of town on the Kisii-Kisumu road. A high cement-rendered brick wall topped with barbwire surrounded a large garden of a thinly grassed lawn, a few tropical flowering shrubs and the odd spindly tree. A previously cultivated vegetable garden lay fallow. A security guard greeted us at the gate and we soon met the cook and houseboy.

Our few days in the field were spent visiting the field camp, Mikei, the five prospect areas and various drill sites. The field camp was a short distance from Migori and east of Lake Victoria but took a bone shattering drive of over forty-five minutes to reach. The gravel road was either unlevelled bedrock or cavernous potholes, capable of tipping a vehicle. Lucas, a local Luo from the region, was a skilful driver and patiently negotiated the track. The field camp was found within a large, fenced, sloping enclosure and included a long whitewashed building housing a mess and quarters for accommodation. A timber-clad core shed was found near the gate with several concrete pads distributed on the north and east side. Halfway between the core shed and the quarters was a roomy timber-clad office. Paul Ogolla, a local Luo farmer, who previously worked on the project with Auvista, oversaw laying out core from several previous diamond holes. Paul's open, honest face and calm demeanour, not to mention familiarity with the project and local community, suggested his contribution to this project would be invaluable. Intercepts of gold mineralisation were barely detected without referring to drill logs, but massive sulphide mineralisation from the Macalder deposit was a standout.

Nyanza, KKM and Gori Maria are three low-grade gold prospects with broader mineralised intercepts and trended ESE through cleared grazing lands. It was likely with increased drilling that these prospects could join up. Most of the drilling was proposed on these targets with the aim of increasing resources. Providing we could locate the old drill collars, this appeared relatively straight forward. Mikei or MK Zone, the fourth prospect, followed a narrow high-grade quartz vein for several hundred metres on the edge of a steep valley. Consistently drilled from surface meant the next phase of drilling was either from the very bottom of the hill or across the valley. Drill site preparation promised to be challenging.

The Macalder zinc-copper-bearing massive sulphide body was located on a prominent hill further west. Mined underground from stopes and shafts, the ore was processed on site with tailings discarded downslope towards a stream. Ubiquitous rain spread the tailings wide, but even more worrying was the presence of local villagers panning the tails and using mercury to extract the gold. Auvista drilled and systematically assayed the tailings dam to include the low-grade gold as a resource. Our task was to redrill Macalder and find more massive sulphides.

Visiting each prospect area highlighted not only the problem of trespassing on numerous African farmed lands, but also the state of the drillhole database. Recorded collar locations were found to show serious inconsistencies and, to make things worse, the error was variable. Fortunately, Paul, the longest serving employee, had excellent recollection of certain hole collars so at least we could work from these. Previous surveying of the holes was of uncertain quality and, compounding the problem, old drill collars were marked by concrete square slabs with details etched into the face. The concrete blocks were a size most convenient and irresistible for use as building blocks so every single drillhole marker was stolen by local villagers. After poring over the interpreted sections and drillhole location plan, I managed to plot five obvious holes in the Gori Maria and KKM areas with reasonable confidence and mark them in the field in preparation for drilling in February 2004.

Next morning, I accompanied Bruce back to Nairobi on an early charter flight and caught a taxi back to the company house. By late afternoon, Basil, ready for a small adventure, bundled us both into a LandCruiser and headed for Nairobi National Park. Established in 1946 as the first of Kenya's national parks, it lies 7km from Nairobi city, next to the airport and remains one of the world's most accessible parks. We drove through typical acacia savannah, a grassy plain, admiring giraffe, eland, Thompsons gazelle, black rhino and hyena. My highlights were the black and white colobus monkeys, which I'd never seen before, and the birds – in particular, some raptors and the crowned crane. We drove back via the Karen Blixen Museum, the Norfolk Hotel and a landmark off the tourist route, Kibera, the world's fourth largest slum area. No measure of independence, capital investment, education or political will has managed to shift this home to so many Kenyans.

Bruce and I parted company at Nairobi airport next morning – he flew back to London while I returned to Gaborone. James and Basil remained behind to seek a suitable Kenyan bank to establish a company account. Christmas of 2003 was looming so, while Africa generally shuts down from mid-December to mid-January, I took the opportunity to return to Australia to visit family.

Migori Drill Programme

I returned to Nairobi by mid-February 2004, delighted to meet Lucas at Nairobi airport. Very businesslike, he announced that when I was ready, we should depart for Migori by road. He suggested it was preferable to leave late this afternoon and we stay overnight to break the journey. We passed Kibera on the way and I enquired about it: "Who lived there? Was it safe?" "My home is in Kibera", he replied, and "Yes, I have a house in a village on the way to Migori but Kibera is my home for when I work. I am not afraid." I got the impression that Lucas was unafraid of anything or anybody.

We set off from Nairobi and I welcomed the chance to drive across Kenya to Lake Victoria. The hardest part was exiting Nairobi for the highway but, soon on the outskirts, we travelled along the rim of the East African Rift Valley with aerial views across to lakes, volcanoes and a patchwork of farmland. Maasai people erected a row of stalls along the highway following the scarp, where they sell traditional red tartan clothing, known as the Shuka. The red colour is believed to ward off lion at a distance.

Our route passed through Naivasha, a wealthy town cultivating and exporting fruit and flowers to Europe, and then Nakuru, a town near Lake Nakuru National Park with its millions of flamingos. By late evening, near Kericho, we turned into a tea plantation which offered accommodation. The view from the hilltop was of neatly pruned green tea bushes downslope and as far as the eye could see. The tea plantations were different compared to those I had seen in the highlands of India, where they are sheltered by tree cover. The teahouse was more or less deserted, but we met at the bar-restaurant and were soon joined by a white Kenyan farmer. We started chatting and soon discovered that he was the brother of Newmont geologist, Brian Levet! Small world.

After leaving the teahouse next morning, we drove through a village, where a crowd gathered on the main road and surrounded a partially covered dead body. Mindful of the seriousness of the tragic accident, Lucas slowed the vehicle and delicately negotiated a route past the on-lookers and the body. We pushed onto Kisii, a sprawling, dirty and chaotic town where we stopped briefly for supplies. Finally, we reached the Migori house where Matthew Wilkinson, a graduate geologist with his bright new laptop, Albert the cook, Johnson the houseboy and Paul Ogolla met us. For the project we had three vehicles; two LandCruiser personnel carriers, previously used in the safari business, and an old white

Land Rover, all vehicles clocking well over 300,000km. We unpacked, reviewed data and made plans. The surveyors were due to arrive on the weekend to locate drillholes so we could get started.

A prominent hill with radio-communication towers, aerials, satellite dishes and mobile phone antennae formed the backdrop to the house and was situated across a valley filled with a patchwork of cultivated fields and mango orchards. I set off in the evening for a walk, crossed a small bridge over a muddy stream and followed a stony path between the fields. Reaching the ridge, I then followed a trail to the top and skirted around the infrastructure to admire the view. Suddenly, I was startled by a man pointing a bow and arrow at me. He quickly put down the bow, I stared at him for a moment and then realised he was our security man! I smiled in relief and he returned a sheepish look. "It is not necessary for you to follow me" I said. "I will be fine. Anyway, I always feel much safer in the country". We walked back to the house together in companionable silence.

On Sunday, with the drilling company's arrival a few days off, I suggested we visit Ruma National Park which is found northeast of Migori and close to Homa Bay of Lake Victoria. Our party included Lucas the safari expert, Matthew, Paul, Johnson, Albert and myself. Once drilling started free time would be rare so this was an opportunity for a break and to see some wildlife. Too often Africans are deprived of the opportunity to access their own national parks. On the way towards Lake Victoria, we passed a huge colony of pink-backed pelicans roosting in flat-top acacia trees – an uncommon sight with pelicans around the world normally found in or close to water. Ruma Park was an open grassy plain, with scattered spiny acacia and swampy areas. While midday is not the best time for a game drive, we nevertheless saw many Rothschild's giraffe, Jackson's hartebeest, buffalo and several roan antelope, a specialty of the park. The only roan antelope I had previously seen were through a security fence to a private reserve of the Pretoria Military Police! The crowned crane was in abundance near the wetlands with numerous other birds.

The Geoserve drilling team arrived from Tanzania at Mikei camp a day later. The crew included ten Swahili-speaking Tanzanians and one

Fanagalo-speaking head driller from South Africa. Bruce Medhurst, both English and Afrikaans-speaking, was the supervisor and manager from Zimbabwe. Later his son, Ryan, with his Afrikaans girlfriend, Anika, took over supervision. Bruce and Ryan were well educated, bush-toughened, highly resilient and capable men. Both fairly small and neat in stature, they enjoyed golf and a lifestyle that hard work rewarded them. Based in Tanzania for several years, they recently moved their team to Kampala, Uganda, so mobilisation to Migori was not unreasonable.

Equipment included a mobile/skid-mounted drill rig with the capacity to drill to 800m, a 6X6 crane support water truck, a 4X4 LandCruiser, drill pipe, downhole survey and orientation devices, a pump, a generator and welding and oxyacetylene equipment. The water truck was an ex-SADF vehicle and was used to manoeuvre the drill rig into position. With three drillers, three drilling assistants, one water truck driver and two cooks, the team was well placed to drill over 24-hour shifts.

Our arrival at Mikei Camp with the drillers suddenly highlighted just how dysfunctional and ill-prepared we were. Paul patiently walked us through the camp pointing out what needed doing. The camp itself lacked water, power, fuel for the cooking fire, bedding and even tables and chairs. Very little shade existed for all the core work, and there were no layout tables and benches for logging, sampling, and core sawing. The submersible pump on the bore was not working, so water supply independent of the driller's water truck was not available. The combination of Paul and his Kenyan farmers and Bruce and his Swahili drilling team swung into action. Teams of Paul's labourers were set to work clearing drill sites and building access tracks. Carpenters appeared under trees knocking up benches and tables while a temporary water supply was rigged up and filled. Food supplies, most importantly, and the first of the bedding turned up on the Geoserve LandCruiser after a trip to Migori.

The Geoserve drill rig was sited over the first hole and work proceeded immediately. Ramani Communications, the Nairobi-based survey company, commenced gathering control points over the region and reviewed the previous grid control of Auvista. Routine surveying of all the drillholes was then in progress for the following two weeks. At last we had a more reliable collar file to work from.

Despite the dreary drive to and from Migori, it appeared both Matthew and I would only stress the situation at the camp until at least water and power were installed. We sourced the only internet 'café' in town, and this provided our main communication to Bruce and the outside world. The office was found at the end of a dark, grimy passageway past a kitchen smelling of cooking, and up a narrow staircase. Eight old desktop computers were sandwiched next to each other on benches. A humourless supervisor stood behind a glassed-in cubicle with a strong-featured, friendly girl, called Julia, to take the fee and allocate the computer. Webmail was available for communication as well as text messaging from a cell phone at the highest point of Mikei camp. Bytes were delivered like snowflakes so while internet speeds were variable in Botswana, outback Kenya was a decade behind.

The start of the program was frantic. Despite an army of help on tap, much training and many jobs needed completion before things would run smoothly. Paul's carpenters soon knocked up benches to lay out core, skillion roofs went up for shade, the generator was hooked up for the core saw and the camp, and the submersible pump was lifted and tinkered with to identify the spare parts needed. We trained local Luo Kenyans, with Paul's help, to set up core for photography, cutting the core and bagging samples. Errors soon discovered in the sampling, meant rematching the core with the samples and starting again after more detailed instructions. Soon the procedure was understood. Both Matthew and Paul were shown how to sight in the drill rig, and I worked closely with Matthew logging core until he was confident to work alone. Oriented drill core came regularly and quickly. Good drilling conditions resulted in excellent core recovery and soon the drilling team settled into a routine of producing at least 100m per day. Long sections were plotted for each area so as many holes ahead were planned as possible.

A week into the program and our first Toyota personnel carrier broke down. Matthew and I were stranded along the road to the camp, while Lucas was busy inspecting the damage. A broken, bleeding axle was not an instant fix. Luckily, Paul was returning from Migori and was the first at the scene. Matthew and I squeezed into the Land Rover and left

Lucas to hitch back to Migori for the second vehicle. Three days later, the second Toyota also broke an axle, so we borrowed the Land Rover from Paul to drive between Migori and the camp. This was the first time two Toyota LandCruisers let me down so badly but with a previous tough life on safari drives, it was not surprising. A day later the old white Land Rover, now with an odometer no longer working, also broke down. Fortunately, Lucas managed to fix the first Landcruiser and we were back on the road, while Paul yet again was forced to commute to his farm on a bicycle.

Ten days after commencing the program, I ran into Geoserve's Bruce at the petrol station in Migori. "Hey Prue, I don't suppose you could put in a word to your boss for an initial payment?" "What amount are we talking about?" "Kansai promised to reimburse me on arrival for all the equipment, core trays, core cutter and gear we brought up from Mwanza plus the mobilisation fee so about US$30,000". I frowned. "Give me a copy of the invoices and I'll contact him tonight". When I reached the camp, I passed on my concern to Paul. He looked at me somewhat grimly as if he already knew and then admitted all his workers hadn't been paid for a fortnight. "Goodness Paul, why didn't you tell me sooner?" He showed me the perilously low bank balance. "My people depend on weekly payments for food and to send their children to school. I assumed their wages were coming so I didn't like to ask". I sent off a stern email to London that evening and waited.

I followed up two mornings later with the drillers in Migori. "Bruce, did you get paid?" "No, we didn't! Things are serious now as the petrol station is threatening to cut me off. We are potentially their biggest customer for the next two months, so they want to see some money first." "Fair enough", I said. My credit card was available but what if Kansai didn't pay me too? Things were looking messy. Anyway, a credit card was also my emergency exit to Botswana and thence to Australia.

We drove to site, where I sent a blunt text message from the hill. "Please pay US$30,000 to Geoserve's account (Kampala) and forward K$20,000 (Migori) immediately. Otherwise tomorrow, drill rig stands down at normal standby rate US$2,000." It worked. The phone eventually rang, and it was Walsham. "I just picked up your email today.

We have a problem getting money into Kenya as we still don't have a Nairobi account". "I can't operate with thirty Kenyans working for a pittance if we can't pay them. The drillers account with the petrol station will close due to non-payment so right now they have enough fuel for two days only. You also promised to pay Geoserve for all the gear Kansai ordered – not my problem Bruce." "Understood", he replied. "Leave it with me". Later in the day I received a text from Basil to say he was booked on a flight the next day to Nairobi from Johannesburg and carrying US dollars to change to Kenyan Schillings. At least he was used to errands like this with Diamond Works. Geoserve finally got paid as did our Luo workforce.

Poverty was widespread in the Migori area and, apart from two American missionaries, Matthew and myself, there were no Europeans in the region. Paid work was rare. Hundreds of Kenyans walked miles to our security gate and submitted their CVs. Every day the security guard would hand us a bunch of letters and shake his head. Albert, Johnson and the security man knew they were privileged employees for this short period of time. I kept the stamps of all those letters to prove that Africans were not lazy but desperately wanted to work. Many walked over 30km to deliver their letters. Our internet girl Julia walked 16km daily to and from her work, as I joined her once walking home: the next time we took the 4WD!

One afternoon I drove into the camp from a drill site and Paul requested an urgent meeting. He drew me into his office and lowered his voice. "There's been an armed hold-up at the Mission". "How far away?" I enquired. "Thirty kilometres", he replied. "A priest was killed and the two nuns working at the mission reported over K$20,000 was stolen". Paul was worried that the Kansai drill program might also present a target. Each week he drew out a large sum of money from the Migori bank to pay the thirty or more workers we hired locally. He felt he might represent a target himself, or get hijacked on the road, especially after regular visits to the bank. "Perhaps I should change the day and time I visit the Migori bank each week?" We agreed it was a sound plan.

At last the first batch of samples was due for shipment to Mwanza.

Lucas loaded the LandCruiser and set off for the border with Tanzania. I felt a little uneasy about assaying at the Humac Laboratory having not checked it out personally but, while Matthew and I were kept busy with photography, logging and marking core for sampling, a visit to Mwanza would have to wait. Auvista previously used the laboratory so it was recommended for the program.

Drilling progressed steadily at KKM (Kakula-Kalange-Munyu) prospects with the first six holes drilled here. While I worked closely with both Paul and Matthew to check drill sites and compensation payments to relevant farmers, to ensure site clearing well in advance of drilling and to line up the rig, in fact this work was soon well covered by both of them. I encouraged Matthew to visit the rig daily and occasionally I accompanied him to a new drill site. One afternoon, after siting three new holes at Gori Maria, we returned to the camp to find a huge crowd gathered outside the gate. "This must be all the local village people for miles around," I thought. "What is the problem?" As we approached the gate, three aggressive, bulky gentlemen wielding sub-machine guns confronted us with loud abusive, shouting. Then I noticed the drillers' water truck, but no Geoserve drillers or offsiders. Confused at the pushy demands and not fully understanding what was happening, I indicated to Matthew to drive through the gate, which was opened cautiously by a security man. The armed men drove off following the water truck leaving a buzz in the crowd.

Paul greeted us near the core layout area and shook his head. "Who are they?" I asked. "What do they want? What are they doing with the drillers' water truck?" The drillers then appeared at the gate and drove in. "I'm sorry but I've had to shut down. I can't drill a metre without water" said Ryan. The programme was going so unbelievably well - so what next? Paul explained "They are local council people and basically they say we owe them money." "Migori Council?" I asked. "Yes, the Migori Council. They want us to pay K$100,000". "But the Council has no jurisdiction over minerals. Minerals are controlled nationally. In Australia, it is state-controlled but I doubt Kenya subdivides this responsibility by state. Kenya would be like Botswana, Zambia and even South Africa with a national mining law. So, Paul, is this legal, this request?" "No", he confessed. "It is a bribe." I added: "We pay compensation to local farmers – have they instigated this? Are the farmers unhappy?" "No, they are very happy with our payments." "Somehow we have to get the driller's water truck back quickly – but first I'll contact Bruce Walsham and let him know the situation". I sent a text from the top of the camp. "While we wait for a response from Kansai, let's return to Migori and perhaps visit the council. We might be too late. At any rate, let's see if we can find the water truck."

By the time we reached Migori Bruce rang: "Phone the Ministry of Petroleum and Mining and speak to the Director. His name is William Kirubi. I met with him on my last trip and he knows about the Migori drill program." I immediately made the call, but it was past 4.30pm. We drove past the council offices and complex, but these were already closed. Nearby, the water truck was incarcerated behind a tall netting fence with a heavy padlocked gate.

Early next morning I rang the Department from the Migori house and quickly explained that I urgently needed to speak to the Director. The woman who answered the phone was understanding and immediately put me through. Mr Kirubi was calm, deliberate and listened carefully. After introducing myself, I referred to a meeting that Bruce held with him a month back, and the Director replied, "Yes, I remember it well. The Migori drill program! So how is it going?" I explained the drilling was progressing well and we were about a third of the way through our program. We maintained good relations locally and awarded compensation to farmers as required. We were also employing 30 local farmers for their labour. However, yesterday three armed personnel from the Migori Council held us up at the camp and ordered us to pay K100,000. Our program has stopped because a vital piece of our drilling equipment (the water truck) was confiscated. To make things worse the three armed gentlemen have incited local villagers to stage a protest outside the camp." There was silence at the other end of the phone. The Director finally responded with a steely, but firm voice, "This is a very serious situation indeed. I am extremely glad you rang me. Please give me at least an hour or two but I will speak to the Mayor of the Migori Council directly. My secretary will phone you when I'm done when I suggest you then call the Mayor and arrange a meeting with him personally to ensure your equipment is returned immediately."

Ryan and his head driller pulled up in their LandCruiser outside the Migori house. "Any news? Have you made any progress?" I told them about the phone call to the Director of Mining and Petroleum and that we needed to wait. "While you are here, I'll show you the next three holes at Gori Maria and would you like a coffee?" The novelty of a coffee break in the middle of a program of steady drilling and peering over plans was welcome, although we all felt anxious. In less than forty minutes, my phone rang. The Director's secretary called to say we should proceed directly to the Migori Shire Council to see the Mayor. I suggested Ryan accompany me, given it was his water truck.

A short drive to the Council chambers was followed by a hesitant walk to the front door and reception. The Mayor was waiting for us, beamed, and ushered us into a large meeting room. He vaguely waved his arms around, talking about land allotments, certain council rights and fees which must be paid, none of which made sense. Ryan and I sat listening not knowing what he was talking about. Finally, given audience, I explained how Kansai held two exploration titles SPL122 and SPL158, which required fees paid to the Department of Mining and Petroleum. I explained we were currently in the middle of a drill program, and any encroachment on locally held farms was duly compensated. Finally, I added, "I believe Mr Kirubi phoned you this morning. Could we please have our water truck back because we have already lost half a day? This will cost our company nearly US$2,000 in standby for reasons out of our control. All caused by the Migori Council". The Mayor was dumbfounded. He squirmed in his chair and turned to one of his support staff, issuing instructions in a low voice. He then turned quickly to us and suggested we accompany the gentleman, who would remove the padlock so we could claim the water truck back. Clearly, he thought we might send an expensive bill to the Migori Council for wasting our time.

Geoserve soon resumed drilling and routines were re-established. The drilling team settled well into the camp, which was now resembling a home. Repairs to the submersible pump allowed hooking up water and hot water showers via a chip heater. Tables and bench seats were fashioned for the mess, all twelve bedrooms were furnished with proper bedding, lights were rigged up for the kitchen, bathroom and bedrooms via the generator used for core cutting, and a large plot was fenced off for growing vegetables. Outside on the grass strolled three turkeys, several chickens and two goats. and every time Geoserve went to town for food shopping, a new animal or bird arrived. 4WD vehicles were parked under the shade of trees outside Paul's office. Matthew and I saved hours of travel during the week by staying at the camp mid-week. Bruce Medhurst left and was relieved now by Ryan and Anika. She watched over the kitchen, kept in touch with her friends via her mobile phone and, in quiet times, studied accountancy. Anika also taught me some tricks with SMS, for which I was forever grateful, given that we depended on this form of communication.

One morning back at Migori, Johnson appeared at breakfast with news that Albert was sick with malaria and couldn't come to work. "Is he all right?", I asked. "He will wait for testing this morning and get medication", Johnson replied. "It is likely he will need one to two days off". Malaria is endemic to the Lake Victoria region and local people are fully aware of the need to seek urgent treatment. "Please reassure Albert that the company will pay for his medical expenses and he must take all the time he needs. Can you cook in the meantime?" Johnson self-consciously responded that it was not his job, but he would manage. Two days later Matthew emerged from his bedroom looking like a ghost. "I feel dreadful", he admitted. He knew we were both under pressure to keep momentum up with logging and sampling. "You must stay home and get well. At least we have a spare vehicle so get Johnson to take you to a clinic for testing and a script. The vehicles are now in good shape so Lucas can be my assistant today." The problem of malaria surfaced after the odd shower of rain and during very late returns to Migori in the 4WDs. After hibernating in the vehicles, mosquitoes come alive in droves at dusk.

Sunday morning, two days later, I awoke feeling like I had 'run into a bus'. Our month-long program with no break was catching up. Matthew was still recuperating from malaria, so I had no option but to keep up with the logging and drill supervision. Lucas did well the previous day and was ready to leave after breakfast. I was quiet for the jolting trip to site. We picked up all the core trays and carried out our routine geological work, including photography of each tray. The sun was shining, we worked steadily and replaced all trays on the benches ready to orient the core, measure recoveries and prepare it ready for logging. By lunchtime, it was becoming steamy and hot and I walked towards the lunch tree, flopping onto the grass like a rag doll. Lucas brought my lunch, but I wasn't hungry. Ryan and his head cook drove up in their white LandCruiser after their weekly shop in Migori and, without hesitation, called through the window: "You've got malaria! Get back to Migori now!" I rose to a sitting position and said that all the core was ready for logging. "Forget about the logging. Go and get tested now, trust me." Lucas nodded in agreement.

The drive back was murder. Every bump, jolt and knock on the door hurt. Malaria has an insidious habit of finding the weakest part of your body and shaking it. My right-hand sciatic nerve – exacerbated by years of long-distance running, took the brunt of it. We drove straight to a group of small wooden buildings in a street parallel to the main road. Opposite were pharmacies housed in similar timber-clad shops. Lucas indicated the clinic, where I might need to wait. I walked into a small gloomy room and sat on a narrow bench, aware that patients were already in consultation. Brochures on HIV AIDS were pinned up on the walls and lying loose on a small table nearby while a hushed conversation from behind a curtain suggested a serious conversation about this very illness. "A malaria diagnosis could be a really good outcome", I thought ruefully. A couple eventually emerged and acknowledged me as they departed. The doctor appeared, invited me into the cubicle and with my brief explanation of symptoms, took a blood sample from my finger, mounted it onto a slide and advised a diagnosis would take 20 minutes.

Plasmodium falciparum is responsible for the majority of malaria deaths globally and is associated with 'cerebral' malaria, particularly prevalent in sub-Saharan Africa. If diagnosed and treated within 24 hours, malaria is generally safely cured. I joined Lucas in the LandCruiser and waited. Right on time a prescription was handed to me and I paid a nominal amount. A walk across the road to the pharmacist and I was soon armed with medication. We drove back to the house and met Matthew. "Oh, not you too," he said. I shook my head with a resigned look. "I'm much better", said Matthew. "Why don't I drive out now and stay the night at the camp? At least I can make a start on logging the hole you've prepared and be ready early tomorrow." It was a good suggestion and, with Matthew taking responsibility and keen to work, even better. I withdrew to my bedroom to rest under a white veil of mosquito netting covering the bed.

My phone rang in the evening and it was James from Nairobi. He was driving up the next afternoon but possibly wouldn't arrive until the following morning. Knowing that I was out of action I told him I was down with malaria to which he responded sympathetically and added, "Tonight you could be very ill. In fact, it is not uncommon for you to feel you want to die. I am warning you because malaria is bad around Migori and can be fatal. You must fight." It was a strange conversation. Ironically, that evening my mind drifted off in a state of suspended calm. Relaxed, at peace and at home with the thought "It wouldn't really matter if I died right now!" No stress about drilling, managing a program and people, or flying back to Australia to see family. Two mornings later, I still felt shattered but gritted my teeth and braced myself, ready to face James and crew. His errand was to deliver a notice from the Department of Mining to help move prospectors away from Macalder Prospect, while we were drilling.

Drilling moved easily between holes from the relatively straightforward Gori Maria and KKM prospects. While movement of the drill rig between holes generally gives geologists a breather and chance to catch up, Geoserve moved so quickly around Gori Maria and KKM that no such additional time was awarded. I wanted to wait for some assays before we continued in these areas, so the drilling team headed for MK Zone. There the target quartz vein lode was narrow but with consistent and high gold grades. The lode was also easily recognised while gold mineralisation at the two former prospects was invisible.

MK Zone was previously explored by St Barbara in 1988 when they reported 360,000 tonnes at a grade of 5.60g/t for 65,000oz. Auvista, in

1997, reported a resource of 136,000 tonnes at a grade of 13.80g/t for 60,000 oz. Incredibly St Barbara started trial mining until 1992 but the operation was abandoned due to insufficient funds. The prospect was clearly insufficiently drilled. A program of six holes was planned, three shallow holes to extend the length of the mineralization and three much deeper holes to test the zone at depth. The shallow holes progressed rapidly with a hole completed every day. The lode followed the line of a crest of a hill and dipped slightly steeper than the slope to the lower valley. Our third deep hole was positioned almost at the base of the hill and accessed from the opposite side of the valley. Early rains arrived so all vehicles slipped and slid along the new tracks to reach the area. This was our most dicey drill site.

The drill rig was trailer-mounted for transport to site and then offloaded to skids. The heavy 6WD truck then towed it into position. At one point I suspected both the drill rig and truck could end up in the creek. Drill team members were on board with lots of shouting directions all at once. The truck manoeuvred the rig awkwardly, but its sheer weight limited any sideways movement. A little coaxing back and forth and the rig fell into position almost exactly. I breathed a sigh of relief. Ryan and his drillers pretended this was business as usual but later agreed it was close! The last thing we needed was an early wet season halfway through the program, with two machines hopelessly bogged in a creek!

I returned to Migori to send an update on drilling progress and expenditure to date. An email from Bruce arrived – first apologising for the lack of funds at the start of the program and, secondly, to let me know we no longer had a logistics manager. Apparently, he was guilty of using substantial company funds to renovate his house in the UK! Bruce suspected that the money allocated for the Nairobi account and drill program was never banked and the Nairobi bank failed to set up an account for Kansai in the first place. Perhaps James used this as an excuse and naively thought Kansai wouldn't miss the funds. "It's my fault", Bruce lamented, "I met him on the train. He was very charming, keen to join us and worked for us in Angola with Diamond Works. He talked about his Kenyan contacts so I thought he could be useful".

A second tranche of samples was ready for delivery to Mwanza. With the drill rig busy for at least two days it was a good opportunity for me to accompany Lucas and visit Humac Laboratory. We loaded the samples and headed off for the border town of Isibania. Accustomed to paperwork accompanying freight, Lucas handed me the forms and directed me into the office for passport control and customs clearance. He then handed me the keys of the LandCruiser. "Where are you going?" I asked. "I don't need to show my passport. I can walk through border control because I am Luo," he grinned. "I'll see you the other side". People and vehicles were massed outside the offices, but the passing traffic was light. As with many African borders, plentiful stalls sold food and cheap goods to passing travellers. We soon hit the road and made good time.

We descended the steep rift-faulted margins as we approached an embayment of Lake Victoria and there were further descents on similar faults south near Bunda. To the east a large flat green grassy plain appeared with silhouettes of large antelope between acacia trees on the horizon. The Serengeti National Park offered a glimpse of its treasures. The four-hour drive went quickly and we arrived at Humac Laboratories. An elderly manager seemed out of place with his broad North England accent, dark sallow complexion and the deep voice of a smoker. A visitor to Humac was a rarity so a twinkle in the eye and some humour soon surfaced. With the presence of coarse gold at MK Zone, we decided to test some intervals with screen fire assays and then he offered a tour of the laboratory. Nothing like a detailed inspection of a laboratory to allay one's fears of the accuracy of their assays.

Mwanza is a bustling metropolis on the southernmost shore of Lake Victoria and sited on the eastern shore of Mwanza Gulf. A distinctive town with white-washed buildings, colourful splashes of red, purple and orange bougainvillea, tall hibiscus and climbing geranium. The streets, which wound around large granite boulders and kopjes, were dotted with date palms, attracting thousands of birds on dusk. Mwanza had been an active slave-trading post and Arabic traders left their mark on both the architecture and gardens. On the way to our hotel, we called by geologist Mark Davey, whom Paul recommended as the only person nearby who worked at Migori and who could answer some of my queries. Mark agreed to meet us for a drink at the hotel later.

Dusk was approaching so after navigating through town, Lucas drove to Hotel Tilapia, a quiet refuge on the edge of the Lake Victoria. I was dropped at the hotel, while Lucas immediately set off to refuel and purchase a few spare parts. After checking into my room, I opened the curtains to look out over the lake. Hitched up to the landing just below was a wooden river boat with a blue and white painted cabin. On the side of the boat was splashed prominently "The African Queen". Right here in the middle of Tanzania remained the props for that crazy Hepburn and Bogart movie. World Wars in colonial Africa resulted in absurd and unnecessary battles in countries where Africans were blind to the cause. No wonder colonial Africa died. The 1951 movie for all its parody was a priceless salute to a miserable era. C S Forester, who wrote the story in 1935, was never too far from the truth.

Mark Davey appeared in the bar and willingly divulged as much as he knew about the Migori prospects. It was good to meet someone who previously worked there and, in fact, he was a useful contact on a Tanzanian property offering a decade later. Mark was a quiet, self-contained and well-grounded English geologist completely at home after living for years in the middle of East Africa. Mwanza was the second largest city after Dar Es Salaam but still remote. Lucas joined us for an Italian meal after which we both retired early.

On the return drive to Kenya we stopped briefly at Lamadai, a tourist destination which is easily accessible from the highway and Serengeti National Park. Lamadai is found on the edge of Speke Gulf of Lake Victoria. We walked to the shoreline and found both wading heron and a few gulls swooping overhead. The expanse of the lake is considerable and neighbouring countries of Uganda and Kenya lie well beyond the horizon. Navigation is dangerous on the lake, which is subject to sudden and dramatic weather changes and many areas lie beyond communication. Lake Victoria is surprisingly shallow at 41m (described as a "puddle" by local geophysicist Phillipa Hutchinson) and lies offset to the east of the much deeper Great African Rift lakes such as Lake Malawi and Lake Tanganyika (570m), the latter second only in size and volume to Lake Baikal, in Russia. Lake Victoria was named after Queen Victoria by explorer John Speke during an expedition with Richard Burton in 1858 to locate the source of the Nile. I recall from a book on their journey a vision of Speke shooting gulls from the deck of their boat as they passed through the Gulf of Aden to Zanzibar to start the expedition.

As we approached the margin of the rift and began to climb, two gleaming, brightly painted buses came hurtling towards us and racing downhill. We passed both before the trailing bus pulled out suddenly to overtake the first at high speed. Lucas shook his head. "They shouldn't be driving", he observed. Otherwise our journey was quiet and the border crossing without freight was easy. Anxious to return to the drill site it was likely the hole, drilled in our absence, was nearly completed.

Nyanza was our next prospect, less well defined with fewer previous drillholes. Gold grades were highly variable. Four holes were planned here, although there was a danger of drilling into workings. This prospect in the end yielded the highest gold grades and intervals of all with 2m@77g/t (35-37m) and 19m@15g/t (102-121m). The first hole appeared highly encouraging with strong alteration, veining, sulphide mineralisation and segments of visible gold. I wondered why the prospect had been neglected. The second short hole, drilled west and along strike from the first, was a problem hitting underground workings. The drillers managed to drill through the first cavity and continue after casing off. A second cavity was struck and the attempt to run a second string of casing failed. Core retrieval suffered and only a few, dismal fragments of core ended up in the tray. However, the hole demonstrated that mineralisation was continuous from the first cavity, so it wasn't entirely wasted. The drill rig moved to a proposed third deep hole and collared between the two shallower holes.

Matthew and I were logging in the yard when Ryan drove into the camp looking frustrated. He jumped out of the LandCruiser with: "I need to make call", he said. "It's not looking good. We've tried welding this part unsuccessfully twice, so I have no choice but to order it. We will be down for at least two days." "Where will the part come from?", I asked. "Johannesburg," replied Ryan. "We will airfreight it, but it will take at least a day to reach Nairobi and another to Migori. I suspect we did the damage yesterday in those workings." "That's bad luck," I replied. "You should take a break", he said and shrugged. "And, anyway my guys need a break as well! We've been keeping apace, with few excuses to stop". A breakdown at just four holes short of finishing the program was a shame. The only relaxation for the drilling crew was playing short games of soccer at the camp.

Matthew stood listening and then said "Why don't you visit the Maasai Mara? I am happy to finish off this hole and stay at Migori." Matthew had taken the odd day off, after contracting malaria, and made some young friends in Migori. He was also planning to stay in Kenya after the program. Lucas wandered over. "How long will it take to reach the Maasai Mara from Migori?" I asked. "About two to three hours" he replied smiling. "We could leave this afternoon and reach easily by nightfall. But I don't know the road." As a safari driver for many years at the Maasai Mara, Lucas knew the game reserve well. Normally Maasai Mara was accessed from the east from Nairobi but we would drive in from the west. "Will we need to book accommodation?" "No, they know me there. Anyway, it is nearly the wet season, so the tourist season gets quiet." "OK, let's finish what we are doing – Matthew, you can take the second 4WD back when you are ready. Do you mind please marking out the core samples for cutting? Sample what you can where there is very poor core recovery– just leave half the material behind. These assays won't be counted. Are you sure you wish to stay?" "I'll be fine," he said.

Another 45-minute jarring journey to Migori and a quick turnaround at the Migori house. A small bag was packed with a change of clothes, a toothbrush and toilet bag, rain jacket, binoculars, bird book, camera and a couple of water containers. The LandCruiser was travelling back to its old stamping ground but needed a shovel, jack and tools of trade for a mechanic to keep it going. While 140 kms is not a great distance, given the state of the road, progress was extremely slow. One creek crossing appeared impassable, but we managed to will our way through. We arrived at the edge of the rift on dusk, a hazy full moon rising from the east, light pinks and a blue sky with a vast rolling plain below us. I pulled out my camera, took one photo and the battery went flat. Lucas pointed out the Mara Safari Lodge in the far distance.

We descended into the rift and entered the park via Oloolo gate before it closed, when suddenly it was dark, and we struggled to see the moon through the cloud. We followed undulating roads with cavernous potholes, puddles and coarse gravel infill. On the plain, Thompson's and Grant's gazelle, topi, zebra and eland came into view with our headlights and soon we saw a spotted hyena. Navigating reserve tracks at night, let alone spotting wildlife, is very different compared to the day, especially with no GPS, no compass nor any moon to guide us. Lucas eventually recognised some trails on the south western side of the Mara Triangle just as a storm broke. We finally located a well-formed gravel road to the Mara Safari Lodge, which is found on a generous loop of the Mara River. Arriving at the hotel in pouring rain after 8pm, and desperate to eat, we checked in, and yes, meals were still being served. I was escorted through a deserted sprawling camp to a large room with crimson curtains, a four-poster bed with mosquito netting, a generous lounge and matching chairs, mirrors, a desk and a chair with a spacious bathroom at the far end. Through the noise of rain on the roof, I became aware of a chorus of grunting noises emanating from the river below. Hippopotami! Plenty of them! A bloat? Suddenly I was excited, despite the long, arduous drive to get there.

I dumped my bag and headed for the bar where two elderly, English-speaking gentlemen were propped. We exchanged pleasantries as apparently they were the only other guests. When Lucas emerged, we took a table to order a meal. "You did well to get here tonight, Lucas". "Yes," he replied, I never guessed how difficult that road from Migori would be. Unfortunately, it is hardly ever used."

Next morning, we met at breakfast at 7.30am and planned to leave at 8.30am. As I walked along the path through the vast garden, suddenly a Maasai warrior stepped out from behind a tree. Incredibly handsome, he was adorned with long coloured beads, earrings, armbands, and anklets, and also beads plaited between strands of thinly braided long hair. Wearing a brightly coloured red and blue tartan shúkà and open sandals, he beckoned, tiptoed a short distance away towards a veranda post and pointed. An Eastern tree hyrax was climbing between the two wooden poles of the post: the left set of claws on one pole and the right set on the other. A comical sight! Hyrax are cuddly little animals also known as dassies, that resemble wombats or koalas but incredibly have a close genetic relation to elephant! I thanked the warrior and he sauntered off through the garden. I met Lucas at reception ready for our morning game drive and we set off soon after through the gate to the camp. I pointed out the warrior, who was striding ahead through bush. Lucas added, "He will work for the camp here and assist with guiding. Because it is quiet today, he will likely return to his village". Soon after, we spotted a woolly stork.

The Mara River forms the boundary between the Maasai Mara National Park and the National Maasai Mara Reserve. Wild game effectively roams across both tracts of land as do the Maasai themselves with their herds of cattle. Increasing herds of cattle and numbers of Maasai do however impinge on the natural environment and affect the livelihood of wild animals. In Tanzania over-grazing has led to desertification resulting in an unsuitable land for both humans and wild animals. However, the parallel existence of both the Maasai and native fauna and flora appears to work in Kenya. Increasing numbers of tourists, camps and lodges is, however, alarming especially when basic infrastructure like access tracks is not maintained.

We spent the morning on the plain of the Mara Triangle. Completely different from the evening before, when visibility was less than 100m, today we could see over vast distances. We picked out herds of gazelle, zebra, impala, eland, blue wildebeest, red hartebeest, topi, giraffe and a herd of elephant, intent on traversing south. Meandering through the maze of tracks, our late departure meant we missed seeing predators, but we did encounter a spotted hyena and black-backed jackal. Before we returned for lunch, Lucas headed for the hills to seek out the last white rhino in the Maasai Mara. We found the rhino under the escort of a security guard, who stood literally metres away supervising the animal from dawn to dusk (and no doubt relieved by another guard for the evening). We stepped out of the 4WD and Lucas chatted to the guard for some minutes. The rhino was as quiet as a domestic cow, but we refrained from venturing too close. Poaching of rhinos for horn is devastating their numbers not only in East Africa but also southern Africa. Zimbabwe, South Africa, Namibia and Botswana each maintain dedicated reserves for black and white rhinos to conserve their numbers.

After lunch I watched hippopotami lazing in the Mara River from the bank, roamed the garden with binoculars searching for small birds and then retreated to my room to read and fall asleep. At four in the afternoon, we left for our last game drive. While we saw the same animals from the morning on the plain, we retreated to a tributary and followed a tree line. We spotted a mongoose, just before two beautiful cheetahs appeared. The male stood tall with his front paws on a rock, peering around for prey, the female skulking around the base of the rock ready to hunt. Evening light picked out their striking, black spots on cream coats and distinctive facial markings. They were easily the finest-looking cheetahs I'd ever seen. We roamed further upstream to its source and then joined the Mara Triangle plain. As we swung back along the opposite side of the stream, two fully grown male lions emerged. Edging closer, these large, golden brown lions were beautiful, nonchalant to our presence and in very fine health. The contrast with Botswana's Kalahari lion, with light sandy coats and rare black manes, was striking.

Dusk approached quickly, when an evening thunderstorm loomed and broke as we followed a less used trail towards the main track. The LandCruiser suddenly sunk into black mud. A quick change into 4WD and then low range only made things worse. We abandoned the vehicle, hurriedly tore off armfuls of dripping vegetation and threw them under the wheels. Luckily, the vehicle drove out cleanly. Safely back on the road, we faced another long drive in pouring rain with limited visibility back to camp. After hot showers and a change of clothes, Lucas spent the evening with his previous employers, while I joined one of the retired Englishmen from the previous night for dinner. He was a keen birder, specialising in warblers and luckily, while no expert, I had also spent hours observing and differentiating warblers in Botswana, so we compared notes. I later concluded that such a pastime in retirement in Africa could be very civilised.

We set off early next morning to return to Migori and, while crossing the Mara Triangle was a lot easier in daylight, the road from Oloolo Gate to Migori was agonisingly slow. After such a wonderful day long adventure in the Maasai Mara, the cold reality of a drilling program awaited us. We arrived after midday at Migori and headed straight out. The deep drillhole number three of Nyanza was finally underway and the fourth hole was already staked, so I left Matthew to carry on with supervision.

I ventured to Macalder for the afternoon as the two remaining holes for the Migori Program were to test for extensions to copper-bearing massive sulphides. Fortunately, with some financial encouragement, the local prospectors had left the abandoned mine and plant, so the site was clear. A scaled plan showing workings existed but geology was lacking. I prowled around the shafts, stopes and outcrops mapping the geology so I could plan some holes. Macalder was riddled with underground workings but, lacking an updated plan of them, presented a difficult target to drill. No drilling had been carried out for many years, except on the tailings dam by Auvista. However, I did manage to site two short holes.

The final four holes of the Migori program for Kansai all ended within a week just as Easter was celebrated. Geoserve packed up their drilling gear, loaded the truck and 4WD and set off to Kampala, on the same afternoon the last hole was completed, just as efficiently as they drilled. The camp fell silent as Matthew, Paul, Lucas and I completed the logging a day after the drilling, sampling and packing the core in the shed. Lucas and Matthew drove the vehicles back to Nairobi as Lucas, being an Elder of his church, wished to be home with his family for the Easter ceremonies and Paul retained the Land Rover on site in case additional core cutting was required.

It was Easter Sunday and with nearly everyone gone, Paul invited me to his church for the Easter service and his home for lunch afterwards. It was a rare social event. He picked me up from Migori as promised and drove straight to the church. Located in a valley beyond the drilled prospects, the church was a large, simply constructed building filled with wooden pews and packed with well dressed, local farmers and their families. While the Christian celebration of Easter Sunday is a solemn affair, it is ultimately a celebration. The strongest, African (Luo) singers found throughout the congregation led the rest of them with confidence and joy. No orderly choir or conductor, no organ, just communal singing at its best.

After the service we drove to Paul's farm and house, a bare wooden construction, with an iron roof and four rooms. On the way, he pointed out piles of manure used to fertilise his crops of corn and cassava. We sat down at a homemade table, with his wife and children, for a delicious roast chicken and vegetables. Paul's work with Auvista ceased at least six years previously so he and his family were totally dependent on his small farm. It was sobering to think the same fate might follow this program for not only Paul but also Albert and Johnson at the house. Vital assistance provided by the local farming community allowed the program to progress smoothly and rapidly and, although the work was casual, it was valuable for both the exploration company and themselves.

Next morning Bruce Walsham arrived on a charter flight from Nairobi on a fleeting visit to review the better intersections from each Migori prospect. Although assays were not available for the remaining four holes, segments of the best-looking core from Nyanza and Macalder were laid out in addition to intercepts from Gori Maria, KKM and MK. The camp was now well established with mown grass, a thriving vegetable garden and workable facilities to log, cut and house core, to accommodate and cook for twelve persons and to enjoy a game of soccer. Only the security guard and Paul remained. After a picnic lunch, we bid our farewells and took one last jarring, slow drive back to Migori, during which conversation usually stopped. We picked up the reports, plans, sections, handwritten drill logs and digital data from the Migori House and caught the charter flight back to Nairobi.

"How about dinner at the Norfolk?" Bruce suggested. "We fared much better at Migori than I expected, so time to relax". On arriving at the Nairobi house, I thanked him, disappeared for a shower and wondered if I possessed any suitable clothes for dinner. We caught a taxi to the hotel, settled with a drink and placed our orders. The seafood rice dish looked tempting and, assured seafood was flown in daily from

Mombasa, I opted for this. Halfway through our meal, I suddenly excused myself and moved swiftly towards the ladies' bathroom. The instant I closed the door, I was uncontrollably ill, and projectile vomited all over the bathroom. "Unbelievable! You really know how to appreciate a dinner in a posh hotel with the boss!" I thought furiously recalling previous WGC colleagues equally ill in India. As funny as it looks with others, vomiting is never really fun! I spent the next 20 minutes cleaning the bathroom with hand towels, toilet paper, soap and water. Luckily, nobody joined me. I quietly returned to the table with a bashful explanation. "I assumed something was wrong", Bruce said. "Maybe the seafood?" "I don't think so – I've never had food poisoning react while I am eating!" Feeling cheated, I finished with a small dessert to sweeten up the evening.

I hailed a cab from the hotel reception back to the house, took a second shower and went to bed. It was a long, tortuous night with not a single moment of sleep; tossing and turning, and aching all over. I rose early, took some breakfast and sorted through some plans in the office when Bruce arrived as cheerful as ever, "So how are you today?" "Not great! I think I have malaria! I didn't sleep all night, so I probably need a test as soon as possible". A clinic was soon found, and a malaria diagnosis confirmed. Armed with my second dose of medication, it was fortunate that I was diagnosed before leaving the country. Later that afternoon, I caught two commercial flights back to Gaborone for a break. We held off final reporting on the program until all the results came back.

Nairobi Office

Both Matthew and I returned to Nairobi at the same time to start compiling the data. While I worked on sections, plans and an up to date interpretation, Matthew wrestled with the MicroMine database and digital sections. He soon discovered that MicroMine couldn't plot drillhole information nearly as well as we did by hand. Whilst in Nairobi, an opportunity to visit geophysicist Phillipa Hutchinson arose so it was a chance to stand back from the drill program and review regional geophysical data. Airborne geophysics won't site a drillhole collar, but it helps for regional targetting.

Each morning I walked around the streets of Nairobi and soon found the Nairobi Arboretum. Perched on a steep conical hill, the area offered numerous trails which varied from extremely steep to gradual either crossing or following the contours of the hill. I donned a pair of running shoes to attempt a return to fitness and easily ran to the entrance near the stream. One look at the steep dirt track ahead and I stopped. A tall Kenyan runner came into view from ahead and rested for a few moments. "Are you training for a marathon?" I asked. A broad smile was his answer. He introduced himself and said that these hills were all part of the training. "I'll look out for you then in future Olympic and Commonwealth Games", I replied, "because I won't be training on these hills!"

Basil and his wife arrived a week later to review the accounts and make a final and more successful trip to a bank to establish a company account. It was ironic that this happened after the drill programme and not before. Whilst taking his job seriously, Basil never relinquished an opportunity to explore Africa so, on the weekend, he organised a Sunday trip to Lake Nakuru National Park. Lake Nakuru is one of the famous soda lakes of the East African Rift with its main drawcard, the spectacle of millions of greater and lesser flamingos, numerous other endangered water birds, and various animals. Black rhino graze around the margins of the lake with various buck and are aggressive if spooked or harassed. Keeping our distance, they were content to graze.

A fortnight later, I needed a break from the office routine and drove myself to the edge of the rift from Nairobi. I parked near a trail to the volcanic crater within Lake Longonot National Park and past the numerous colourful Maasai stalls on the side of the road. Without any known predators and only occasional bush birds, I safely walked around the rim of the volcano overlooking Lake Naivasha to the north east. After reaching a third the way, I realised it would be nightfall if I kept going so had to back track! The state of the trail meant slow walking and suggested this national park was reserved for geologists only, not tourists.

When we commenced the program at the Migori Project, total

indicated and inferred resources from the five prospect areas using a cut-off of 0.75g/t was around 7.3Mt grading 2.37g/t for 550koz. KKM held the greatest resource after Gori Maria and MK while Nyanza and Macalder offered the least. Of the twenty-two holes drilled, at least fifteen returned significant gold intersections, four anomalous mineralisation, two were abandoned in workings and one was barren.

Kansai undertook resource calculations after the program but neither Matthew nor I were involved – a sad indictment of our industry today when field geologists rarely participate in resource estimation. Red Rock Resources purchased the deposit from Kansai Corporation around 2010, and the most updated resource for Migori is now 30Mt grading 1.26 g/t for 1.2Moz. Just how many ounces were added with a change in cut-off grade is not understood. Further drilling and ground geophysics (EM) in the Macalder area certainly took place after our programme in 2004.

****
****

In 2006 the Commonwealth Games were held in Melbourne. My husband and I moved to Victoria on business and timed a visit to Melbourne during the games. We picked up tickets for a day of track and field and watched an outstanding display of athletics. The long distance running from mainly Kenya was a standout. I couldn't help but touch base with both Paul and Lucas from the MCG via texting! They were equally aware of the games in progress and just as thrilled as us.

Paul Ogolla rang me a few years later to report that Lucas passed away. Lucas was a fine leader of the Luo people but lived with an ongoing liver complaint, which was expensive and sadly difficult to treat. Paul most likely continued working with Kansai and Red Rock Resources on further exploration programs. Geoserve remain based in Kampala and undoubtedly are drilling as efficiently as they did at Migori. The vulnerability of the many farmers who worked for us, and of all those applicants, who walked miles seeking work, will no doubt continue with population pressure, drought and deforestation. Equally endangered are African game and birdlife from human pressure, poaching, farming and drought. I was extremely fortunate to experience a vibrant, pro-active Kenya with enthusiastic help from the local Luo people, not to mention the blow-ins from Botswana, South Africa, Zimbabwe, Tanzania, Uganda and the UK!!

THE MEDITERRANEAN

AND

THE MIDDLE EAST

LETTERS HOME FROM IRAN

TONY GATES
IRAN 1977

Tony Gates completed his BSc at the University of Sydney in 1963 and has been a practising geologist since then. He initially worked for companies such as Aquitaine Petroleum in the NT, Queensland and WA and followed by overseas postings in France, New Zealand and Papua New Guinea. His petroleum work in PNG established excellent relationships with tribal groups, which persist to the present. He then worked for Geopeko in Darwin searching for metals and uranium with successful results.

He moved to Perth at the beginning of the nickel exploration boom and has been based there since. Originally he was involved in nickel exploration for the American company Kennecott.

In 1970 he started his own consulting companies, Tony Gates and Associates and Pacific Exploration Consultants, which see him still exploring and acquiring mineral assets in many countries.

(Editor's note: The following letters by Tony from the Iranian outback were submitted by his wife, Pauline, who has saved them, and they are reproduced verbatim.)

December 1977

So they – ie. The driver and English speaker – Name of Nardar, left a day early. The driver had to take Nardar because of the danger! said Nardar.

First day great consternation, when I met them by helicopter, long trip – slept at the Mayors house of a village – don't know where we are going, where is the gendarmes (cops)....etc. Finished the first days work with a severe case of the craps (figuratively only) with the worst driver in Iran, at a place called Zaboli – a small village of mud houses of about 200 people and a radio transmitter that transmits Iranian Central Government propaganda to the Baluchi people throughout SE Iran & Pakistan. This transmitter is above a huge balloon, about 100ft long tethered to a 5000ft cable. So as we are 4000ft above sea level its quite effective

Then out of nowhere comes two Americans – one a paramedic, the other the site manager of the balloon site. Dinner that night consisted of prawns, French fries, with Mushroom Flapjacks followed by Thousand Island Salad and then apple pie like Mama made; beautiful ground coffee and a colour movie (TV cassette) of Ann Margaret and her boobs bouncing and a quick look at the latest American Football championships.

They had no sleeping room, so we drove down the road and slept off in the bush – ie. In the gravel – there are NO bushes or trees or grass – nothing but rocks and gravel.

Our brave Nardar slept in the car – animals etc might eat him! So we began an uneasy night – what with the balloon across the road going up and down with motors and generators roaring and the whole place lit up like Luna Park. But night got the best of us and a night in -2deg.C was had.

2nd day we left Zaboli to drive 70km east through about 6 villages, but the last vehicle this way was in 1901, since then camels and donkeys have made a mess of the track. It took us nearly all day to travel here – what with pleasant discussions with the military police – "bayonets fixed" identification card etc. Etc and a complete lack of knowledge by the locals of their area beside 10km around their villages and an equally unbelievable way of confusing simple questions into major United Nation type discussions with 10 people all giving directions in 2 languages pointing in 8 directions at the same time.

So second day out and I've spent 1/2hr collecting twigs out of the creek bed to boil some tea in a baking dish; never heard of billys, eat on our beds (no table or chairs) no tin opener, eggs, salt, bread, meat etc. – shit are they hopeless – of course Nardar blamed the driver "I told him to pack them" – the driver blamed the cook etc. Etc., eaten a tin of ham (it had a key) and settled down to a cold night under a clear crisp sky in a clean beautifully gaunt and ruthless environment. The Beluchi have been friendly, helpful and lovely people – the Tehranians are spivs – that's my first uncharitable intolerant consideration.

Last night on television at the Balloon we saw that Tehran had 2 metres of snow, closed the schools and airport so no mail at camp yet out here the days are warm and sunny, beautiful conditions with below 0 deg.C at night but little wind so quite tolerable (in my tracksuit, socks, jumper & beanie and hands on my cock) also three blankets and a sleeping bag.

Tomorrow I must complete about 48km of traverse then drive north to a place called Caravan to meet the chopper – I'm afraid I won't make it in time, so if the chopper doesn't fly down the road, it's another night in the "bush....with the animals".

Until then my love

Your loving husband Tony

16 December 1977

Friday 2.40 wrong, 4.40 and we are at least 3 1/2hrs from the helicopter pick up. So it means we miss it tonight and another night out under the stars, but I'll bet the young spoilt smart arse will want to sleep in some tavern or such, anyway from our instructions last night at the police we were only three hours from our destination, it's taken us 4 hrs so far and another 3 ½ and that's at breakneck speed. The roads are rough and rocky and although the Government has the 3rd biggest army in the world outside of USA and Russia & spends 50% of its GNP on arms, it has never heard of a grader to help these poor farmers in this area. At least that helps them remain independent from the central "Imperial Government of Iran".

We have travelled through many small villages where the water supply still remains a hole in the ground and the women drop a bucket into the hole and pull up a 10 gallon bucket. These villages have no power or sewer or water supplies and each house depends on the small fuel lamp – food is cooked over the most hard to get wood, so there's little hot food instantly. Another small interruption ....

17 December 1977

Each kilometre travelled seemed a hundred kilometres - eventually lights came into view and we entered the dusty town of Caravan – a larger bigger Meekatharra run mainly for Government Instrumentalities.

After asking the cops the whereabouts of the best hotel in town we finished up at a concrete modern blockhouse, two stories high, three beds per room ie. Per square cell, cold showers only, no food and a toilet that consists of the typical hole in the ground squat type. The small squat hole stank to high heaven and the water didn't flow. But the rooms were clean and the proprietor very friendly. Although the hotel belonged to the Government we were the only ones patronising it. We then travelled through the town to the best restaurant and thank goodness at 7.30 it was closed, as the kitchen must have been the filthiest I've ever seen. Around the corner in the next mud-straw & brick building was a bar – smoke filled with about 10 heavily drunk men. We bought 1/2doz Carlsberg cans (@ $1.00 each) and bought 50c worth of bread – like huge pizza bases and sat in the city square and drank and ate our evening meal. The square was lit by neon type signs upholding the Iranian Revolution, such as "hand for the people" "equal pay for equal work" "water belongs to the people" Nationalising big industry" "51% shares to the workers" etc. – quite radical and well meaning phrases that the people think is actually happening. However when Sampey Exploration could make $2million from this contract it makes one wonder.

After dinner, we went to the local Mosque for Muharram – the month of mourning for the death of the 2nd & 3rd prophet during the religious wars of about the 13th centaury. During this service I sat cross legged, shoes off, on a dusty old carpet and listened to an Imam (or priest) who spoke in classical Persian about these deaths – the whole speech was broadcast over the amplifiers to the vacant streets of the town. The whole thing built up into a singing crescendo, until the lights in the mosque were turned off and everyone cried and wept and howled for the prophets and the lost close ones. During the middle of this break we snuck or is it sneaked out and back to the hotel.

At six this morning we were up, packed and off to the public bath house, a swelling, damp, steamy collection of six closets about bathroom size & showers – mine of course rained hot for 2 minutes, so I de-soaped in cold water.

Now I sit, sad and wistful on the 44 gal fuel drum of helicopter fuel awaiting the chopper as the sun streams down, the myriad of motor cycles stir up a cloud of dust from here to the horizon and the stark mountains turn hazy blue in the thickening dust.

I sit here thinking of my love for my beautiful family – Pauleen I love you, kids I love you both so much, love each other, be happy and good.

Love Dad and Tony

xxxx,xxxx,xxxx

A GREEK TRAGEDY

JOHN NETHERY
GREECE 1987

John Nethery was born in 1946 and raised in country New South Wales, before attending UNSW in Sydney to graduate in Science (Geology) and with a Diploma in Education in 1969. He joined AIG Corporation, rising to General Manager – Minerals of AOG Minerals Ltd, exploring throughout Australia, Oceania and west Africa for gold, base metals, uranium and diamonds. Since 1997, as a consultant, he has explored, mainly for gold, throughout Australia and elsewhere in Oceania, South-East Asia, the Mediterranean region and briefly in western USA. In 1993, after 16 years living in Sydney, he moved to Chillagoe, North Queensland, and is still exploring for gold and base metals at this time.

In mid-1987 I had recently left the corporate scene when I had been General Manager – Minerals running AOG Minerals Limited for parent company Australian Oil and Gas Corporation, where I had spent my entire career of eighteen years. The scene at AOG had changed markedly with the takeover by Industrial Equity Limited which had more plans for AOG other than oil exploration. AOG Minerals Ltd was considered a strange anomaly to the newcomers: "Why explore for mineral deposits? Let someone else do that and just take them over if they are successful". Well that was not to my liking as an obsessive explorationist, so when the company was merged and rebadged as Australmin Holdings Ltd yours truly departed the scene soon after to become a consultant.

Geoff Loudon and Gavin Thomas at Niugini Mining Ltd were restless in 1987. Their world-beating discovery on Lihir Island in PNG was stalled in interminable politicking and corporate manoeuvres. They wanted to use their skills, honed by discovery and definition of a large bulk tonnage volcanic related epithermal gold deposit, to further advantage elsewhere.

The idea of finding massive deposits of "invisible" microscopic gold elsewhere was rather appealing. Ian Plimer was pretty much focused in academia at the time, but was constantly in and out of consulting exercises and had done some early research at Lihir and spent a goodly part of his frenetic lifestyle consulting to industry. I had found several much smaller pristine volcanic-related gold deposits in north Queensland with AOG, and was looking for new horizons. We were all good mates and kindred spirits and commonly discussed potential dreams as one does (often with the creative assistance of food and alcohol). Mediterranean volcanic arcs were one such dream.

Ian, sponsored by Niugini Mining, had recently spent a weekend detour on his way to Europe on the island of Milos in the Aegean Volcanic Arc, attracted there by the record of active hydrothermal activity, recent volcanism and published data on extensive bentonite, kaolinite, barite and perlite mining. His frantic running around on a motorbike resulted in a return to Oz with few clothes but a bag full of rock samples. Only one of these produced a detectable gold assay of 0.4g/t from an historic abandoned barite and silver digging. Plimer and I determined to interrogate these disappointing results over copious quantities of alcoholic beverages at the quaint old *Wallarah Arms Hotel* at Catherine Hill Bay, south of Newcastle. On viewing the full laboratory report I had become quite enthusiastic about the wide areal extent of enhanced volatile elements such as arsenic, antimony and mercury, throughout a section of this large island, which indicated the potential for the discovery of large deposits. We agreed that more substantial reconnaissance was required.

A joint venture was formed whereby Niugini Mining would fund the two of us to research and examine potential areas of interest throughout the Mediterranean. The first priority was to revisit Milos. It was on this second visit that we found float of low sulphidation epithermal quartz veins in a gully which we followed to outcrop and roughly defined a vein-bearing zone several kilometres long. We knew then we were onto something. Further reconnaissance revealed sinters and hydrothermal breccias. I volunteered to climb the mountain, Profitis Ilias, working on the old adage that the most interesting rocks are on the tops of hills. I found extensive stockwork veining. The rough silicified outcrop at the very top was occupied by a large, hairy majestic and obstinate billy-goat, who was rather put out by this intruder trying to enter his domain. I told him he could keep his perch as long as I could chip some rock from around the periphery, which I proceeded to do, hurried along by the occasional nudge of horns. In one of those simple twists of exploration fate we learned later that if I had approached the mountain from the opposite side I could not have missed several of the bonanza grade veins. After a reconnaissance of some of the surrounding Cyclades islands in a hired fishing boat we returned to Athens to find that global stock markets had crashed: October 1987.

Ongoing low-key reconnaissance defined interesting epithermal type gold mineralisation elsewhere, including in one area on a large island close to Turkey. The latter resulted in an abrupt early morning invitation to the local police station where I spent an interesting few hours trying to explain my interest in nature and geology to a couple of fairly heavy grey-suited characters with assistance from a bilingual lady-friend of one who had been cajoled to interpret. They were not in the least interested in volcanics and were suspicious of anyone who was: "Why were you taking photographs of an army camp hidden in a heavily wooded circular valley?" "Well I didn't know those buildings were an army camp and actually I was photographing the interesting volcanic caldera that is not shown as such on the government geological map". "Well if it is not on the government map then it does not exist. This proves the point that you were photographing the army camp and therefore you are obviously spying for that country just over the water that you can see from the window". Brilliant deduction my dear Watson!

I never did find out what the lady thought of my explanation of interest in the volcanic structures of the northern Aegean. I felt she was a little sorry for my predicament but she probably thought a couple of weeks in the slammer might get me back on the straight and narrow. My motorbike was confiscated and it was suggested that I head back to Athens pronto. I spent the next morning strolling along the beach, waiting for the bus to the airport and, to my surprise, found a nice piece of epithermal quartz vein and traced the trail back to a creek draining onto the beach. Something to come back to – which I did several years later.

The return of reasonable gold assays from Milos presented a quandary as to what to do next. The discovery of invisible gold on an island where glassy obsidian had been mined for spear and arrow heads since 9000 years ago; and where extensive open pit clay mining had been continuous since Minoan times, 4000 years ago, through to the present day, would surely set off alarm bells. Milos is the largest perlite producer on the planet. The operative phrase was "proceed cautiously". A family connection of a Niugini Mining board member was a very astute young Athens lawyer who initiated the plan hatched to apply for exploration and mining tenure for "all metals". Element 79 (gold!) should not be mentioned specifically as its occurrence in Greece was not common knowledge so it might create a nationalist stir. Discussion at a high level with the relevant minister and senior bureaucracy established that a foreign company was welcome to apply for mineral rights on condition that a Greek company would eventually be established if the search was successful. Not a problem! No drama! Three long years of negotiation ensued. We found that discussion and debate are a major part of Greek culture. It is the debate that is important not so much the outcome. It is a way of life. Aspects of Australian mining law were a big part of this interminable talk-fest.

To cut a long story short the Ministry in its wisdom finally decided to offer the mineral rights for the area for public tender. *Epharisto poli!* Thank you very much! It just so happened that our bid was just pipped at the post by a large Greek industrial minerals company, Silver and Baryte Mines, that had substantial industrial mining interests, including perlite, on the island. Several generations previously that company had pitted a small barite and silver deposit on the edge of our area of interest. Consequently we offered a joint venture arrangement should they feel the need for some experienced explorers and foreign funding.

A series of joint ventures involving Greek, Australian, French, Canadian and USA companies ensued over the next decade which saw a number of very classy veins and breccia systems defined at substantial cost and the gold ounces content exceeding seven digits. Eventually the inevitable Greek tragedy saw the provincial government offside on some trivial issue, northern European Green activists take a sudden interest in the rare threatened snake species of Milos Viper, and a claim made, incorrectly, that this rare animal just happened to nest in the middle of our gold vein system. Our Environmental Authority was cancelled and exploration terminated.

If one visits there today one sees an array of wind turbines beautifying the higher points on the ridge where the quartz veins crop out and one can take comfort that the geology of the turbine foundations was mapped in detail by an Australian geologist. Wind turbines obviously don't disturb vipers. *C'est la vie* or as Greeks say *Afti einai I zoi.*

YOUNG TURKS

IAN PLIMER
TURKEY 1987 - 1992

Professor Ian Plimer is Australia's best-known geologist. He is Emeritus Professor of Earth Sciences at the University of Melbourne, and has served as Professor and Head of Geology at the University of Newcastle, as Professor of Mining Geology at the University of Adelaide and German Research Foundation Professor at the Ludwig Maximilians University at Munich. He has published more than 120 scientific papers on geology and was one of the trinity of editors for the five-volume Encyclopedia of Geology.

He has been granted numerous Australian and international medals and awards for his scientific and educational works. He has been an advisor to governments and is a regular public broadcaster.

He has written and spoken extensively and fearlessly on climate change matters, with emphasis on the need for scientific rigour and veracity, common sense, and the perspective and relevance to be gained from the long-term inputs from the geological timescale.

He began his geological career in Broken Hill, and has extensive exploration and mining experience. After leaving academia, he has served on the boards of several listed and unlisted mining and exploration companies.

In the late 1980s and early 1990s, a gold exploration program was run in Turkey for Niugini Mining Ltd by John Nethery. There were no maps and so a group of geologists comprised of he (Neddy), Mike Barr (Kiwi Mike) and I, together with Turkish geologists Ömer and Okan, scoured the country looking at areas of interest.

These are stories of some of the Turkish children we met.

Aleşehir

Western Turkey has experienced stretching which resulted in the sliding down of huge uplifted blocks from the basement. Continued stretching has produced a series of seven flat-bottomed graben valleys separated by rugged mountains of older basement rock.

The graben contains a great thickness of water-worn sediment and braided streams that are continually changing course. The valleys are in the agricultural heart of western Turkey and provide fruit, wine, grain and nuts. The edges of the graben are active faults characterised by frequent earthquakes, geysers, hot springs and boiling mud pools. Sulphurous gases fill the air. The valley walls are rocks, steep and forested. The hills and thyme-covered slopes are grazed by sheep that provide milk, yoghurt and scented Işkender kebab, all with a distinct thyme taste.

Hot sulphur-rich geothermal fluids can carry large amounts of metals in solution. The most commonly carried metals carried are gold, antimony, arsenic and mercury. If these fluids are suddenly boiled or chemically react with limestone, coal or salts, then a mineral deposit may form. Such areas are geologically attractive for gold and silver.

I was attracted to these graben because, in the long ago in this area, King Croesus built a fabulously wealthy empire based on gold. His city of Sardis is still well preserved. The King's gold derived from the alluvial sediments on the floor of the valley and elaborate sluicing treatment works had kept thousands of slaves employed, fed and watered. In those times, rainfall and temperature was higher than now and there was more water for sluicing. There was presumably little gold left in the valleys as slave power is very cheap, very efficient and, in relative terms, the price of gold has not changed for thousands of years. A troy ounce of gold for the last thousand years has been worth a week of a carpenter's time. The gold mined by King Croesus undoubtedly originally came from the hills - but which hills?

After some investigation, it became clear that the gold derived from dissected terraces of older partially consolidated gravels draped half-way up the walls of the graben. These gravels were once flat lying sediments on the valley floor, now they were steeply tilted well above the valley floor as a result of very rapid land rise.

In tectonically active areas, the geological processes are very rapid and the land rises and falls quickly. For example, on the southern coast of Turkey, subsidence has drowned ancient Lydian cities that are now a few metres underwater, whereas the famous port city of Efeses is 15 kilometres inland as a result of uplift. There can be no understanding of sea level change without an understanding of land level changes, which can be very rapid.

The historically mined gold had come from the source area of gravel, and the cobbles and boulders in the gravel immediately told me where to look for the primary gold sources. They were fractured limestone masses perched high above the valley floor at the boundary between the graben and the basement rocks. It was here that hot acid gold-bearing fluids had chemically reacted with limestone and precipitated gold in rocks. This gold is normally invisible and grows into larger grains in alluvial sediments.

I climbed up donkey tracks to the part of the scarp that I thought was shedding gold into the valley during aeons of uplift, weathering and erosion. I knew I had come to the right place because there were hundreds of old pits, an abandoned crushing and mercury distillation plant, and large dumps of waste that had been roasted to volatilise the mercury which was used to recover the gold as an amalgam. A village had been built on the area flattened out by the old mercury mining and distillation activities. For some decades last century, the village was able to feed itself as a result of the employment of its inhabitants in the mercury mines. The village probably provided the mercury needed to detonate the shells for the Turkish defence of Gallipoli.

The village dwellings were made of mercury-bearing stone. The tracks were cobbled with mercury-bearing tailings and the village well was an old shaft used for extracting the mercury-bearing ore. The village had its normal quota of ferocious-looking cowardly wolf dogs, elderly men sitting in the sun, old humped women carrying the firewood and doing the heavy work, deep mud and dung, and pens constructed from sticks and stones for the goats and sheep.

As per usual, one must seek an interview with the village mufti and explain why one is in the area and what will be done. After the necessary niceties, the village mufti instructed a young boy, no older than eight years old, to be my guide. No guide was necessary as I was looking at rocks in a five square kilometre area around the village but my arrival presented a commercial opportunity for the village.

My guide was a charming child. He was a typical village boy with a shaved head, ear-to-ear smile, ill-fitting mud-caked hand-me-downs and plastic shoes with no socks. He knew every one of the old mercury mines in his playground and he took me to every shaft, pit, costean, ore dump and prominent outcrop. He knew his patch well.

My child guide had severe brain damage, typical of congenital mercury poisoning. He was constantly shaking, had slurred unintelligible speech, deformed digits and ears, and constant salivation. He showed passing interest in my maps, compass and satellite-based ground positioning system and was absolutely fascinated with my Estwing geological pick. He would keenly watch how I would select partially silicified limestones, break off a number of chips, place these rock specimens in a cloth sample bag and label the bag with an ink pen.

He indicated that he would like to collect my samples for me; I gave him my hammer and, after attacking every rock within sight, I managed to coerce him into sampling only the rocks which I considered were of geological interest. Far too many samples were collected and, as it transpired after analysis of the samples, none of the specimens were of interest. This is not unusual in the exploration business.

Some of the mountain streams were fascinating. Long tapering ice crystals had grown in from the banks to the centre of the stream. Water flowed under the ice and only in the centre of the stream. Because I had often seen the same texture in quartz veins which formed from silica-saturated hot fluids flowing along fractures in rocks, I took a number of photographs.

My little guide then took me to every icicle, every ice waterfall, every ice-laden stream and indicated that these sites were far better for photography. I wanted to photograph an analogue of a quartz vein in rock and he thought I was interested in photographing ice. He was a lovely kid. We had lunch together next to an ice-laden stream in the sun and out of the wind. He was highly complimented when I took his photograph.

When back at the village, I paid the mufti for the boy's services and gave my little guide the same amount of money. Custom required me to have a çay with the mufti and senior village men using what was mercurial water. Boiling water to make tea does not get rid of mercury. The young boy was dismissed. Upon my departure after just one cup of tea, I asked the mufti to find my guide for me. He was again presented to me, I offered him my geology pick as a present which he took with a high-pitched yelp of joy and ran off down the village road swinging my pick around his head.

He was happy. The village will continue to look after my little guide for the duration of his shortened life.

Kücükyenice

As Ömer and I drove up the track from the river, we could see a solitary miner working in an open pit at the top of a hill. Our rather poor maps recorded no mines in the area but we had been attracted to the area because of its apparent geology, also derived from poor maps. The sandy, limey and salty sediments of an old lakebed had been cut by a sequence of 20 million-year old volcanic rocks. The ancient lake had formed by the stretching and subsequent sinking of the crust. Sinking occurred in fault-bounded blocks that went from Balya to Bergama in western Anatolia. It was along these faults at Balya that lead and silver had been mined since antiquity and gold had recently been re-discovered near Bergama.

An old miner was very happy to stop working and let us look at the rocks in the pit face. He was mining yellow antimony oxide on a tribute basis for an antimony smelter near Gönen. The antimony oxide mineral cervantite just looks like a mixture of yellow iron oxides and clays which are common in all weathered rocks. It was clear that we were dealing with an experienced miner who knew his stuff. He had to know his stuff because there was no old age pension, social security or medical benefits in Turkey. He may have been 50 or he may have been 70. It's hard to estimate the age of someone who has had a hard life in the open doing back-breaking physical work. Even at his age, he had to work to live; all he knew was mining and, if he didn't physically exert himself in an open pit, then he didn't eat. Mustafa was poor and proud.

I was interested in the host rocks. They were altered limestones. Alteration had occurred from hot antimony-bearing fluids passing through the porous permeable reactive limestone and depositing opaline silica, clays and the bladed silvery-grey antimony sulphide called stibnite. It is a very easy mineral to identify in the bush. If a match is rubbed along the face of a stibnite crystal, the match catches alight. No wonder it is used on the sides of matchboxes.

Mustafa the miner had a collection of picks, pelican shovels, sledgehammers and drill rods. It was all hammer and tap for rock drilling and no explosives were used. Boiling water was produced from a fire and a billy-full of it was poured into the drill hole rock to try to make the rock crack. This was one of the mining techniques used by the Romans. There were wooden wheelbarrows and leather bags for collecting ore. He was dressed in thick tattered muddy clothes for working on the northern side of the hill in winter. No hard hat. No safety boots. No gloves. No mask to filter out the deathly silica dust. Just a green woollen cloth cap and a pair of cut down Wellington boots. The boots probably leaked. It could have been a scene from the goldfields of eastern Australia in the 1850s. I took photographs of this living history and offered to send prints to Mustafa. I asked him to write his name and address in my field book for the despatch of the photographs. He changed his mind and said he really did not want the photographs after all. It was then I realised he was illiterate. I should have been more sensitive and I lobotomised myself for this stupidity.

We were given a conducted tour of all the small pits, shafts and open cuts in the district. We went to inspect çakmak hill, the original reason for our visit. Çakmak is the Turkish word for flint that was used by the locals to describe a great diversity of silica-rich rocks. Because only some of these varieties were of interest to us, all çakmak locations in volcanic settings had to be inspected. The most interesting locations of çakmak were normally steep rocky pinnacles; old place names such as gümüs (silver), altin (gold), şap (alum) and bakir (copper) gave clues about what might have been mined at that place in antiquity.

The collection of silica-rich rocks I sampled suggested that we were in a place where very acid hot springs had repeatedly and explosively boiled. At times, the hot springs had exploded and the çakmak had been broken into sharp angular pieces, slivers and needles which were re-cemented by silica from a new influx of silica-rich waters. This jigsaw of fragments could be put back together showing that the fragments had only moved centimetres. This also showed that boiling was explosive, the energy to convert water into steam came from the superheated water itself which then instantaneously cooled and precipitated previously dissolved silica. Some geological processes take aeons whereas what I was looking at took less than a second. I could see remnants of an old land surface that told me that the hot springs had boiled close to the surface and, from experimental studies and work on the New Zealand hot spring systems, I deduced that any gold would have been precipitated at a depth of about 200 metres. Traces of gold, silver, mercury, antimony and arsenic should occur at the old land surface. This was confirmed by the texture of the opaline silica in the çakmak, the presence of ruby red crystals of the arsenic mineral realgar and a large surrounding bleached zone where acid steam had converted the hot volcanic rock into flinty clay. I was getting geologically excited which, as all exploration geologists know, can be dangerous and expensive.

My experience and knowledge of natural systems gave me all the clues I needed but I needed rock chemistry to validate my observations. I knew that I would be back in the area. All the rocks in the area had that smell of gold. Sometime in the future, I had to work nearby. By then, all samples would have been processed and chemically analysed. I could then be armed with greater knowledge and have another brief look at the area as part of planning the logistics for the next summer season's exploration.

The assay sample results were, not surprisingly, encouraging and I passed through Kücükyenice on my way to another field area to the north. My intention was to do a brief compass and tape survey of the antimony mines before starting the exploration season in earnest. My survey started on a ridge where the rocks had been slightly hardened by silica hence underwent less weathering and erosion and stood out as high ground. There were a number of old open pits, an old two-compartment shaft and a disused adit. One open pit had recently been reopened and the same Mustafa the miner was extracting stibnite ore. When he caught sight of me, he downed tools and sprinted out of the pit. Mustafa hugged me, thanked me for sending the photographs and wanted to stop work and take me to the çay house.

Prints of the photographs I had taken were enlarged and sent from Australia and I just addressed the airmail package to Mustafa at Kücükyenice village in western Anatolia. Very few village people receive post, mainly because most village folk are illiterate, and the Turkish postal system must have thought that this delivery from Australia was very important. It was a marvel of the Turkish postal system that he actually received the photographs that had in turn created great interest in the village. Not only was the status of Mustafa the miner greatly enhanced, but also an air of intrigue had developed about the mystery photographer from far away. In order to finish my survey, we agreed that I would meet Mustafa at the çay house at dusk.

When I came down from the hills to the village, it was clear that Mustafa the miner had reached the çay house well before me. Business was as usual for the village women who had their coloured washed clothes spread out on rocks and hawthorn hedges to dry. Other women were cooking the daily bread in the village's communal oven and others were sitting around in a circle on the ground grinding wheat by hand into flour.

A group of children stood on the road that separated the village school from the bread-making activity. They would not let me past unless I took group and individual photographs. These were absolutely delightful kids. The boys with crew cuts, ill-fitting hand-me-downs, open plastic shoes and no socks. Some had a few front teeth missing which made their smiles even more beautiful. The girls were too young to wear chadors and were draped in cotton print pantaloons, pinafores, blouses and jackets. All smiles and a lot of noise but they kept their distance from this stranger with white hair and blue eyes. A number of the children had rounded faces, copper-coloured hair and aqua blue–greenish eyes. These are Turkmenistani features. One particular girl stood out from the crowd. She was taller than the rest, had a long shock of almost iridescent copper-coloured hair and piercing aqua blue-greenish eyes. I took a few rolls of film of the group and individuals and promised the children that I would send the photographs to the village school, the kücück eskola.

At the çay house, there were the normal pall of smoke and numerous idle, smoking, talking, tea-drinking men. In most villages the women do the hard physical work as well as bearing children, cooking and washing. At the çay house, I was warmly welcomed individually by every man and was directed to sit down at the head table. Everyone wanted to honour the guest with a glass of tea and, after about ten small glasses of the sweet tannin-rich çay, I laid my small thin aluminium spoon horizontally across the top of the glass indicating that I wanted no more. I explained who I was and what I wanted to do in the summer season in the hills overlooking the village. The village was to be my base, I would employ a few chain men to help me and I would employ Mustafa the miner to clean out the abandoned adit so I could go underground and map the rocks.

The teacher from the one-teacher kücück esloka, Ibrahim, insisted that I talk English with him. In the field, I always carried an English-Turkish/Turkish-English dictionary with me, despite having a working knowledge of Turkish. Ibrahim had brought with him a tattered dictionary from his school. He needed it as much as I needed mine. We conversed using nouns separated by the frantic searching for words in our dictionaries. Our conversation was very slow, had many misunderstandings and amused the onlookers crowding around our table. The teacher challenged me, in English, to a game of backgammon. It was on.

Backgammon is played everywhere in Turkey, especially in the çay houses. In the big cities, it is tea drinking, smoking and backgammon that are the main activities in bazaars, back streets and shops. It is a game I knew well from my previous geological work in Iran and from social occasions in Australia. I accepted Ibrahim's challenge because I thought that I was a better than average backgammon player. It is a Persian game that was given to the Raj of India. On a state visit, the Raj of India gave the game of chess to the Shah of Persia. For the return visit, the Shah asked his intelligentsia to invent a game that, unlike chess, was a game of both skill and chance. A game with no complicated pieces. A game more like the game of life. A game that anyone could play, even an Australian in Turkey. The game invented and presented to the Raj on the return visit was backgammon. Even Persian prisoner slaves from the Punic Wars chained underground in the Lavrion silver mines of Greece had cut backgammon boards out of the limestone and made dice and counters from clay.

It is an international game thousands of years old. It is a game I enjoy. A fast game is a good game. I started to play this game of chance and skill against Ibrahim and, much to the enjoyment of the onlookers, Ibrahim's skill was far greater than any chance given to me by throwing the dice. I could not win a single game. I tried playing skilfully, recklessly, conservatively and randomly and just could not win a game. The crowd discussed my every move and, after a while, realised that I needed some help. When I went to move a counter, an anonymous gnarled guiding hand from the encircling crowd would stretch out over the board, grip my hand and make sure that my counter ended up in the correct spot. I then started to have a few wins. We must have played thirty games before I threw in the towel. I was looking forward to coming back for a summer field season.

Enlarged photographs of the children sent from Australia preceded my next visit. I drove slowly from Ivrindi to Kücükyenice along the battered dirt roads in order to understand the lie of the land and to get to know the geology better. Road cuttings, local road base, stone walls, stone buildings and piles of rubble in the fields are invaluable clues to the local geology. No one is going to cart stone from miles away to build a stone wall around a field. These walls were obviously made of stone collected from the field to make ploughing easier. As I drove up the track from the river, I saw a new pile of fresh rocks. These clearly had not come from a new mine as the rocks were rounded ones of fresh volcanics that had been carried up the hill from the river. I stopped and saw that a new well was being constructed for the village.

The well was more than twenty metres deep and access was down some very flimsy frayed rope and spiralling wooden ladders. Materials were hauled using a counter-balanced battered bucket and steel kibble with a somewhat worn rope over a pulley. If the load shifted or fell, anyone underneath would have been killed. Safety is a luxury we wealthy westerners enjoy. The shaft had been dug by hand to considerably beneath the water table. During digging, the shaft had to be kept dry by continual baling out of the inflowing water using the bucket and kibble. The well was now being filled with large round boulders from river gravels to stop the walls collapsing and to make the water accessible. At the bottom was my long-lost friend, Mustafa the miner. I greeted him and came down because I was interested to look at the fresh rock in the bottom of the well.

He hugged me, we sat at the bottom of the well and ate some bread. It appeared that once the new village well was finished, Mustafa had no work because the London Metal Exchange price of antimony had slumped due to the dumping of antimony on the market by the Chinese. The excess antimony on the market probably came from the Dachang tin-lead-antimony mines near Guilin that I had visited a few years earlier. Mustafa could no longer eke out a living as a tribute miner and he was at the mercy of the international metals and money market beyond his ken and so far away. He was an innocent bystander who ate or starved depending on matters beyond his control in far-off lands. Mustafa was relying on my return such that he could eat over summer.

After we made arrangements for the dewatering and cleaning out of rubble from the old two-compartment shaft and the connecting adit in the hills behind the village, I made the dangerous climb to daylight and my vehicle. I continued the drive up to the village and, at the edge of the village, was met by the normal mangy dogs, the odd free-range goat and old women walking back to the village doubled up under a load of firewood that they'd just cut. I stopped in the town square in the late afternoon sun between the çay house and the mosque and was immediately surrounded by a crowd of children who appeared out of nowhere. They clearly were very pleased to have received photographs of themselves and, for most of the children, it would have been the first time they had seen their own photograph. I was later to learn that, in some of the village huts, my photographs took pride of place on the wall.

The children were dancing around me and squealing with excitement. It was like a scene of the encircling Willis' midnight dance from Adam's *Giselle*. These will-o'-the-wisps were silhouetted against the mosque, swarming in the dusty hazy light like butterflies and encircling me with flashing cabalistic colours. School had just finished, we gave the children some boiled lollies and tried to quell their excitement. A few lolly papers were dropped in the town square, my backgammon-playing teacher Ibrahim appeared, chastised the children and instructed them to clean the whole town square. This they did. The teacher placed the oldest child, the tall girl with copper-coloured hair and the aqua blue-green eyes, in charge of cleaning the square. This was the way Australia was when we were poor.

My work kept me in the village for two months. A routine developed. An early morning gathering of the chain men and miners I employed would disperse to our various chores, we would all meet on the dumps of the old mine for lunch where someone was employed to keep a fire going and another person was employed to make tea and, at sunset, we would drift back down to the village çay house. I now had a special table for the plotting of maps, downloading of data and organising the next day of activities. Every morning when I left the village to work in the hills, the copper-haired girl was waiting for me. Every evening I returned, she would be waiting for me outside the school that was on the outskirts of the village. She would warmly greet me with smiling eyes. What was she thinking?

Every night at the çay house, Ibrahim the teacher would ask me to come to his school the next morning to have a cup of coffee. Every invitation I refused because my field work in the hills was not finished, the first snow falls had come and I was aware of how long it took to partake in the simple act of drinking coffee or çay in Turkey. Maybe the Protestant work ethic some of us have in Western democracies is a blight on our lives?

At the end of the field season on my last day in Kücükyenice, I agreed to come to the school at 9 am for a cup of coffee. When I arrived at the village school on the slope overlooking the village, there was a school assembly in the snow-covered school yard. I was the honoured guest,

Ibrahim the teacher formally welcomed me in front of the assembly and the children saluted me. It reminded me of my post-war primary school days when straight lines, silence, undivided attention, marching, the singing of a hymn and the saluting of the flag was the normal morning ritual at assembly. I'm sure it did me no harm.

There was only one school room and one teacher for all the village children of all ages. A barn-like room with a worn knotted oak plank floor, a wall slate, a central potbelly stove, benches for all the children and a front desk for the teacher. It was entry into a 19th Century Australian bush schoolhouse. As soon as I entered the classroom, I knew what was meant by coming up to the school and having a cup of coffee. On the slate was drawn the outline of the continent of Australia. Tasmania, as per usual, had been expunged. It was clear to me that I was to give a lesson about the land I came from and that the children had been primed.

These are the challenges I enjoy. I spent the morning in the classroom providing an illustrated lesson on the geography and history of Australia, including Tasmania. I drew and labelled in Turkish the mountains, cities, agricultural and mining regions on the wall slate board. The children furiously copied everything I drew and wrote on their own smaller slate boards. As a result of my one and only lesson at the kücük eskola at Kücükyenice, whenever I now work in impoverished countries, I bring along scores of pens, stolen from hotels, and notebooks to distribute at the local school. A heart-starting Turkish coffee was served at recess and a cleaned slate allowed me to draw aspects of the vegetation, climate, wildlife, fisheries and reefs.

I didn't mention the war. I wonder if these children knew that we Australians invaded their country in 1915 during World War I? I wonder if these children knew that the first act of Islamic terrorist killing in Australia was done by two "Turks" and took place in Broken Hill on 1st January 1915? I didn't mention the Ottoman genocide of Armenians in 1915 and 1923. These were not matters for children of this age, especially when linguistic difficulties could easily lead to misunderstandings. During my time in Turkey, older people had sometimes come up to me and told me that they are looking after our young men buried in their country. These people never told me that Turkish casualties in 1915 were four times those of the Allies.

It was then question time and this involved the rapid use of my Turkish dictionary. The first question was from a little crew-cutted boy who wanted to know the religion of Australia. This was an Allah-given opportunity to demonstrate the multicultural nature of Australia and the fact that we have Christians of numerous denominations, Jews, atheists, Hindis, Shiites and Sunnis all co-existing in peaceful harmony. I hope he understood that Australia had no state religion and that it is a secular society with all institutions underpinned by the basics of Christianity. Other questions indicated that the children had been stimulated by this foreigner from a far-off land. As expected, there were questions about crocodiles, snakes, sharks and spiders and surprise that we didn't have wolves and bears. I hoped that one day, if there is a great change and Turkey becomes a fundamentalist theocracy, that these people might remember the foreigner who gave them a lesson. He was not all that bad, was he? Maybe they won't accept xenophobia, maybe they might develop some religious tolerance, maybe they might even try to understand other cultures or maybe they might not take up arms for some spurious nationalistic, religious or ethnic cause.

The copper-haired girl was the class monitor. She was given the jobs of making and serving coffee, adding wood to the pot belly stove, disciplining and organising the smaller children, cleaning the classroom, providing me with chalk and cleaning the board. She was being trained to be a village wife and mother. This she will do well but it was just perpetuating the past. She probably will spend the rest of her life in Kücükyenice or an adjacent village doing what her mother and grandmother did.

I wondered if she had her appetite whetted about the world beyond or whether her mind had been opened?

I will never know, but I tried.

The flautist

The Anatolian Fault has been active for more than 100 million years. It is an east-west structure that can be traced from the Pyrenees to Kazakhstan. The sliding of one block of the crust past another has produced a long history of earthquake and volcanic activity in Anatolia and has led to the opening of the Marmara and Black Seas. Major earthquakes occur roughly every 300 years and multi-storey buildings are re-built along the fault scarp to replace those destroyed by previous earthquakes.

A recent rotation of stresses changed fault movement from a sliding motion to stretching which led to crustal extension and the pulling apart of the crust. Great blocks of the crust were thinned, broke off from the surrounding rocks and sunk with the inevitable associated earthquakes, volcanism and geothermal activity.

It is this stretching of the crust that allowed the intrusion of lavas into limestone at Sebepli. Sediments and small coal layers were deposited over the blocks of crust that had sunk. The cooling molten rocks which intruded during thinning released huge amounts of hot acid waters. These geothermal fluids moved along fracture systems and chemically reacted with limestone. Limestone is basic and reacts with acid. It also contains hydrocarbons hence can chemically reduce hot fluids. When geothermal fluids hit limestone, all hell breaks loose and chemical reactions take place very quickly.

Limestone contains 44% carbon dioxide by weight; reaction between acid and limestone produces carbon dioxide that is released to the atmosphere, much of the limestone is dissolved away and replaced by quartz. Cavernous quartz-rich rocks that were once limestone still contained brachiopod and crinoid fossils that had been totally silicified. The material leached out of limestone by hot acid fluids was dumped at the surface by hot springs as a sinter which may contain the tell-tale traces of antimony, arsenic and mercury and, in some places, gold and silver. There are huge changes in rock volume, density and chemistry - these alter the solubility of gold in geothermal fluids and gold is commonly rapidly precipitated. Our interest in this area was because of the yellow metal and such settings elsewhere in the world such as Carlin (Nevada) we had studied and visited.

In summertime, the line of sight in the oak-forested mountains of Sebepli was less than twenty metres so we decided to put in a grid in late winter after all the leaves had fallen. Oaks are one of the last deciduous plants to lose their leaves in winter. Once we had established a grid, we could use the spring season for mapping and sampling and, if targets of interest were delineated for a three-dimensional look, then drilling could be undertaken in summer.

Wintertime is certainly not the best time to be doing field work in alpine Turkey. The thick snow and biting wind made conditions somewhat unpleasant and there are just not enough hours of daylight to achieve a good day's work. But we decided to work in winter anyway because we thought we would have a greater line of sight for surveying.

Our plans for detailed investigation of the site were devised in Australia. Like all well made plans, they came unstuck by an unexpected fundamental flaw. Sebepli was at the ridge crest between the Biga Peninsula and the Marmara Sea. In winter the cold land mass met warmer moisture-laden air driven in from the sea and the area was shrouded in a pea-souper fog day and night. The line of sight was only ten metres rather than the twenty metres in summer.

Our days that winter were spent blindly stumbling on the precipitous slopes over jagged rocky outcrops of quartz, silicified limestone and limestone. We were most aware that our presence disturbed the villagers' activities in the mountain forests. There was the distant chopping of wood for cooking and heating fires, the tinkle of sheep bells and the occasional shepherd's shout.

Crib time was spent at the datum peg on the rocky peak of Sebepli Mountain. Crib is an old-fashioned underground miners' abbreviation for the 17th Century English board game of cribbage. It was played on a board with match sticks in semi-darkness in underground openings cut out to serve as the miners' lunchroom. The miners carried their crib (or lunch) down underground in a sealed metal tin (crib tin) to prevent rats enjoying their lunch. In the crib room, tea was drunk, cribbage was played, lies were told and, after half an hour, work started again. A routine at Sebepli was established. My job was to try to get the saturated oak to burn, Kiwi Mike cooked a steaming hot stew for lunch, Neddy prepared the billy of tea and Okan did the housekeeping with the downloading of data.

Our lunch spot was used to store pegs, equipment, excess clothing, cooking equipment, food and water. Like all isolated country areas, whether Turkey or elsewhere, items left in the bush are neither interfered with nor pilfered.

Every lunchtime, a few shepherds just happened to drift in from the thick fog to stand around our fire and warm themselves. They carried their thin ekmek wrapped in newspaper for lunch and they never rejected our invitation to drink piping hot çay with us. They enjoyed the company of strangers and our conversational attempts were greeted by toothless smiles.

One little shepherd boy, no older than twelve years old, was too shy to come in close to the fire. He was poorly clothed, poorly shod and often had no food. He always greatly admired our field boots. We would regularly give him a hot lunch which he would then take away from the fireside-populated comfort and eat alone. We never heard or saw him disappear back into the foggy forest to tend to his sheep.

One lunch, Neddy asked whether we had heard the sound of a flute drifting up the mountain through the fog. The three of us had been at the other end of the grid and had heard nothing. We suggested that maybe Neddy should drink less raki, a potent local firewater which occasionally made him hallucinate.

The next day, Kiwi Mike, Okan and I also heard the flautist's melodies. Upon questioning, the shepherds told us that the flautist was the little shepherd boy who comforted his sheep with music. In the thick fog, the sheep could not see their guardian shepherd. The boy's sheep kept within the safe earshot of the flute and the shepherd could also hear his sheep's distinctive bells. We tried everything to try to get the boy shepherd flautist to play for us over lunch. He was too shy and too self-conscious to play in front of an audience. He was clearly pleased that we enjoyed his music. We did notice that he now spent much more time herding his sheep in proximity to the grid which enabled us to hear music all day. The music was evocative and hauntingly beautiful. The melodies drifted in and out of tonality and were all an unfamiliar blend of Asian and European phonics. Thick snow, defoliated forest, interesting rocks and simple melodies drifting through the fog over Sebepli Mountain gave us an unforgettable experience of interesting geological field work set to music.

At times, one of us needed to drive into town to despatch our samples and get provisions. On one trip, Kiwi Mike made some purchases and presented the boy with a new flute and a pair of field boots. The boy took the gifts and quickly vanished into the foggy forest. We couldn't see him but he thanked us by serenading us with his whole repertoire during our lunches at the datum peg at Sebepli.

Our young flautist had never been to school. He had never had formal music training. He probably couldn't read or write. Why would he need to? He had never been more than five kilometres from his home village. He had been entrusted with the villagers' sheep and probably will have a stable fulfilled life comforting the village sheep in his care with music from his flute. His sheep would never know how lucky they were. He will never experience a prosperous or totally destitute life. He will do the same thing year in, year out all his life. It will be a simple life. Will the 21st Century even have an impact on him? Will he ever hear a symphony orchestra? Will he ever hear the hauntingly beautiful flute music of Mozart - such as the passionate andantino movement of the Concerto in C, K.299 or *Die Zauberflöte*? I suspect not but I hope I'm wrong.

The last time I was in that neck of the woods a few years later, I made a diversion to pass through Sebepli *en route* from Izmir to Istanbul. I went to Sebepli Mountain and tied my field boots to the top of the datum peg.

He'll know where the boots came from.

Şapdağ girl

She was well aware that, in a matter of years, she would die a slow horrible death as others had before her. All her childhood had been spent knapping ultrapure silica into cobbles for use as grinding blocks in ceramic works.

Agricola knew that working in an environment of silica dust produces irreparable lung damage within months, coughing, silicosis and death at a very early age. Sharp spicules of silica pierce the lungs to produce bleeding, lithification of lung tissue, loss of lung expansion and a painful death by drowning in your own blood. Agricola wrote:"...*some mines are so dry that they are entirely devoid of water and this dryness causes the workmen even more harm, for the dust, which is stirred and beaten up by digging, penetrates into the windpipe and lungs and produces difficulty in breathing...it eats away the lungs and implants consumption in the body.*"

Even if she had read Agricola's 1556 AD book *De Re Metallica*, it was too late. She probably couldn't read anyway and she certainly would not have had access to this classic volume in Latin or English. Herbert Hoover and his wife translated the book from Latin to English. He worked as a mining engineer at the mines in Broken Hill (NSW), Gwalia (WA), various places in China and Bawdwin (Burma). He later became President of the USA and, although not an outstanding President, he could read, understand and translate Latin - unlike Presidents who followed. She would not have been aware of Calvert's Holland's 1843 AD description of the working conditions of cutlery grinders in Sheffield who used silica for grinding and polishing:

"*The dust, which is thus every moment inhaled, undermines the vigour of the constitution, and produces permanent disease of the lungs, accompanied by difficulty of breathing, cough, and a wasting of the animal frame, often at an early age of twenty five. Such is the destructive tendency of the occupation that...many sick clubs have an especial rule against the admission of dry grinders, as they would draw largely on the funds from frequent and long-continued sickness...Grinders' asthma in its advanced stages admits neither of cure nor of any material alleviation.*"

Large volumes of silica-laden steam some 20 million years ago moved up fractures in the volcanic rocks of Western Anatolia. The fluid stripped material out of the volcanic rocks and left pipes of porous silica behind. Small amounts of alum (şap) and gold-bearing crystalline silica were also deposited. The silica was so pure, so hard, so even-textured and of such ultrafine grainsize that it was prized as the grinding blocks for clay in ball mills at ceramic factories.

The silica mines were perched beneath the alum cliffs of the mountain called Şapdağ. Mining was by drilling and blasting, and sizing by sledgehammer. This was done by men. Drilling used water to cool the drill bit. This water also served to keep down the dust and, as a result, the men did not inhale silica dust. A group of teenage girls and women sat in a circle on boulders on the edge of the cliff underneath a bamboo and reed gunyah knapping the sized boulders down to small cubic blocks with a hammer and chisel. Knapping produced dust and sharp fragment of excess silica. We went to the pine slab hut and had çay with the foreman. The hut had a pot-belly stove in the centre for cooking, making tea and warmth, and around the pot-belly stove were a number of four-gallon drums for seats, some wooden benches and a warped trestle table. Pinned to the wall were three home comforts. The statutory photograph of Kemal Ataturk, last year's calendar from a ceramic company in Bursa and a large faded colour photograph of all the mine workers.

One face on the photograph had been cut out. There was just an oval hole in place of the face of a woman. It appears that she was a young married woman who had the misfortune to be looked at by another man. That man was not her husband. Her head was lifted high enough for another man to look into her eyes. She now no longer works at the mine and all memory of this sinful woman had been erased. Another face in the photograph attracted my attention. It was that of a strikingly beautiful teenage girl, maybe eighteen years old who, even in a photograph, had a sense of presence. Her cotton print pantaloons were tied at the waist and feet, she had a floral blouse and was partially covered by robes. In my mind, I called her the Şapdağ girl. Her head was covered with a chador which made her even more attractive.

After a few cups of sweet çay, we climbed the cliff to inspect the geology of the silica mine. I asked the foreman if I could take photographs and the men in their dirty ragged work clothes posed for a still-life photograph of them swinging sledgehammers. Such photographs could have been taken in the goldfields of California or Victoria in the mid-late 19th Century. The miner's clothing, footwear and lack of safety equipment were no different. No steel-capped boots, no hard hat, no gloves, no dust masks, no eye protection and no spats for leg protection.

The women looked on while the photographs were taken, talked, laughed and tried to avoid eye contact. Many of the knapping women had a rasping cough and I noticed that the Şapdağ girl was among them. She too had the cough. She was still in the same clothes from the photograph in the hut, she was a little taller and older, maybe in her late teens or early twenties. She was even more beautiful than in the photograph and her poise suggested that she was just not aware of how beautiful she was. She was confident and challenging and her brownish skin was covered in grey silica dust. A young mineworker only had eyes for the Şapdağ girl but she didn't seem to, or didn't want to, see him.

To my surprise, the foreman suggested that I photograph the women. One of them must have quietly asked him and, because I was no threat to the social stability of the village and their women, he gave me permission. Some of them didn't want to be photographed and the rest wanted a group photograph. The Şapdağ girl not only wanted to be photographed with the group but, to my astonishment, asked to be photographed alone. She removed her headscarf for the photographs and actually looked into my eyes. She smiled. Her greenish-brown eyes were piercing me, they followed me round like the eyes of the Mona Lisa. They never left me. I wondered: what she was thinking? Did she realise that she was doomed? It was fortunate that there was some employment for her within walking distance of her village and the money would have kept her extended family fed. The Şapdağ girl had to work otherwise a few people would not eat. Marriage would have been arranged but not to the young mineworker who admired her and, if she didn't die in childbirth, then the silica dust in her lungs would leave young orphans to repeat the cycle. She probably knew this and accepted that this was her future and a portrait photograph would be her immortality.

I returned to the Şapdağ silica mines after an absence of some years. The photographs I took of the workers were also hanging in the çay hut and the pit had advanced one more bench into the cliff. There were still young women knapping silica into cobbles. Some of them had the same rasping cough I had heard on my previous visit.

The Şapdağ girl was not there.

She would have been, if a watery mist for dust suppression was used or if she had worn a cheap dust mask.

Gümüsler

In the springtime, the nomads bring their flocks of sheep from the valleys to the emerging grasslands of the high Taurus Mountains. They travel long distances to graze their ancestral pastures. They always travel the same route, graze the same pastures and camp at the same springs. Wooden picket sheep enclosures are constructed at the camps for protection at nighttime and their dogs have metal-spiked collars to prevent them having their throats ripped out in a fight with a wolf. Each day the shepherds bring the flocks from the pens to the grasslands and back.

After the departure of the shepherds last autumn, we surveyed in a very large grid. Each point was defined by a marked wooden peg. Such pegs are ideal for the construction of sheep pens, for firewood or carving into crooks. We had experienced this before and were in the habit of asking the locals not to remove our grid pegs until we had finished our surveying work and the winter snows had melted. Once they knew what we were doing, then the only pegs that were disturbed were those pushed over by sheep trying to scratch themselves. We kept track of the shepherds as they came through Niğde to Gümüsler. The Turkish word for silver is gümüs and, because the maps were so poor, we had to use ancient place names to give us a clue about the geology and old mines. We knew why we were in the area but did the shepherds know? As the caravan of shepherds and their flocks ascended through the gridded areas, it was explained what we were doing and why we needed the wooden pegs in the ground over summer. They were happy to have company on their ancestral mountains and the chief invited me to visit them once they had established their alpine camp.

The Taurus Mountains snake their way through southern Turkey. They formed as a result of the twisting and pushing associated with the opening of the Mediterranean Sea as northward-moving Africa collided with Europe. Large kilometre-thick slices of older rock had been pushed or thrust over younger rocks. These thrusts once contained broken rocks that were now cemented together with quartz. These thrusts were at altitude and draped themselves over hill and dale. It was here that we found the gold. Well, not really. It had been discovered in pre-Roman times.

We certainly were not the first to be attracted to the area. The thrusts had numerous small pits which had been sunk for gold. The workings were minor, very little tonnage had been removed and they were possibly of Hittite age. The miners knew what they were doing. The pits were at the intersection of fractures in the thrust where the gold had been remobilised and grew as coarser grains. Glacial gravels shed from the high peaks had been mined on a massive scale on the grassed alpine plateaux and Roman relics were still present. It was fortunate that at that time the planet was considerably warmer than at present and the Romans were probably able to work their mines all year round. Unlike the Carthaginians, the Romans did not explore for minerals and invasion was commonly to acquire areas that had previously been mined for lead, silver, copper, tin and gold. So too at Gümüsler. The silver and gold were probably originally discovered and mined by the Hittites which gave the Romans good enough reason for invasion. The home of the Hittites, Cappadocia, was not too far away. Hittites were miners and it was quite possible that they worked the high Taurus Mountains in times when the climate was far more hospitable than it is now.

High in the Taurus Mountains, I wondered whether the average suburban Australian was aware of the influence that Hittite miners has on their lives. Although not an expert on the culture of nanology, I am aware that the humble garden gnome was used in the underground iron and copper mines of Cappadocia. In this land of phallic hills, strange monuments and cave dwellers, the garden gnome at a cave dwelling or underground mine entrance was a necessary protection against evil spirits and devils.

Like many good ideas, it was stolen by others. The gnome entered Teutonic mythology with the exploitation of metals from the mines of central Europe in the Middle Ages, of which the best-known characters are Snow White and the seven vertically-challenged. The area of ancient mines in Saxony and Bohemia centred around the Erzgebirge also required gnomes to ward off evil spirits and devils, despite the deeply Christian population. There are some in this life who have been called to be gnome specialists. The purist gnome specialists tell us that a gnome can be no greater than 68 centimetres high, must have a grey-white beard and must be dressed in a red bonnet (*Zipfelmütze*), a green apron and over-sized shoes. Next time you look at a humble garden gnome, please treat it with far more respect as it is a relic from thousands of years of superstition. And they work. No one in suburban Australia with garden gnomes can prove that they've ever been visited by evil spirits and devils!

Traverses through the alpine grasslands and across the spine of the Taurus Mountains were necessary to map the shape of the thrust surfaces. On one such traverse, I came across the shepherds' camp. Their tents were circled around a spring and the water ran down slope across the fresh alpine grasses into a number of picket sheep pens. The white tents were a hybrid between a yurt and a tepee. Most of the shepherds were out in the hills tending their flocks in the alpine meadows and the chief came out to welcome me.

A visitor is a gift from Allah. A large red carpet was laid out on the grass in the sunshine for the chief and his visitor. A circle of onlookers developed and we were waited on by a number of women. My field boots were removed, I was given an embroidered cushion to sit on and I was welcomed with a drink of cold curdled sheep's milk. This was followed by another drink, eyran, made from diluted yoghurt and some ekmek (a local bread). The chief discussed the quality of the grazing lands this summer and I was able to explain in detail about our exploration activities.

The chief was inquisitive about the other lands I had worked in and a number of times commented upon my white hair and blue eyes. The onlookers pressed closer for a full facial view of their visitor. It was clear from the quality of my clothes, my field vehicle and the equipment that I used that I came from affluent circumstances. The chief asked me about my homelands.

Where was your homeland? How did people live? What did they eat? Was there enough food? How did people dress? Did the young go to school? Did the snow also cover the grazing lands for half the year? What was the religion of my homeland? Who was my leader? Why did I need to leave my homelands? Did I have any sons? Explanations were not constrained by language but by the size of one's world. I pulled out my field book and drew a map of the world for the chief. I showed where his lands were and where my lands were. I was able to tell him that I lived a long distance from him, more than 1,000 days walk or 5,000 days walk with a flock of sheep. I drew a map of Australia and, with various coloured pencils, drew the mountains, cities, the various agricultural activities, the mining centres, the forests and the deserts. I showed him the population distribution, the breakdown of the various religions in Australia and drew a picture of the average modest suburban house with its 2.2 children. I tried to explain that children went to school and that my land had adequate resources for its population.

He was interested in Australia's sheep. By using *Kalabity*, a 778 square mile property in outback northeastern South Australia that I know well, I sketched out what a large outback sheep station looked like. Because of the fences and the lack of wolves, he recognised that shepherds were not necessary but couldn't understand why we didn't use the milk from sheep. He was not aware that in the 19th Century before pastoralists could afford to fence outback properties, shepherds looked after flocks and protected them from dingoes and Aborigines.

After two hours of paying my respects. I decided that it was time to get moving and bash a few rocks. The chief quietly spoke to one of the older women who returned with a young woman. The chief asked whether I would like this woman and whether I could take her back to my land to work for me. The chief was a pragmatic altruist. The plump mongoloid teenager was a burden to nomadic shepherds living on the bread line and she would certainly have a far better life in those strange lands where they were so affluent that sheep's milk from huge outback stations was not used.

It was difficult to explain that the chief's proposal was just not possible. I did not attempt to explain the basics of immigration and visa conditions. I did not attempt to explain that in my far-off land there are many forgotten abused kids and mentally disturbed people on our city streets. I did not try to explain that the young woman would be far better off with her own family rather than transposed into an alien culture and family. It was just all too delicate and difficult. I stated that my chief only allows his own in my lands and that the chief's daughter would not be allowed to enter my chief's lands. At that time, Bob Hawke was Prime Minister of Australia.

As I was driving away from the camp, I looked in the mirror and saw that the young mongoloid woman was running after the vehicle, pleading and crying.

TURKISH DELIGHTS

ROGER CRADDOCK
TURKEY
1989 – 1995

Roger Craddock is a Mining Engineer. He graduated from the Camborne School of Mines in 1968 first working in South African gold mines before returning to the UK to gain an MSc in Mineral Production Management from the Royal School of Mines, Imperial College, London. He then worked in base metals in Eire and tungsten/tin in the UK before moving to New Zealand and then Turkey. Later, while based in France, he covered southern Europe, large parts of Africa and the 'stans. "He spent his working life travelling the world at other peoples' expense and loved every minute of it (well most of them!)".

It all started with a 'phone call.

I had an interesting, challenging time helping to take New Zealand's first new modern gold mine through from concept to production. This included Feasibility and extensive Environmental Studies, long and involved permitting procedures, financing, construction and operation. However after 18 months or so, it dawned on me that I no longer wanted to be Production/Mine Manager but yearned for something else new and exciting. I had made this known to my boss in the Australian parent company and was in the process of negotiating a transfer to Australia for my whole family so that I could take up a new and interesting position.

Then there was the 'phone call from my Australian boss, "have you heard from David yet?" "No". "OK, call me back after you've spoken with him."

Half an hour later the telephone rang again, "Hi, it's David – we've just had the first assay results from our drilling at Ovacik in Turkey." He summarised the results that were very promising and showed extremely good to high grades. We had a brief discussion about the find and its potential and then he got to the point. "We'd like you to visit Turkey to review the country and the project with a view to taking it on and basically repeating what you did in New Zealand!"

I spent the next few weeks in Turkey, learning about the project and the country and, in particular, trying to understand the processes required to permit, build and run a mining project. Turkey had recently been opened up to foreign investment and this allowed a foreign company to explore, develop and mine minerals, including gold, with no requirement for a local partner or part local ownership or investment. However, all of this was new and, although there were several other foreign companies actively exploring in other parts of the country, none of them had progressed enough to be even close to the stage of developing a new mine.

It all seemed "possible" which is exactly what I reported on my return through Australia. This resulted in the next stage. "Take your wife with you next time and investigate possible accommodation and schooling for the kids". That proved a little more contentious and problematic.

We were to be based in Izmir on the Aegean coast, Turkey's third most populous city after Istanbul and Ankara. Set in a beautiful location around a large bay, I believe Winston Churchill once referred to Izmir as the 'Pearl of the Aegean'; but, unfortunately, at the beginning of the 1990s the head of the bay, surrounded on three sides by Izmir, was heavily polluted; smelled badly, particularly on a hot summer's day; and then there was the regularly-told story about the waters burning! Presumably ignited methane from the accumulated rubbish in the shallows. Izmir was a very Turkish city, but it was also home to a major NATO headquarters and therefore had a large resident US armed forces community, along with a much smaller UK one. For this reason there was one English-speaking school in the city which was the US DODDS (Dependents of the Department for Defense) school located in a large old warehouse that was surrounded by guards and large concrete crash barriers. My wife was not impressed!

Always up for a challenge and perhaps relying too much on the good old Kiwi/Australian principles of "she'll be right" we moved to Turkey!

All went generally according to plan. Diamond drilling of the deposit was completed, although we had to bring in an offshore company to do the bulk of the work as the local company regularly failed to give acceptable core recoveries. We worked up a small but very high grade resource and completed a feasibility study that indicated a very viable open pit and underground mining operation. Permitting for the project went surprisingly well in some areas and we produced what we believed was a Turkish first, a very high standard Environmental Impact Study, and presented it to a very new Department for the Environment in Ankara for approval and Environmental Permitting.

While all this was progressing we experienced a few more "interesting" problems.

We managed to fall foul of the local town Mayor who decided that he was going to oppose our activities. We later discovered that, in the past, he had used other "causes" to keep his name in the press and so on in front of the local population. He was very good at it and, of course, the local and national tabloid press loved him: he always gave them a good story. I remember quite vividly seeing pictures of supposedly expiring locals who were dying from "cyanide poisoning", and stories about how our project would affect pregnant women who would give birth to babies with two heads and, at that time, we hadn't even started work on site, except drilling. This might all seem rather ridiculous and even amusing but when you personally, and the company, are on the receiving end, one's sense of humour tends to evaporate! At one meeting with him, where I attempted to explain what we were doing, using the best methods and technology to protect the environment, he listened for a few minutes and then quite openly told me that he understood what I was saying but that really he wasn't interested and didn't want to hear any more as he intended to continue to oppose whatever we did. It's no fun being on the receiving end of most of the Turkish tabloid newspapers that seemed to have no interest in talking to us about what we were really doing despite numerous attempts by us to engage with them.

However, it wasn't all bad. It became evident that to develop an open pit at one end of the deposit we needed to relocate a small village graveyard. Graveyard is perhaps a rather grand term for it. It was really a cluster of shallow graves around the side of the low hill that formed part of the outcrop of the gold deposit. The graves were very shallow because of the very hard rocky ground and so rocks were placed on top of them to give sufficient cover over the bodies. Needless to say, we thought this might be a real Red Flag that would stop the project in its tracks and we approached it with great trepidation.

Fortunately we found that one of the local MPs was prepared to assist us and, with his help and guidance, we started to get the required permissions, including those from the Ministry of Health, a local Mufti – the religious leader, and the Izmir Health Director and his deputy. I really can't remember how many "permissions" we obtained but it just seemed to happen. We purchased a piece of land that had been approved as suitable and approximately 100 new graves were prepared. The most difficult part was getting the permission of every affected family to relocate the bodies. Some of our local staff, together with the local Mufti, successfully undertook this delicate task. Each relocation was accompanied by the religious formalities requested by each family.

As for the actual relocation of the bodies, two local Romany Gypsies were hired; the Izmir Health people supplied all equipment, including the latest in sanitary wear (now better known as PPE). With the guidance of the local town's hospital Head Doctor, the village Headman, the Mufti and two volunteers from our staff, the job was quickly and efficiently completed. Obviously the company bought the land and paid all expenses, i.e. the labour and equipment to complete the task. The only other cost incurred was for an inexpensive car that we were asked to "donate" to the local hospital. The only problem resulting from this incredible job was that one of our volunteers suffered nightmares and vomiting from the experience: thankfully he made a full recovery!

This sounds like a simple story but it has always summarised in my mind the Turkey I experienced in the early 1990s when what one would have thought was a simple task could take forever to accomplish, while what seemed like a total Red Flag event could be completed quickly and relatively easily. Another interesting task was to acquire the land for the project. Any arable land in the area was used for cotton or tobacco farming and most of the plots were quite small and there appeared to be few records of ownership. When the owner of a plot of land died his ownership was split between his relatives. If the plot was large enough it could be subdivided, if not, then the number of part-owners increased. Obviously when yet another owner died his relatives got their share, and so it went on, and therefore it is easy to see that a small plot of land could end up having a large number of part-owners. Finding all these people and getting their agreement to sell the land proved to be a major undertaking.

I recall meeting a man tilling a plot of land that we hoped to purchase. I approached him and, through my trusted interpreter, asked who the owners were. We were given a long list of people, including one of his brothers who lived in Australia, so to contact him and get his permission was going to be a problem. All of this was dutifully relayed back to me and for some reason I asked where his brother lived in Australia and was told "Melbourne". I remembered a few trips around Melbourne in taxis with what always seemed to be Turkish drivers who never seemed to know their way around or, more particularly, the address I was looking for! So rather flippantly I asked my translator to ask if the farmer's brother was a taxi driver. My translator queried why I should ask but then posed the question. The farmer's face lit up in a big smile and he said: "Oh, you know my brother"!

Another interesting experience associated with purchasing land for the project was the actual closing of the deal. Once one had located all the part owners, and agreed a price, one had to arrange for all of them to be in the local town at the same time in order that all parties could meet at the equivalent of a Notary Public's office to sign off the transaction. Then we had to pay each of the landowners their share of the purchase price. Getting them all in the same place at the same time was not the easiest of tasks, but it was definitely necessary so that each could see and understand that none of the part-owners got more than their fair share.

We discovered that none of the locals trusted the banks, which meant they would only accept payment in cash. Therefore, on the appointed day, I would collect the total price of the land in cash from our bank in Izmir. Unfortunately I can't remember the dollar to Turkish lira exchange rate in those days, sufficient to say that there were a lot of lira to the dollar. That resulted in my carrying a very large bag full of cash. This was hidden under a seat in my vehicle and I drove (accompanied by one of my local colleagues) for approximately an hour to the meeting at the Notary's office.

Once settled in the office, the Notary, with great ceremony, would take over and organize the signing of all the paperwork (there were usually more X's on the documents than signatures). Once this was completed, all the part-landowners would take seats around the walls of his large office and the Notary counted out the money into separate piles, one for each landowner. These piles were then placed on the floor in front of each of them so that they could all clearly see what each of them was getting. When all this was completed, a signal was given and they all literally fell onto their pile of money, pushing it into their pockets/bags and then they were gone. Done deal!!!

There were so many land deals that I became accustomed to these transactions but I still remember my total amazement after witnessing the first: it was almost unbelievable.

The trip between Izmir and the project area was at least a weekly event for me and it was not unusual to be stopped for "speeding". This usually happened towards the end of a month, presumably when the local police needed the funds to help pay the wages. I proved to be an obvious target as I drove a "Blue Plated" imported vehicle! The usual procedure, after being stopped, was to approach the police car parked at the side of the road and pay the "instant cash fine" (no speeding tickets issued) straight into the sergeant's hat that had been strategically placed on the front passenger seat. Job done, I was free to continue my journey. On one occasion, when I was once again pulled over, I had one of our Mr. Fixit locals with me. He insisted that I let him go and talk to the sergeant to see if he could sort it out. He returned a few minutes later with a grin and said: "Let's go!" We left with a wave and a smile to the local police while he told me that he had carefully explained to the sergeant that I was in charge of the gold mining project in the area. It seems that the sergeant had been impressed and told my colleague to tell me that in future I could drive at whatever speed I liked through his area and I wouldn't be stopped again.

I don't think I was ever stopped again on his patch!

As we progressed our main project, we decided to take another look at the surrounding area, particularly the hills behind the project site, as there appeared to be evidence of ancient mine workings for gold. This investigation inevitably led to a plan for a few diamond drill holes to test the theory. During this drilling programme I received a telephone call, in our Izmir office, from the drillers who said that local villagers had surrounded the drill site and consequently they had been forced to stop work. The villagers were threatening to burn the rig because they claimed we were polluting the village water supply, which came from a spring, with diesel. I knew the village in question and also knew that it was quite a distance from the drilling site that was in a separate valley. It seemed very unlikely, but technically not impossible, that we were polluting the water supply.

Once again I summoned my local Mr. Fixit and we headed to the village, rather like the cavalry, riding to save the drill rig! Arriving in the village we were immediately surrounded by a crowd of noisy, maybe angry, women and their children. This was very unusual because the women normally stayed well clear of strange men, even of non-related males in the same village, and any formal dealing we would have with a village would always be with the Headman and a representative group of male villagers. However, the Village Headman did appear and led us up through the village to the water supply followed very closely by the very noisy, arm waving group of women and children. I duly made a big thing of tasting the water. It was pure, clean and tasted very good. There was no evidence whatsoever of any contamination by diesel. This was conveyed to the Headman and the assembled crowd.

The situation then started to deteriorate quite rapidly, the crowd got very much louder, lots of shouting and arm waving and even sticks being waved in the air. My Mr. Fixit was busy trying to talk with the Village Headman but he slowly pulled away from the crowd, came over to me, grabbed my arm and steered me back towards my vehicle saying: "Come, it's time to go." I resisted, as we really hadn't resolved anything, but when you don't speak the language it's always better to take the advice of your local right-hand man. Once we were safely back in the car and heading out of the village I obviously wanted to know what had happened. He told me: "you really don't want to know what they were threatening to do to you," so I naturally pushed further and was told "they were going to beat you with their sticks." I think there was more but that was probably all I needed to know. Thinking back I recall that this wasn't the only time that one of my minders guided me out of a meeting that had turned noisy and potentially nasty. Anyway, on this particular occasion, we circled around the back of the village in the hills to the drill rig, told them to stop the hole and move on to the next one, which was even further away from the village. No point in fighting a battle that you are not going to win. We had no more problems with our drilling or with that village!

During this time when we were having "such fun" at work Saddam Hussein invaded Kuwait. Now you may wonder what this has to do with Turkey. Well, Turkey has for centuries been the meeting point of the east and the west. Most of Turkey is in Asia while its major city, Istanbul, straddles the Bosphorus between west and east and its northwestern part can be considered to be in Europe.

As previously mentioned, Turkey is a NATO member and the latter has a large base in Izmir, while the US has a large task force at the Incirlik Air Base in South Eastern Turkey that is within easy striking distance of Iraq and Kuwait. For these reasons, Turkey was very much involved, if only through association, with the First Gulf War in 1991. Not everybody in Turkey agreed with the US position in the war which caused problems around the NATO and US bases.

It was about one in the morning in January 1991, and my wife, children and I were asleep in our apartment in Izmir when the telephone rang. It was the headmaster of the DODDS School: "Morning Roger, sorry to disturb but just wanted you to know that 'the boys are going in'. There will be no school tomorrow, we'll let you know when we will open again: got to go, bye." His words exactly and those few words "the boys are going in" will forever stay in my memory!

So, one in the morning, just got the news, albeit not totally unexpected, that Operation Desert Storm was about to commence: what do you do now? Answer, go make a cup of tea, get back in bed and turn on the

BBC World Service and find out what they know! No more sleep that night!

Well, the school did reopen, but not for long as soon afterwards there were a few bomb blasts in that particular area of Izmir. We heard the noise from across the bay. No one was hurt and there was very little property damage; the locals called them "suitcase bombs, just for noise and effect". Nevertheless the US community went into lockdown and an obvious part of that involved closing the school. We lived with that for two or three weeks before realising that the school was not going to reopen anytime soon and that we still had to educate our four children. Answer, find a school in England that would take them and move wife and kids to the UK for the duration. Easier said than done but I will say here and now that it would have been a lot harder without a wife who could manage the problems of the move, hotels, house leasing, new schooling, etc, and without a company that was prepared to help and to meet the costs.

We eventually got the environmental permit but it required us to incorporate a cyanide treatment plant. We had been forced to agree to this addition despite that we had already agreed to a fully lined tailings pond, which meant that the cyanide in the tailings was isolated, and didn't need treatment. One got the feeling that they didn't really understand our proposals: they simply asked for everything they could come up with. After many long months we finally got the last of the permits required to build the plant and operate the mine, which involved approval of all of the detailed layout and construction drawings.

This frustrating period of seemingly endless negotiations and meetings finally enabled us to start construction of the mine and begin open pit and underground development. The construction went reasonably well and was completed on time and on budget. However, our construction engineers needed to constantly check for quality of work and materials, more so than on other projects in which I had been involved, but we always managed to achieve the required specifications. Our main construction contractor was an Ankara-based company and it generally performed well. But, when we really needed them to go up a gear to get the job finished on schedule, it took a trip to Ankara for me to meet the owner of the company and to form a personal relationship/friendship with him to get the accelerated progress we needed to complete on time. Once we had established that link, then it all happened quite quickly and efficiently; in fact, we were amazed at the speed with which they could make things happen.

In order to start production, our permit required that the plant be inspected to confirm that it had been built as per the construction drawings that had previously been approved by the authorities. We all thought this was merely a bureaucratic formality but were duly set upon by large numbers of suited inspectors armed with tape measures and copies of the "approved" plans. They proceeded to crawl over every inch of the plant before disappearing back to Izmir to compare notes!

We anxiously awaited our final permit to operate but weren't particularly worried because as far as we were concerned the plant had been built to the approved plans.

However we did not receive the expected good news as our final approval had been refused! Investigation revealed that the inspectors had found one small chemical storage tank that wasn't quite the same size as the dimensions shown on the "approved" plans. Sounds ridiculous, and it was: we tried everything to overturn the decision, but all to no avail. Someone or some group didn't want a foreign company to own and run Turkey's first modern gold mine. We thought there was nothing we could do about it, certainly in the short term.

But in fact we did do a landmark something. While all this "inspecting" had been happening we had dry-commissioned the plant. We arranged a shipment of cyanide to be landed on the other side of the country and to be trucked very quietly across country to the site, arriving in the very early hours of the morning when no one was around. We were well prepared as our commissioning crew had already run the front end of the plant to fill the tanks with ground ore slurry: all we needed was the cyanide, which we fed in as soon as it arrived. We ran the plant, including the cyanide treatment plant, for long enough to at least partially load the carbon with enough gold to give us a small bar of bullion.

Very late one night we stripped the carbon and "did a small smelt". I still have an old picture of myself in the mine office with a can of beer in one hand and a small bar of bullion in the other. It's the only gold that I ever saw come out of the plant that I had worked so hard for over those approximately six years.

I suppose I should at least try to excuse our cavalier response to the Turkish authorities in having effectively run the plant illegally, but we did have a plan to use the exercise to illustrate that the plant could be run in an environmentally sound manner with zero cyanided waste being discharged to the ponds. Despite that we did it on the quiet, we did it properly, sampling and testing all of the plant's discharges so that we had this data. We still perhaps believed that environmental issue were the ones that we needed to solve whereas, in truth, the issues were political.

The saga of this project became long and boring as it continued to go through many phases of attempting to get the operating permit reinstated. The project was sold on to others who thought they could solve the problems, but failed. A Turkish company, who had no previous mining experiences but had very "close" ties to the then current government, ultimately acquired it and brought it into production in 2001.

Many years after I had ceased to be involved, I received a message from an old contact/friend, an Ankara based businessman, who had found my name through LinkedIn and wanted to re-establish contact. He had been the company's insurance broker in those early 1990s. After his greeting, he updated me with news of the project and said that: "We had got it right, but that our problem was one of timing; we had simply been there too early in Turkey's westernisation phase, and that the Turkish company who finally acquired it had made a lot of money working it".

I suppose that gave me some satisfaction, but not much!

SAUDPLAY -
THE CRADLE OF GOLD MINING

RALPH STAGG
SAUDI ARABIA 2005 – 2011

Ralph Stagg graduated as a geologist from the University of the Witwatersrand in 1970 and subsequently completed a Masters degree in Mineral Exploration at the Royal School of Mines, University of London, in 1977. He began his career as a geologist with Falconbridge exploring for nickel in Western Australia, returning to Southern Africa and working mainly in base metals for Cominco. In the 1980s he became involved in mining geology and then engineering consulting for a number of different companies in Australia, South East Asia and the Pacific. In the mid-1990s he spent a number of years in China working for a private investment group.

In the early 2000's he became involved in the Middle East and subsequently founded Citadel Resource Group, a copper and gold company in the Kingdom of Saudi Arabia; and later Celamin Exploration, a phosphate and base metals company in Tunisia and Algeria. He is currently a director of listed companies exploring in Chile and the Murray Basin of Australia.

The first time I went to Saudi Arabia was a few months after the attack on the American Consulate in Jeddah. The people at the Ministry told me they watched a huge firefight that went on for some hours from the roof of the Ministry. Later when we drove past the Consulate, which was a through route that we used regularly, it was never without a thought that this might happen again, in Jeddah or elsewhere. Happily, it never did during the time I was there. Saudi security was very tight.

Every building or compound that one entered had armed security (usually semi-automatics) and larger weaponry discreetly hidden. Any vehicle was searched front and back and inspected underneath using hand-held mirrors. No exceptions. After a while, you got used to it.

I first flew into Jeddah in early 2005 from Bahrain after travelling there from Australia. At that time Gulf Air (the Bahrain National Airline) flew from Sydney to Bahrain via Singapore, about 13-15 hours flying time depending which way you were going. My colleagues were based in Bahrain so it was a logical route for us use and Gulf was remarkably inexpensive. So I flew to Bahrain and had meetings there for a couple of days and then on to Jeddah.

I had met the colleagues in Bahrain whilst working on a copper project in the Sultanate of Oman. We had formed an informal relationship that focussed on finding projects and prospects in Arab-world countries in the Middle East with a view to ultimately forming a company listed on a Stock Exchange. My expertise was evaluating the ones that came to our attention and determining whether they fitted our commercial criteria. Their expertise was local and they were already working in the Kingdom of Saudi Arabia on the Al Jalamid phosphate project. Because of that, it came to their attention that the Saudis had changed and modernised their Mining Code, with the most important changes being that a foreign company could hold 100% of an asset and that foreign companies could invest in gold mineralisation.

We therefore decided to fund a reconnaissance in the country to determine whether there was any prospectivity additional to that which we already knew about informally. In 1978, I had worked as a geologist for Consolidated Goldfields PLC (CGF) in the City of London and had become marginally involved in Saudi Arabia. CGF had an exclusive right over the famous Madh Adh Dhahab ('Cradle of Gold') gold mine that has a very long history. Madh is rumoured to be one of "King Solomon's Mines", with some supporting archaeology, but when CGF was working there, it was to discover extensions to the known epithermal gold lodes. This was very successful; subsequent mining has produced more than two million ounces of gold at very close to a very high 30g/t Au grade from the new orebodies it discovered.

At that time I never did any fieldwork in Saudi for CGF, my role was desktop analysis in Head Office. We were also aware in 2005 that there was a then government-owned mining company called Ma'aden that had amongst its assets five operating gold mines including Madh Adh Dhahab. To be accurate Dhahab translates more as metal rather than gold, however general usage of the name is the latter.

I had another piece of luck when we were planning the reconnaissance in Saudi. One day I met a long-time friend who runs an industrial mineral practice and travels extensively as part of this business. I mentioned to him my plans to go to Saudi and he immediately offered the loan of book he had been given at an Industrial Minerals Conference a few years before entitled "Mineral Resources of Saudi Arabia – Not Including Oil, Natural Gas and Sulphur" and published in 1994 by the Directorate General of Mineral Resources. This extraordinary piece of good fortune enabled me to concentrate on areas of potential that I could examine in more detail during my visit and allowed early focus during what would always be a time-limited first visit.

The Kingdom is far-sighted and commenced mapping and surveying the Arabian Shield within its borders from the early 1960's and employed such luminaries as the French BRGM and the American USGS as contractors to do specific surveys that are all summarised in the book. Anyone who has worked internationally is aware of the competence of these organisations which were held in high regard for many years. This meant the information provided was competent and credible. The added advantage for me was that everything is written in English. That meant, at least for the Technical Reports, early translating from Arabic was not required.

Therefore, with this background and after detailed discussions and planning with my colleagues in Bahrain, I flew from Bahrain to Jeddah on a Friday in order to commence my studies at the Ministry early on Saturday (the start of the week in this part of the world). Jeddah airport at the time was extremely large as it is the pilgrim airport for Muslims from all over the world who are heading to Makkah (Mecca) as part of their religious pilgrimage (Hadj). The most important title and duty of the King of Saudi Arabia is "the keeper of the two holy cities" which are Makkah and Madinah (Medina). The airport handles large numbers of pilgrims although they enter through their own terminal. Members of the Royal family have their own terminal too. At the time, we ordinary folk were bused to the terminal: there being no air bridges there.

The business class passengers were disembarked before the economy ones giving them an advantage in the immigration queue. Therefore, it was with some trepidation (will my visa work?) that I arrived at the terminal and was pleased to find a vacant immigration booth. The fiercely bearded immigration office studied the document for a while and then shouted "KART!" I did not understand so he explained by shouting (somewhat louder – obviously hoping I was deaf) "KART!" Luckily then the penny dropped and I realised I needed an "immigration card" which had helpfully not been supplied by anybody either on the plane or in the terminal. There was a small queue behind me, but hoping to pre-empt the rush of economy passengers, I left my documents on the counter and went in search of an immigration card. Somewhat surprisingly, I found some scattered helpfully on the floor and filled one in. By then the queue was much longer and I thought I would just go to the same counter where I had been previously. So with a few "excuse mes" I slid back to it and handed documentation to the, as I thought then, "grumpy" immigration officer. With a flourish, he stamped everything that did not move and then grimaced and said "Welcome to Saudi Arabia" and I had made it!

Just to backtrack a bit, my mention of a visa is because it is a significant issue in Saudi. Initially I received a single entry visa valid for a month. Later, the best I ever achieved was a six-month multi- entry visa. This system is different for people who choose to come to Saudi to work for a Saudi-based organisation. Then you have to surrender your passport to your employer and operate on a local identity document system, which I found untenable as, under it, if you wish to leave the country you have to ask your employer to return your passport. This may not be approved. Therefore, I stuck with the visa system as I did not have a Saudi employer (nor did I want one!). Leaving Saudi can be equally challenging, as you have to keep in mind that Saudi does not operate on a Gregorian calendar. So your month may not be their month and if you have overstayed your visa – you stay. I once got a phone call from one of our geologists in this predicament at 11.30pm one night. I was forced to contact our visa specialist who was in the middle of his wedding (usually a three-day affair in Saudi) and he went to the airport and sorted out the problem.

In order to get a visa from the Embassy in Canberra you have to provide an invitation letter from a Saudi organisation (in my case the Ministry did this) and a letter from your foreign employer (my own consulting company sufficed) that provide guarantees and is 'legalised' using approved organisations in Australia. The first time I did all of this I thought it would speed matters up by going to Canberra taking my application with me and effectively getting a visa 'on the spot'. No such luck! A very nice man explained that I had wasted my time but they would process it and return my passport in due course. Of course, many of my geological and other colleagues have been through this sort of process and our understanding government allows people to hold two passports in these circumstances. Visas to Bahrain on the other hand are purchased from Immigration on arrival, all you need is some cash (preferably dinars), and in you go. Other currencies are subjected to unfavourable exchange rates!

Therefore, I had finally arrived after all the preparation and rigmarole, found my pre-ordered car and driver, and headed to the hotel recommended by the Ministry. I entered the maelstrom that they call "traffic" in Jeddah. My baptism was a quiet day, as Friday is equivalent to our Sunday but I had my first experience of very expensive sports cars overtaking at "warp speed" attempting to force themselves into spaces that did not really exist. A regular feature of Jeddah traffic. All this takes place on the right hand side of the road, naturally. It took me more than a year to pluck up the courage, or perhaps I should say stupidity, to begin to drive myself. By then I knew Jeddah reasonably well geographically.

At least on that first day the Ministry was within walking distance although strolling the badly paved sidewalks of Jeddah is not for the faint-hearted, and don't be tempted to step into the traffic. The driver to the hotel was moonlighting from his regular job as a professional soccer player. He was from West Africa and had not quite made the grade so was struggling somewhat in a work, cultural and passport limbo. There are some beggars on the streets of Jeddah, people who have come on Hadj and/or overstayed their visas and are trying to eke out some sort of survival. It seems strange that in such a rich country with significant infrastructure, and that provides generously to indigent Muslims who wish to partake in Hadj, yet the beggars are always there.

The next morning, suited up, I walked to the Ministry for my appointment with the Deputy Minister's attaché (an English geologist at the time) and, after meeting with him, I was taken meet to the Deputy Minister of Petroleum and Mineral Resources. Here it was explained that a Reconnaissance License had to be issued before I could examine any data. We had discussed this in Bahrain but there was a hurdle. I was not an officer of the Bahraini Company that my colleagues operated and hence not authorised to sign any documents on their behalf. After I explained this, we compromised and agreed that, if an authority could be transmitted from Bahrain, then we could go ahead. This took a little while and by the time this had been completed my colleagues' company had the licence covering the entire Saudi Arabian Shield. This entitled us to look at Open-file, or public domain, data throughout this geological area.

I then asked if a tenement map of this area on the same scale could be provided. This would enable me to rapidly screen Projects and Prospects that were clearly unavailable as they were obviously held by somebody else. This is Desktop Prospecting – 101 and any professional geologist would have asked the same question. Eventually, after much discussion, I was given three maps at different scales covering most of the area: a compromise that I had to accept. Another compromise was that I really had no idea how up to date they were! By then, as it was approaching 2pm I was invited to lunch with the Deputy Minister, his attaché and a number of functionaries in the Ministry whom I came to know well over the next few years as I had occasion to work closely with them. Lunch at 2.30pm is the end of the working day as I was to discover. Generally, the day starts around 8.15am and don't try to talk to anyone later than about 1.45pm as they are heading home or usually heading to a school to collect the children as their wives were at that time not allowed to drive.

Lunch was very entertaining, the discussion revolved around the first elections to be held in Jeddah since the Kingdom was founded. Both men and women were enfranchised. The election was for the equivalent of the Jeddah Town Council. The council was to have 50% elected representatives and 50% Government appointees – a first foray into democracy. Candidates had to set up tents in a park in one part of town and then try to persuade anybody who came to the area that they were indeed worthy candidates. There were also female candidates, but in a separate area. A few days later, a fierce dust storm came through Jeddah. I later found out some of the tents blew down and those candidates had to withdraw (shamefully). The Saudis are very proud of their nomadic desert culture and if you cannot get your tent to stand up during a dust storm then you are not worthy of that culture.

A geological digression. The area of interest for our geological investigations was the Arabian Shield, a large area of Proterozoic, that is relatively older, rocks that forms the Western part of the Arabian Peninsula. This shield outcrops over an area of 1,500 kilometres north south and about 800 kilometres east west. Phanerozoic rocks, or relatively younger ones, flank the shield in the eastern and northern parts of the Kingdom and underlie the Red Sea coastal plain. The vast oil and gas reserves in Saudi are hosted in those younger rocks in the Eastern part of the Kingdom along the Arabian Gulf. Another part of the shield terrane outcrops on the western side of the Red Sea rift but we never pursued any prospecting in that area because of the difficulties of holding tenure and operating. At the time we entered exploration in the shield it was very under-explored. As an example, in the 'Mineral Resources of Saudi Arabia' volume mentioned earlier they claim that about 400,000m of drilling had been undertaken in total up to 1994 whereas, in comparison, at that time more than 300,000m of drilling is undertaken annually in Canada in similarly prospective geological terrane. This relative lack of drilling gives some idea of the remaining prospectivity of Saudi. Notwithstanding this relative paucity of drilling there are more than 6,000 known mineral occurrences and more than 1,000 ancient copper and gold mining sites in the Arabian Shield.

I spent four days (about six hours a day) in the Library and Archives of the Ministry researching projects I thought may be of interest using my knowledge and experience gained in my 40 years in the Industry. That relatively short space of time meant all the investigations were necessarily superficial. I returned to Bahrain with a series of recommendations to my colleagues there. In short, I believed we had sufficient prospects to allow us to apply for nine or ten Exploration Licenses that would give us a portfolio of precious and base metal projects that could form the basis for a small exploration company. The curve ball was one project that, although it had no tenement marked over it, the head of that department told me verbally, was already taken by a third party. That project, named JS, had a large resource of copper already delineated by nearly 50,000m of drilling and 4,000m of underground development. My belief was, if we could acquire JS in some way, then we could build a mid-tier mining house in Saudi.

On my return to Bahrain, we discussed my findings and the suggested portfolio for applications and decided to go ahead with applying for the proposed exploration licences, to see if we could find out who had the rights to JS, and to find out if we could acquire it. This took some time but ultimately we found a family company in Al Khobar who had the rights to it and agreed to deal with us for a number of reasons, but mainly because of our combined expertise. It took more than two years from that first visit before we finally got all the projects listed ultimately on the ASX.

Al Khobar is on the eastern side of the Kingdom and is the hub of the Saudi oil industry. From here, a major bridge system called the Causeway, and which includes customs and immigration, leads to Bahrain. Therefore instead of going through immigration in Jeddah, an alternative is to cross this causeway by vehicle and then fly from nearby Dammam airport to Jeddah on a domestic flight. This flight was always operated by Saudi Arabian airlines or 'Saudia', usually in a 747. These flights were fine so long as you travelled up the front of the plane. I rapidly learned that on arrival at Jeddah rich locals had their vehicles waiting on the tarmac to collect them and whisk them off, unlike the rest of us who were subjected to a hot bus and collection of baggage in the terminal.

I was offered a lift by one of the locals one day but had to decline, as I was unsure if I would ever find my baggage again. Through our proposed drilling company, which had a member of the Royal family as its Chairman, we arranged for my female colleague on her first visit to Jeddah to be picked up off the plane in such a way (as a surprise).

She was very amused by the comment of another passenger: "Look at her – she must be married to a Prince!" I usually arrived in Jeddah in the normal mundane way – no Prince's chauffeurs for me.

Our second visit to Saudi was to iron out issues that had arisen with our Exploration Licence applications but we also needed to visit and start making contact with the Saudi Geological Survey (SGS). This organisation is autonomous and separate to the Ministry and has the archives of all geological work undertaken in the Kingdom. It has an uncooperative relationship with the Ministry but it was important for us to make contact and assess the depth of information and also the skill sets held by this organisation. Whilst the Ministry is located close to the business centre of Jeddah, the SGS is located on the outskirts near to the Goat Market. It is housed in a very large compound and, unlike the Ministry, is quite difficult to access. Once we had made contact, I spent many hours in its Library and Archives although it was sometimes difficult to find out where certain materials were housed and in which department. The bureaucracy was very compartmentalised. As an example, it took us more than a year to find the digital records of the BRGM's drilling at JS. We knew they existed from other documents we had found, and had even resorted to visiting the BRGM headquarters in France (to no avail), and finally unearthed them in the 'computer' department at SGS purely by persistence.

Getting around Jeddah was a challenge in the early days. We started using taxis, which was very idiosyncratic, and at times downright dangerous, until we met Abdul who could usually (at least in the early days) be found lurking on the cab rank of the hotel that we used close to the Ministry. Abdul was a very large Yemeni who had a disability caused by falling asleep as a child in the desert and rolling into the fire. His medical treatment was rudimentary. Abdul rapidly became our driver of choice and also guide and ultimately friend and mentor. He was very proud when we finally opened our offices in Jeddah and he could proclaim he was the first employee. In the early days, he would drive us around town and, when we were seeking equipment, he was invaluable. Often suppliers in Jeddah had obscure locations and, as a long-time taxi driver, he either knew the location (street signs only in Arabic!) or he would call them and be able to discriminate rapidly whether they were able to help or not. If in doubt, we hopped in Abdul's cab, went, and investigated in more detail.

The geology business needs access to drilling equipment and, just as important, reliable and accredited assay laboratories. In Jeddah, we were lucky; there was an existing drilling company as well a highly recommended and very competent laboratory. Naturally, we ran our checks and balances before using the latter, and they proved their worth throughout all our exploration campaigns in Saudi. We were not forced to export large quantities of samples to internationally accredited organisations. Sadly, after a promising start with the drilling company, with which I made contact during my first trip to Jeddah, their standards were lacking and we were forced to get a new drilling company started in Dubai and import the rigs to Saudi. Later we brought a selected contractor for underground drilling in from Kalgoorlie. Their competence was excellent.

Accommodation was another issue in that we could not run an exploration program from a hotel room or even a series of them. Therefore, we rented a villa in a compound to start with and used one room for an office. Later, when we set up an office in a high-rise in downtown Jeddah and as our needs expanded, we rented a number of villas in the same compound in northern Jeddah near the corniche. This area was chosen because initially the exploration manager of Ma'aden, an Aussie, was domiciled there and he was able to provide us with considerable assistance. He allowed me (after-hours) access to his archives for our projects so I could gather further information that was not available elsewhere.

He also introduced me to 'hooch'. Saudi is alcohol-free, however the compounds are largely unsupervised and many people make varying quality wine and beer. Huge containers of 'grape juice' are readily available in the supermarkets and home-brew kits are brought in from abroad. When driving over the causeway it has been known for gin or other white spirit to migrate to a water bottle under the seat and arrive unscathed through customs. In all compounds ethyl (we hope) alcohol can be bought if you know where to go. All of these have a risk, not so much being caught by the authorities and deported, but a health one in the quality of the alcohol. I experimented for a time but ultimately decided that as I was generally only there for 3-4 weeks at a time I could adopt a 'liver-cleansing' diet!

Communications were generally very good in Jeddah. Mobile phones worked and fast internet abounded. In the field, it was a different matter. At JS and other sites, mobile phones had little to no signal. If we climbed to the top of Lode 1 or around the back of Lode 3 sometimes a signal was possible. Later we set up internet at JS and that helped considerably. Nevertheless, at the start of our work, we needed emergency contact in case of an accident or other troubles. It was not possible to get radio licences as the government restricted these. We found this out when we tried to put together an electrical geophysical survey at JS. These types of surveys require two-way radio communication but we were unable to get licences for this so the survey was never undertaken.

However, satellite telephones were readily available and reasonably inexpensive to operate. With Abdul's assistance, we found an outlet in Jeddah and purchased a number of them. One was deployed at JS. Now if you have ever used these before you will know that they only operate line of sight to the satellite. If the latter goes below the earth's horizon or an artificial horizon e.g. under a roof, then you have no signal and no communication. At JS, we found a place at the window of our main room in the camp where the phone could receive a signal and thus calls from outside.

We were extremely lucky at JS that we inherited a 30-cabin camp from the original work in the 1970's and 1980's. This camp was fully fitted and the site included an airstrip and water supply. However, it was dilapidated because no maintenance had been done for 25-30 years. The cabins were imaginatively decorated: one I recall with Swedish Alpine scenes, somewhat bizarre in the middle of the desert. Later we were required to upgrade the camp for security purposes by the local army but initially it served its purposes. In the early days, we employed two Czech geologists to manage the drilling program that I had planned to verify the work done in the 1970's and 1980's. They were competent operators but with relatively poor English language skills.

When we started the drilling program, I explained to one of them about the sat phone. I had to leave site for a week in order to undertake other matters so I told him I would call every night for a safety check and progress report. On the first night, I could not get through, the same on the second and third nights. By then I was sufficiently concerned that I ceased my business in Jeddah and drove six hours to site. My geologist was fit and well and somewhat puzzled about my early return. I looked on the windowsill – no phone. "Do you have the sat phone with you?" I asked. This would have been a reasonable safety precaution for him and I would not have been annoyed except that I tried to call during the day as well so I was not optimistic that he had it. He looked puzzled then crossed the room to our locked storage cabinet, opened it and proudly produced the phone – switched off. "Very secure!" he said. Clearly, my briefing had been lost in translation!

The first time I took this person into Saudi we drove across the causeway and through Saudi customs to Dammam airport. We were then planning to fly domestically to Jeddah. I had not long arrived from Australia and was still jet-lagged. We were told that tickets would be available for collection at the ticket counter. I told him to wait and I found the counter and quizzed the official about our tickets. He said no tickets in our names. I went back where my colleague was patiently waiting trying to work out what we did next. After a few minutes I realised that I did not have my wallet which contained everything: passport, visa, cash, credit cards etc. A mild panic ensued. I searched every bag and eventually retraced my steps to the ticket counter where my wallet appeared sitting on the counter while a long queue of people had been politely ignoring it whilst they did their business. This was a salutary lesson that nobody steals in Saudi. The penalties are very severe.

The Ninth Arab-World Mining Conference took place in Jeddah whilst we were there. It was a grand affair opened by the King of Saudi Arabia. This was the first time I personally had attended a conference like this opened by such a luminary. We had a stand, which he visited, although as only two people were allowed to be there at a time, I was not chosen. It seemed prudent for our Arab partners to be there.

All our tenement applications had to go through a site visit process, which was largely attended by officials from the Ministry, in particular one official – Mahmoud. I talked to him about issues that arose in these meetings. For example on one site, local people were concerned that we would knock down trees in a wadi, which was the main access to the area. These were small and stunted – this is desert country after all. I assured Mahmoud that we would be prepared to sign a document guaranteeing that those trees would never be touched by us. This and water were the main concerns raised by the locals in these site meetings. I chose to accompany Mahmoud to one of these visits for two applications we had made near a town called Hail which was four hours driving time north of Madinah. Mahmoud advised me to fly there and he would meet me at the airport. He personally drove everywhere for these meetings and did not like to fly. So, I flew to Hail and was duly met and we drove a couple of hours into the desert to the nearest community. A group of representatives came to the sites and we discussed their issues, which were not serious. We then returned to their town for hospitality: in this case, a traditional feast sitting on the ground and serving yourself with the correct hand.

We then set off to drive back to Madinah and Jeddah. A small problem was that I, as an infidel, could not go to Madinah or Makkah without a special permit that are hard to get and I did not have. Mahmoud and his companions explained that, as we were going around Madinah on the ring road and I was not entering the town, then there was no issue. And so it proved. We stopped on the ring road at a roadhouse for the usual ablutions and I asked the guys if they wanted to go and pray. I was used to this in Jeddah where Abdul the taxi driver regularly left us to perform his religious observances. Mahmoud said they did not need to as, whilst travelling, they had an exemption from their observances. So on we travelled into the night. The road from Madinah to Jeddah is a two-lane sealed highway in both directions fenced to keep out cattle, wild animals (there are baboons on the escarpment) and people (although that does not deter them). That night we travelled at 160-180 kph down the road on sidelights, which occasionally and seemingly unpredictably were flashed onto high beam briefly. I was extremely concerned that we would hit something and was never as happy as when I was dropped safely at my compound gate at midnight. I did not volunteer for any more of these visits after that.

Mahmoud and the Ministry were very helpful and, because in the early days we did not have a vehicle, they drove us to a number of our sites so we could undertake a field inspection. These were my first trips out of Jeddah and so extremely interesting both culturally and geologically. We had a number of projects with two or three Exploration Licence applications close together; one such group was on the escarpment from the coastal plain. As you leave Jeddah, the road is very flat but eventually you start seeing some hills of the western escarpment and ultimately you rise nearly 1,000m onto the high plateau. One of our groups of projects, which was for gold, was in the breakaway country on the edge of the escarpment and was reached on dirt tracks along the wadis. The road into it was bewildering as there is no marked track as such, and everyone drives as they feel like along the wadis so it easy to take a wrong turn. Anyway, Mahmoud knew where he was going.

We then continued to Mahd Adh Dhahab, which is a quite reasonably sized town with the gold mine and plant attached. Later we arranged to stay at the camp there and for our personnel to stay there until we fixed the camp up at JS. We had a gold project close by and in the same belt of volcanic rocks, and then our lead JS project was about an hour's drive further on. Travelling through this country for the first time is most impressive. There is little humidity, in winter it gets down to single figures at night but, in summer, it can reach 50 degrees C. We had to work through all of this.

Later when I drove myself to the exploration sites, the security police nearly always stopped me on the main highway out of Jeddah. They also stopped Mahmoud on the occasions when I was with him and he had to present his papers as did all the other Arabs with us, but I was never asked to. On my own, that incident was always interesting, as my Arabic is very poor to non-existent and they had similar English skills. The trick I found was to say "Al Mahd" which was my destination to all intents and purposes. The first part of the drive to Al Mahd is on the Madinah road and I worked out that they were concerned that I was driving up to look at the forbidden Madinah and thus potentially breaking the law.

Dumb foreigner! Driving in the Kingdom was always fraught, especially in Jeddah, and the penalties were high. There were many stories of foreigners being arrested and sent to prison after they had been in a collision with a Saudi, whether at fault or not. These places I am told are not for the faint-hearted and it could take a number of days, or even weeks, even if you are innocent, to get free. In the meantime, the facilities are very poor. Woe betide you if you are in a bad accident or, worse, manage to kill someone. You are locked up until you can negotiate 'blood money' and this can take years. Therefore, I was always a very cautious driver and luckily survived unscathed.

One of my friends at SGS was named Sami and he accompanied our exploration manager and me on a site visit to our nickel and gold projects near the port of Yanbu. Ultimately, Yanbu became the destination for copper concentrates from JS but that was not the main purpose of our first trip. We hired a car and stayed the night at Yanbu where Sami introduced us to an interesting and very different restaurant. We then drove and inspected both projects, never an easy task when you are trying to navigate by GPS and there are many intertwining tracks. We were particularly interested in finding a drill site at one of them but could not locate it in the maze of small tight valleys.

Sami and I used to go to restaurants in Jeddah but, because we were two men, we had to dine in the mens' section and not the family section. Sami, who had a good sense of humour, tried to argue that we had families - they were just at home! The doorman did not appreciate this, but it amused us. Many of the restaurants provided facilities for smoking shisha. It is smoked in a hookah and has many flavours. This could be annoying as there is no such thing as 'non-smoking' sections in the restaurants. At one point in time the Jeddah Council banned shisha smoking within a certain radius of Jeddah but that caused a furore and proved almost impossible to police.

So, where did this lead? It took six years, but the small start-up that we hopefuls initiated in 2005 was taken over not once but twice. The first time was for AUD $1.3 billion and then six months later the resultant company was again taken over at a 30% premium. We had taken JS to a point where we had been granted a mining lease and raised debt and equity funding to build the Project. After the second takeover, the project was constructed but was unable to gain the permits necessary for it to be put into production. Eventually 50% was sold to Ma'aden and then mysteriously all the permits appeared! Glad we got out early and I ended my Saudi odyssey.

A BALKAN ODYSSEY

FRITZ FITTON
GREECE,
MACEDONIA AND ALBANIA 2008

Fritz Fitton graduated from the University of Leeds in 1970 with an Honours degree in Earth Sciences. He migrated to Australia in the Nickel Boom and worked in the Pilbara for five years. He subsequently successfully explored for base metals in New South Wales and Queensland for EZ Corporation, before returning to Western Australia and working in the Northeast Goldfields.

In 1987, he founded a geological and mining consultancy. He was involved in several mining floats, including the very successful Jubilee Mines and Athena Resources and, as a consultant, assisted numerous others.

Background

Upon graduation at the height of the Nickel Boom I already had a job lined up in Australia with US-based Utah Development Corporation. Before I took it up, I was lucky to get offered paid work for a couple of months in Cyprus as an exploration geologist for Cyprus Mines Corp mapping in the rugged Troodos Mountains in the north west of the island. It was summer and very hot. The job could be quite dangerous as the Greeks and Turks were constantly at war with other and there were snipers everywhere. I had a lucky escape one day and got winged by a bullet whilst mapping a ridge top where there were several dug outs. I was carrying a clip board and had a beard so I reckon it must have been a Turkish sniper who mistook me for a Greek spy as the Turks only grew moustaches.

On arriving in West Australia, Utah sent me up to the Pilbara, firstly to explore for copper - zinc deposits around Whim Creek and then a bit further south to look for nickel around the historic Pilbara mining centre where there were many prospective ultramafic rocks. After five years I was retrenched and then worked around Cobar, then Charters Towers.

I didn't really like the North Queensland humid climate so moved back to WA where I was head hunted by Esso Australia as a senior geo in early 1977. To start with, I was mainly based in Leonora looking for the next Teutonic Bore base metal mine. I rose up to become joint head of their 'Regional Studies' group, and travelled much of the world. All good things come to an end though and Esso decided to get out of metal exploration in 1986 so once again I was retrenched.

After a couple of months of concentrating on my artwork and selling my homemade pickles and sauces in Fremantle Markets, I got a phone call from my old lawyer mate Kerry Harmarnis who told me to get off my hippy arse and help him float an exploration company. I reluctantly agreed as he said he wanted me to pick up tenements around my old stamping ground of Leonora of which he knew I had extensive knowledge. So Jubilee Mines was born. After a shaky start it went onto become the stock market darling after the discovery of the ultrahigh-grade Cosmos nickel deposit near Leinster. From a penny dreadful, Jubilee got bought out by international company Xstrata for $23.50 per share in 2007 in a deal worth $3 billion.

I did very well out of this as I'd done the IPO for shares and had "kept the faith". In 1987 I founded a geological and mining consultancy, Maprock Pty Ltd, specializing in high quality geological mapping. Over the next 20 years I was asked to write many prospectuses and technical reports involving numerous commodities. Over the years I had always kept a keen eye on nickel and in 2006 helped float a company called Athena Resources which had assembled a portfolio of prospects around Ravensthorpe near the south coast of WA. The nickel price had reached the dizzy height of $35,000 per tonne at the time and BHP-Billiton were constructing a large lateritic Ni-Co mine down there after doing a deal with Tony Cooper's Comet Resources. Athena had plenty of potential for further Ni-Co lateritic deposits on its ground and I became their chief consultant.

One of Athena's main backers was a well-known Perth businessman called Tom Galopolous who had a Greek Macedonian background. Tom had a finger in many pies including shopping centres, restaurants and a winery down at Margaret River. He had made a pile from investing in uranium explorer Bannerman Resources a few years earlier and was keen to do it again. We had met when I helped float Trafford Resources the previous year and I found him very convivial. He said that he was soon going back to Greece to visit his extended family and that his cousin Socrates had told him they were mining nickel not far from Kastoria, Tom's hometown. After doing a bit of homework I found out that the partly Greek government-owned company Larco was mining ferro-nickel laterites to produce nickel pig iron that was being treated near Athens and was mainly exported to China and Ukraine. I suggested that when he went back he should do a bit of snooping around and collect a few likely looking rocks. He'd been down to Ravensthorpe several times prospecting with me where I'd shown him typical laterites. I said to look for brown stuff! You might even find rocks on the side of the road that have fallen out of ore trucks. When you get back to Perth, we'll have them analysed.

After a bit more research, I realised that the nickel in northern Greece is associated with a huge ophiolite belt that extended further west into Albania. In fact my work showed me there was more nickel potential in Albania than in Greece and that there were several significant historic laterite mines. When Tom returned from his reconnaissance trip, he produced a sack containing around thirty samples he had collected. They all looked very interesting to me. I said we'll send get them assayed for nickel, cobalt, iron and sulphur. The results we got back were encouraging with most samples assaying between 0.75 and 1.5% Ni and 500 to 1500ppm Co with high iron numbers. The Albanian samples were the best.

He said he had managed to get into Albania quite easily as it was no longer under a communist regime and everyone in that part of the country regarded themselves as Macedonian like him. Tom didn't mess about and went straight to Korce, the main town of Korce Province, to speak to the top government officials about exploration rights in the country. The Provinces in Albania are the equivalent of our Australian

States with Korce Province being the most mineral-rich, like WA is. He told me he'd had an audience with the 'Prefect', the head of government, like our Premiers. I asked how did you get on with him? With a broad grin he replied: "it's a she and she's a bit of a looker!" He said she was called Elfrida Zefi and that she had a degree in Earth Sciences from the US. She wanted to know whether he had any exploration people working for him. In typical Tom hype he told her that of course he had, and that he had the best nickel geo in Australia on his team.

To the Balkans

So a few weeks later we decided to head to Albania. I met Tom at Perth Airport where he panicked as he couldn't find his passport! He was with his wife who reckoned he'd left it at home. They lived in City Beach, not exactly close to the airport. Fortunately there was someone in the house so they phoned them and they made an emergency dash to the airport, luckily just in time! After a brief stop in Singapore we flew to Athens where we were met by Tom's cousin Socrates and by well-known Perth mining identity Shane Sadleir, MD of Athena Resources. After a superb lunch of spit-roasted lamb and Greek salad (naturally), Socrates drove us to Kastoria in Greek Macedonia. We went via the east coast past Thermopylae, site of the famous battle where it is said 300 Spartan warriors defeated a huge Parthian army led by Xerxes. They made a film called The 300 not long ago which was apparently banned in Iran. I saw it and can sort of see why; some of the special effects were spectacular to say the least. The huge creature with lobster-like claws capable of severing a human head was awesome!

The modern Greek road system is excellent (thanks to EU subsidies I was told) as were Socrates' driving skills. He was a fast driver and usually exceeded the speed limit, so I asked whether he was scared of getting pinged by one of these speed cameras, as there are a lot of them? He laughed and with a laugh said: "no, this is Greece, the government can't afford film for them!" Just north of Thermopylae we headed west through lush green fairly flat terrane on motorways carrying very little traffic. We passed a few large coal-fired power stations and arrived in Kastoria late that afternoon. It turned out to be a very pretty little town sitting by the side of a lake. I noticed several snow-capped peaks in the distance to the west. It was early May; we had arrived in the Balkans. Tom stayed in the old family home with an aunt while Shane and I were put up in a very pleasant little hotel overlooking the lake: perfect!

That night we all had dinner in the hotel dining room. On the next table were around eight other diners of both sexes, one of the men was dressed like an Orthodox Greek priest with archetypal tall black pillar box hat. We exchanged greetings and naturally they were curious as to what we were doing in Kastoria. Tom told them we were exploring for minerals and on our way to Albania on a trade mission. The priest said they were fur traders and attending a convention in town. Then he told us he was actually the Archbishop of Kiev in Ukraine and that he also traded in armaments! We were all astonished, well archbishops don't lie, do they? Umm not sure, after all I'd worked in Cyprus during the era of Makarios. He said he was particularly looking for a reliable supply of ferro-nickel. Tom looked at me knowingly and then replied to him that we were specifically targeting nickel iron laterites in the Balkans. Cards were exchanged and glasses clinked: Yamas!

The following day we set off for Albania, only around one hour away from Kastoria to the border. The country was now a lot hillier and well wooded. We passed an old pig herder who had a long crook and was tending a mob of black and brown striped young wild boar. Had to take a photo of this. He didn't seem to mind and waved cheerily. The border was reached around midday and there was quite a queue of assorted vehicles. Passports were shown and the guards waved us through. No visas required. They seemed to like the fact we had Australian passports. They did ask us the purpose of our visit though, standard practice everywhere now, of course. We said we had trading business to discuss with the government. Think they were impressed and passports were duly stamped. There was a very good duty free shop where Tom bought an expensive top of the range bottle of Johnny Walker Blue Label Whisky to give as a gift to the 'First Lady'. He always liked to show off a bit. We also had a carton or two of his Margaret River Driftwood wines.

The first thing you notice about Albania is that it's nothing like modern Greece. It's more like the Greece and Cyprus I first visited in

1969. Donkeys and horses and carts abound everywhere on the roads and in the fields. People tilling the land by hand. Bit like going back to the Middle Ages; very photogenic. I had a ball! Just over the border we saw our first nickel mine where a few dump trucks were parked. Across the road was quite a large excavation where we observed a dark brown flat-lying seam of iron-rich laterite about seven metres thick overlain by a steep hill of pure white limestone. We bagged several samples of the laterite for assay purposes. Interesting geology!

We set off again for Korce. The road was fairly narrow with many potholes, unlike Greece. There were many horses and carts. It was obvious we were in a very poor country: actually Albania had the reputation for being the poorest country in Europe at the time; it may well still be. A bit further on we saw a large mine on the right of the road. A sign proclaimed "Feni-Korce, Betincka Fe-Ni/Ni-Si Mine". There didn't appear to be much activity but we noticed quite a few big waste dumps. After another forty-five minutes we arrived in Kastoria where Tom's cousin had booked us all into a hotel in the centre of town just across the road from Parliament House and from a splendid looking Orthodox church. We parked on the street next to the hotel entrance as there didn't appear to be any parking restrictions. By now it was around 5.30pm so, after checking into the hotel, which was quite modern, we went to have a walk around town. Quite an experience; lots of big old Mercs, but more horses and carts often followed by large dogs and, a sight I'll never forget, two swarthy, mean-looking youngish guys leading a huge brown bear by way of a short chain attached to a ring through its nose. I managed to sneak a quick photo but they didn't like it. Thought they might have knifed me. "Roms" said Tom's cousin, "be careful they can be very dangerous!"

We had a great feed that night and, after breakfast the following morning, made a phone call to government house to get an appointment to see the Prefect of Korce, Elfrida Zefi. We were told to come immediately , so took a short walk across the town square then up the steps to the front door. I had expected to see a lot of security guards but couldn't spot any obvious ones. We were ushered into a smallish office where there were several soberly suited male officials and, at the end of a large desk, sitting next to a flag displaying the double-headed Albanian eagle, was the striking looking 'First Lady', Elfrida herself. I felt very underdressed as I was wearing typical Aussie field gear rather than suit and tie. She had already met Tom who introduced Shane and me. She shook everyone's hand and turning to me said that you look like a geologist who likes to get his hands dirty, and I am very glad you are not wearing a suit! I was impressed. She told us that she had a degree in Earth Sciences from Nebraska University in the US. I thought this is someone I can deal with on an equal level, at least technically.

She said Tom had told her I was a world-renowned nickel geologist and I replied that I hope I live up to your expectations. "We are a very poor country'" she explained, "I want you to help me to make it richer". I responded that if your government gives us permission to explore, I will do my best. She said I am the Prefect of Korce Province, so I can approve things here, but not in other Provinces. I told her that Australia has similar State regulations and jurisdictions. She went on to say that since the fall of communism in the early 1990s, there was virtually no metal mining anywhere in Albania anymore. I said I thought that there was plenty of potential for nickel-cobalt laterite deposits in Korce which could be profitable at the present high nickel price and with the new pressure acid leach extraction technology. I told her Western Australia already had one of them to the northeast of Kalgoorlie which was operating in a large way at a place called called Murrin Murrin, with a second one being built by BHP at a town called Ravensthorpe near the south coast.

Elfrida introduced us to her personal assistant, a youngish lady called Marissa. She had long lank black hair and was dressed like a bikie in tight black jeans and leather jacket. We were told she would be our guide and interpreter during our time in Albania. She spoke good English with what sounded like a Birmingham accent. She said she been to university in the UK to study English. I asked whether she studied in Birmingham? She looked surprised and asked how I knew? Your accent, I replied. I told her that I originally came from Liverpool before emigrating to Australia and visited the UK every year. Then she advised she used to visit friends in Nottingham and loved English beer. I replied that Nottingham has some great pubs and I used to go drinking with my friends from there quite often. She said her favourite pubs were in a district called Canning Circus. I said so you must know the Red Lion, the Running Horse, the Robin Hood and the Falcon then? They were my favourites she replied. A promising introduction to our new guide I thought!

Next we met the Minister for Mines, a very pleasant man called Milo. He gave us some maps showing all the old mines in Albania and wished us good luck. All very civilised! Before we left Parliament House, Elfrida invited us all to join her for dinner that night in her favourite restaurant and told us we should go and visit the nearby Betincka nickel mine that afternoon. She gave us a diplomatic pass to show to police if they stopped us on the road. She said the police were not very well paid and often handed out spot fines for spurious reasons. This pass was to prove very useful over the next week or two.

The Betincka mine site was easy to get into and there didn't appear to be anyone around. There were several large excavations into a dark brown 7 to 9m thick flat-lying seam of iron- rich laterite overlain by a hill of white limestone. The footwall was a brown rock criss-crossed by numerous thin beige-coloured veins. I guessed it was weathered serpentinite. We took many samples for assaying in Australia. We went back to our hotel in Korce to clean ourselves up before dinner and have a few beers. Rock chip sampling is thirsty work!

We had arranged to meet the Prefect and her party in the lobby of Parliament House. The restaurant was only a short walk away. It proved to be cosy and unpretentious and a small open fire was burning next to our table. There was a rack full of assorted wines as you went through the door and a stuffed sad-looking fox. My sort of place. I was seated next to Elfrida at the head of the table and she insisted we toasted each other with raki, the Albanian national drink. "Nostrovia," she exclaimed as she raised her glass. Tom toasted everyone with his usual "Yassou". We had some great traditional local food cooked on a hinged grill device over the embers of the open fire. The lamb chops were sensational and you could tell the grilled vegies were very fresh. We drank Albanian beer and wine - also very good. I was beginning to like the country. When we had finished the meal, the fun started. Several musicians came in with a variety of instruments, including a bouzouki, and started playing some wild traditional Balkan tunes and gypsy music. Being a folk musician myself I was familiar with this sort of music and really enjoy it. Elfrida proved to be a very feisty woman and got up to dance. She was dressed in a peasant skirt which she kept lifting up when she twirled around. Hard to imagine this sort of thing happening in Australia though Bob Hawke did have his moments!

Back at the table she turned to me and said Tom has told me you are a good singer. I nodded and told her I had been a singer in an Aussie band called 'Rocky River" and that we had toured Europe and the US several times. Then you must sing for me she commanded, what sort of music do you do? I like songs of the sea I said, especially sea shanties. "Good, drinking songs", she exclaimed raising another glass of raki.

I said I had been born in Liverpool, England's greatest seaport and the home of sea shanties. I got up and launched into a hearty rendition of 'The Leaving of Liverpool'. She said Tom was right and that I sang very well. I thanked her and told her my band had once played in front of Princess Anne at a big maritime festival in Portsmouth. "And now you play for royalty again", she joked. She then announced that we are now in for another treat, as her cousin is coming to sing us some opera, and that he has a great voice. A few minutes later I noticed a small shabby-looking oldish man coming into the restaurant. He was wearing non-too-clean wellington boots. I thought we had been gatecrashed by a beggar! Elfrida looked round and said good, here is my cousin now; he has just come in from the fields. He started to sing. She was quite right about his voice, it was magnificent. After a couple of arias, the dance music started up again. This time the table was cleared and incredibly Elfrida leapt onto it and strutted her stuff again. Reckon she must have been pretty well oiled by then as she always seemed to have a glass of raki in hand. I thought to myself this is getting to be like a Borat film. We soon called it a day, certainly one I'll never forget. Before leaving we were told to go to her government office the following morning before doing any more field work.

When we got there she said she'd been talking to Tirana and that we should go there to meet the federal Minister for Mines after visiting several more old mines on the way. She said she thought we should register a mining exploration company there as it would further our credibility in the country. Tom nodded and told her that was already what he had planned to do. We set off on the road north out of Korce with Merissa as our guide and interpreter. The road was narrow and in bad repair, Merissa said it was being upgraded. Well, we did see a few road gangs with picks and shovels and the occasional bit of earthmoving gear; it looked like the upgrade would take a long time. There were increasing numbers of horse and donkey carts which I kept photographing, much to Tom's amusement.

Eventually we arrived in a town called Pogrodec. It looked pretty shabby. You could smell the poverty, there were quite a few people missing limbs, one old legless man was shuffling around on a type of skateboard; very distressing to my privileged western eyes. Melissa said it was the result of the Albanian civil war when the country was extensively planted with land mines. I remember seeing footage of refugees trying to get out of Albania at the time, some even attempting to cross the Adriatic to Italy on airbeds; they must have been truly desperate. I was now seeing a side of Albania that was pretty grim and could now understand why people back home warned me about getting shot in the Balkans! Well, as I said earlier, I did get shot in the arse in Cyprus in 1969 but getting shot in the Balkans sounds a lot more painful, ouch! We set off to look at our first mine of the day just north of town, unsurprisingly also called Pogrodec. It had obviously been a substantial affair with a large, but now derelict, processing plant and it was adjacent to a railway line which we were told still had a few operating passenger trains per week to the capital, Tirana. Marissa told us that during the communist era the mine was operated by China, and then by North Korea as the Albanian dictator, Enver Hoxha, had a special relationship with the Kim regime. She said the mine workers were used as virtual slave labour and, when the mine closed in the early 90s, they went on a rampage and stole and sold all the scrap metal components to Romany gypsies. Well, you can't blame them I replied, they were probably starving. She nodded her head. We carried out our usual sampling and once more noted the two distinct ore types. We headed north again in the direction of Tirana along the western side of the beautiful Lake Ohrid. It is one of the deepest and oldest lakes in Europe with a unique fauna. We kept passing fishermen holding up eels for sale by the roadside.

We stopped for a snack at a place overlooking the lake. On the far side were the mountains of north Macedonia, formerly part of communist Yugoslavia. The highest peaks still had snow on them. It was mid-May and there were many spring flowers around and the occasional shepherd tending his flocks. All very biblical; just about every small village we passed had a mosque, unsurprising as, until 1912, Albania was part of the Ottoman Empire. At the northern end of the lake there was a road junction, the right fork pointed to Macedonia and Skopje and the left one to Tirana and Durres. We turned left; there were scores of ominous looking giant grey concrete mushrooms dotting the landscape. Pillboxes from the communist era, said Marissa. Rounding a bend we were pulled over by police, we weren't speeding or breaking any road rules we could think of. Their English wasn't good. They asked to see our passports but seemed happy enough with them as they'd been properly stamped at the border. It was obvious they wanted us to pay an on the spot fine for some trumped-up charge. We had been told to expect this sooner or later. Marissa told them we were on important government business and I showed them my diplomatic pass bearing the Prefect's name. Their attitude immediately changed and they became quite apologetic, obsequious even, and saluted us asking if we needed any special help. Marissa thanked them but told them she didn't think so. Nice to be able to pull rank sometimes!

We descended the mountain and found ourselves in a fertile looking valley. I noticed water buffalo in the fields and some were being used to plough the land. Some of the people behind the plough were wearing Vietnamese-style conical straw hats. We could have been in Asia! Suddenly a buffalo came onto the side of the road being led, believe it not, by a tall blonde girl wearing long plaits in her hair and wearing a conical hat. She was so close I could see she had bright blue eyes. Tom and Shane were incredulous. I have always been interested in history and told them she was probably a descendent of Vikings some of whom settled in Albania and parts of Greece after failing to take Byzantium off the Turks around

1000 years ago. Before that they had founded Kiev after sailing from the Baltic up the rivers as far as possible before hitting high ground; they then hauled their long boats over the mountains and down to the Danube, then into the Black Sea and through the Bosphorus to defeat. They were probably knackered by this time! I reckon they must have waited for winter to haul their boats over the snow like sledges. After Albania they sailed west and conquered Sicily. Their first king there was called Guillaume de Milio otherwise known as 'Bill of the Mill'. He was a Yorkshireman as, by that time, England had become Norman French after Harold copped an arrow in the eye at the Battle of Hastings.

Where was I? History lesson over, (for now). Ah yes, we were off to Pyrrenjas to visit our next old nickel mine. We stopped for lunch at a small restaurant next to the railway line. It was empty and they welcomed us in. As usual there was an open fireplace in which they lit a fire using wood from smashed up pine crates. There was no menu as such. Didn't need to be; after ordering a round of local beer, insisted on by our guide Marissa (she loved her beer), our hosts brought out a pile of lamb chops and went out the back to pick home-grown vegies, mainly red and green paprikas. When the fire had died down to embers, the chops and vegies were put between a hinged grilling device like the one we'd seen before in Korce. Once again a great meal, very simple but just the sort of food I like best. A few years later I bought one of these grillers in Sydney and often cook with it on my big open fire in Fremantle.

After lunch we went to have a look at the railway. I was gobsmacked; there was row upon row of abandoned diesel locos, I counted over fifty. The makers' plates indicated most were built in Czechoslovakia, and a few in East Germany. They were obviously used to haul ore trains when the nickel mines were operating. It was quite a surreal sight. I had to photograph it. Then onto the Pyrennjas nickel mine which was up in the mountains. We passed another derelict processing plant which had a tall chimney. There was a veritable forest of those grey mushrooms on the way up. We looked inside one and it was very claustrophobic. The old mine was the usual story, a flat-lying dark brown seam of laterite under limestone hills. There were several adits and we again took samples for assay purposes. The geological story wasn't hard to understand. The

African tectonic plate pushing north towards the Eurasian plate and thrusting the ancient Tethyan ocean floor over younger rocks. It's how the Alps were formed.

We pressed on towards Tirana through more rugged country and came to the town of Elbasan where there was a large sprawling industrial complex. Smoke was coming out of some of the several tall chimney stacks but, otherwise, the place looked half-derelict. Our guide told us it was a chromium refinery and that there were still a few operating chromite mines in the district. I took several photos from a high vantage point when we turned right onto the Tirana road. That road was narrow and winding so we made slow progress. After staying the night in a fairly ordinary hotel close to the Mines Department, the next morning we presented ourselves to the relevant Ministers. Things were quite civil but a lot more formal than we'd found in Korce. As expected, we were told we should set up a locally based company, preferably in Tirana. We chose the name 'Alban Resources' and registered it that day. I still have a couple of shirts brandishing this name that we had made when we returned to WA. They told us that they would liaise with the Korce provincial government and give approval to our exploration activities in their jurisdiction now that we had registered an Albanian-based company.

We looked around Tirana, the main street where the Department was had a large open storm water drain running through its centre: not very appealing. In the main city square there was a huge statue of a fierce-looking warrior and freedom fighter called Skanderbeg who it is said drove out the ruling Turks in 1450. The Turks came back though again in 1478 and Albania became part of the Ottoman Empire until 1912. We went to an alfresco German bistro for lunch where they served litre-sized steins of Bitburger lager. They were very heavy to lift. I suggested that after lunch we should go and have a look at Durres, the main port of Albania, to investigate its capacity to export minerals should we be successful in our quest. It was only 35 Kms away and everyone agreed it was a good idea.

Durres was quite big and was a good-looking town. It had a decent harbour where we saw a few medium-sized cargo vessels. Another plus was the fact it was connected to the Albanian rail network. It is now a popular port of call for cruise ships – or it was until Covid-19 came along. We stayed at a great hotel on the beach and could look out to the Greek island of Corfu not far offshore. The Albanian Adriatic coast is fabulous and ripe for development, especially in that part of the country which has wide, clean, sandy beaches. That evening we went to explore downtown. It was not at all like what I had expected. Many of the buildings were new and very avant garde. Apparently it had been redesigned by architectural students from Tirana University after damage caused by the civil war. There seemed to be a thriving night life with plenty of clubs and restaurants; there were lots of young people around. Durres used to be part of the Venetian Empire during the Middle Ages and the restaurant menus reflected this heritage. We had pizza and spaghetti marinara washed down by red wine from Tuscany, very nice.

Next day we headed back to Korce via Elbasan and Pogrodec where I took a photo of a family of five riding on the same small moped, very economical! The road from Pogrodec to Korce was, if anything, even worse than ones travelled before; we nicknamed it "the goat track". We stayed in the same hotel as we had previously and informed the authorities we were back from a successful trip to Tirana. They already knew; they seemed to have a good relationship with the central government; always a good sign. Next day we were unsurprisingly summoned back to Parliament House. Elfrida seemed well pleased with our efforts to date and asked if we were interested in other commodities such as copper, gold and coal and we replied that we were. She confided she was having problems with a couple of Turkish companies who had been operating in Korce Province but were not fulfilling their fiduciary legal obligations. Her Mines Minister, Milo, gave us details of the locations and how to get there.

So we first went to investigate a few abandoned coal mines but I wasn't impressed, they were too small and low grade. At one of them we saw a string of donkeys walking along and a little old man struggling past with a huge sack on his back full of some sort of greenery. Tom felt sorry for him and gave him a bit of money. There was a donkey on its back giving itself a dirt bath, it seemed happy enough. At another old mine a family of three, including a very young girl, were scratching in the dirt with picks presumably fossicking for scraps of coal. They all looked undernourished. It was obvious that this was certainly not a rich country. Fourth World would be the best description.

The countryside however was beautiful; snow-covered peaks in the distance, spring flowers and many shepherds tending flocks, not to mention whitewater rivers. We saw no tourists anywhere. Next we visited an old copper-gold mine. This looked a bit more promising as there were trays of diamond drill core on site. We saw a man tending a huge pink pig nearby, it seemed very content. We returned to Korce, which wasn't far. Elfrida had told me to report back to her without getting changed. She must have had a Napoleon complex: he had given Josephine the same instructions! After yet another dinner in her favourite restaurant and more clinking of glasses of raki, she said she wanted to talk to me privately in her private apartment that night to discuss important matters of state. Sounded a bit ominous but I felt I had no choice.

She took me on a short walk across the square from Parliament House to a rundown apartment block modelled on typical East German-style brutalist architecture. The brickwork was shocking. My old man had taught bricklaying at a technical college in Liverpool when I was growing up and he would have been appalled. I must say I was more than a bit apprehensive. There were a couple of what looked like security guards around the door. I didn't notice any obvious guns though. Bet they had concealed ones somewhere! All very James Bond. They saluted us: reassuring. We entered the front door and I saw the decor was palatial. The outside was just a decoy, clever! She opened the door to her private quarters and I was gobsmacked, utter luxury but, I have to say, in good taste. Albania, the land of contrasts. There were only the two of us inside. She said before we get down to business we have to drink raki, it is the custom here; I nodded. Raki is drunk like vodka, in one hit; I was starting to get a taste for it but I must say I prefer wine. Anyway when in Rome as they say... Nostrovia! She said: "I like your style, I think you are doing a good job in my country, you are also a good looking man" Shit, was she putting the "hard word" on me? I have to say she was a striking-looking woman; red hair, 49 years old and very forceful with a great figure. A bit like Margaret Thatcher on steroids! By this time we were both a bit pissed. She asked if I liked her, and I replied that of course I did and said that you are a strong leader, and you must have a very difficult job here. She agreed.

She told me that she fancied me, and asked would I like to stay with her here tonight? The ultimate 'Catch 22'! How are you supposed to answer such a question? Was this a test of my moral integrity? I told her that last week we had dinner with you and your husband and your two boys, so I don't think it would be a good idea. She said she thought he was fat and boring. I said that I was eleven years older than her and was married with grandkids. Slight exaggeration! Had I passed the integrity test? Could I have been arrested and shot at dawn? To this day, I'll never know.

The mood changed, she said I need you to do something to help me and my country. "We are very poor, as you must have noticed, and I think you are the man to help me to change this". I replied that before this can happen our new company, Alban Resources, must be granted proper mineral rights to explore and mine like we do in Australia. "We call it 'Sovereign Risk'". She nodded, I'm sure she understood. She confided she had ambitions for Albania to become a full member of the EU but that it all depended on the state of the Albanian economy. I said I thought that the mineral potential of the country, especially in her Province of Korce and particularly for nickel and cobalt, was very promising. "I hope you are right", she replied. I told her I thought that the geology was favourable and said that, as you must know, in the past you had several producing nickel mines. She said I want you to write a report on what you think, as an international expert, that I can take to the EU in Brussels. Good idea, I thought. Well I said, your personal assistant Marissa told me when we were in Durres that Albania is now a member of NATO which must be a good thing for strategic reasons, to which she nodded. She advised that Marissa will help me to write my report and will put it on her laptop. She said she might even take it as far as the UN: very ambitious, I reckoned, but didn't disagree. Could I be helping to change world politics? If so, hopefully for the better.

I told her it would probably take me around a day to write the report.

I said I would do it in the hotel lounge and that Tom and the rest of our party should meet her in her office with the Minister for Mines to discuss legal matters to do with exploration rights; she agreed. "I will send Marissa with her laptop to help you", she said. Next day, with the help of a few beers, my report was finished. Tom successfully negotiated the political and legal side of things but I pointed out that we really needed assay results from our rock chip sampling of the old mine sites before making any further binding commitments. Elfrida understood: she was a canny woman!

Our field work and time in Albania were now just about over and, after a farewell dinner that night, we left for Greece the following morning. Tom said we should stay in Kastoria that night after sorting out our samples to send back to Australia. I thought it a good idea, and he said his cousin Socrates would arrange the shipment. The following day we drove back to Athens via the famous town of Meteora, a surreal place where there are many ancient monasteries perched on top of towering rock pinnacles. God knows how they built them. It has long been a photographer's paradise.

We were now back in the "real world" driving on good highways instead of goat tracks. Think I preferred Albania in many ways; it was completely different to anywhere else I'd ever been, yet it was actually now a centre-right democracy in Europe. Despite being told before we went that I'd get shot or robbed by bandits it seemed to me to be a pretty safe place. Back in WA we sent off our samples for assay. Most of them came back much as expected: nickel between 0.75 and 1.25% and cobalt in the range 500 to 1300ppm, very similar to Murrin Murrin grades; we were very encouraged. Tom was already talking about a return visit. The following year we did go back although things were about to radically change: it was the start of the GFC. The nickel price crashed and our friend and supporter Elfrida Zefi, the Prefect of Korce, lost the election. All this is the subject of another story!

RUSSIA AND THE FORMER
SOVIET UNION

VODKA ON THE ROCKS

DEREK FISHER
SIBERIA 1992 – 1998

Derek Fisher studied geology at the University of New England on an NSW Geological Survey scholarship, graduating with honours in 1969. Following a stint in WA and the Solomon Islands with INCO, he moved to Canada where he completed a PhD at the University of Toronto.

His experience in Canada introduced him to the junior mining scene, and since his return to Australia in the early 1980s, he has been instrumental in creating, listing and managing numerous companies on both Australian and Canadian stock exchanges. These companies have had exploration projects on all continents, except Antarctica, and developed mining or mineral processing operations in Australia, Indonesia, Mongolia, Armenia and the Congo. Most recently were Moly Mines with the Spinifex Ridge molybdenum deposit and iron ore mine, and Salt Lake Mining which resurrected the Beta Hunt nickel/gold mine at Kambalda.

He joined the Association of Mining and Exploration Companies shortly after it was formed and served on the Council for many years, four as President.

In recent years he has expanded into farming with the family farm supply a significant portion of WA's blueberry production.

We first went to Russia in August 1992. There were four of us: Paul Morgan, a Queensland-trained geologist, then living in the UK, with previous experience in Kazakhstan and Russia; Peter Hannah, a Sydney stockbroker, mining analyst and metallurgist; my Perth business partner, and lawyer, John Hopkins; and myself, an Australian exploration geologist with nearly 20 years Canadian experience including time in the Arctic.

Paul had a strong contact in Moscow, Youri Koulagin, who had spent several years during the 1980s running a USSR trade commission office in Melbourne and had facilitated business in Russia for Greenbushes Tin and in Kazakhstan for Paul.

We were sufficiently encouraged from our meetings in Moscow for me to return in March 1993, at which time I first ventured into Siberia. After our first visit we had left Youri with a wish-list for him to target including gold tailings and rich alluvial gold deposits.

Gold tailings were of particular interest as they tend not to be technically challenging and there was not a strong history in the USSR of using cyanide for gold processing: in Australia a number of successful mining companies had their origins in reprocessing old gold mine tailings in the 80s. Alluvial gold had also caught our attention as on our previous visit deposits with grades measured in the grams per cubic metre were mentioned, grades not seen in Australia since the 19th Century. Australians with their metallurgical skills in processing mineral sands deposits could bring new technology to bear on these deposits.

The projects Youri had identified were in the Chita Oblast (Region) of eastern Siberia. This Oblast, named after its principal city, Chita, is east of Lake Baikal and borders China and the eastern tip of Mongolia. It has a long mining history going back 200 years with the region first being settled by European Russians in the 1600s.

Chita is the gateway into China from Russia with the main railway south branching off the Trans Siberian at this point. The city was and still is a military city and had been closed to westerners from the 1930s. The Russian military clashes with China in the late 60s occurred along the Amur River Valley to the east while the border to the south of Chita along the Argun River was also disputed. It is on the same time zone as Perth and is six hours in time zones and about 6000 km east of Moscow.

It is a long way and it usually involved an overnight flight with a stop in Omsk for refuelling. These stops usually occurred in the middle of the night and we always had to deplane while it was refuelled. In winter it was invariably significantly sub-zero and involved crowding into a freezing, unheated bus and waiting in the drab, concrete, Soviet-era airport building for an hour or so. Fortunately, having spent many years in Canada I was well prepared for the cold.

On that visit to Chita we had numerous meetings with officials from various government departments, including the head of the geological survey but, importantly, with a number of mining companies. Under communism the USSR had put an enormous effort into mineral exploration. Considering that it covered about one sixth of the world's landmass this required a very large exploration workforce. To put this into context we were told that at that time there were approximately one million geologists in Russia and its satellites! There was even an official national Geologist's Day, celebrated every year in April.

Greenfields exploration was not undertaken by the mining companies but by teams or organisations that covered particular areas and these were termed "Expeditions" (a translation of the Russian). In the new Russia, capitalism was rapidly advancing and many of the Expeditions had been privatised. New companies had sprung up from these organisations, usually called "Artels" which loosely translated means cooperatives. They were groupings of former members of the teams and they saw opportunities to exploit many of the deposits they had discovered and worked on.

We met in Chita with the management of several Artels and quickly identified opportunities that on paper looked outstanding. The first was in the Baley District about 300 km by road south east of Chita. It was the site of the famous (in the USSR) and infamous (many Japanese prisoners-of-war perished there in WWII), high grade gold mine, Taseevskoe, which, as an underground mine, produced 17 million tonnes averaging 14 gm Au per tonne. The tailings of that mine had to be attractive as they had never been cyanided and, assuming 90% recovery of the gold, they should be carrying in excess of 1 gm per tonne; a very attractive grade for a reprocessing operation. There was also significant alluvial gold associated with the hard-rock mineralisation with the local Artel working ground averaging 1.5 gm per tonne.

The other attractive area was 600km north-east of Chita near the remote town of Chara. This settlement was built to service the large Udokan copper deposit, one of the largest undeveloped copper deposits in the world. Discovered in the 1950's, this large, flat-lying, sedex copper deposit is finally under development, with a planned initial 12 million tonnes per annum mining rate, expanding to 48 million tonnes per year.

We were shown the data from a number of high grade alluvial gold projects located south of Chara (and south of Udokan) with grades of 1 to 5 gm per cubic metre, with some isolated sampling up to 10 gm per cubic metre. These spectacular numbers certainly sent my brain into overdrive, having worked on deposits of 0.1 gm per cubic metre with 0.5 gm per cubic metre considered high grade.

It was decided that a visit to these areas was the next requirement, and that that should be undertaken in the coming summer (Chara's average winter maximum temperature was about -20 degrees Celsius, while summer was more like 20 degrees Celsius).

MOUs were prepared and executed in preparation for a June visit and then it was back to Perth.

John Hopkins and I returned in June following a hectic couple of months. Companies had to be organised to hold the projects, requiring research into country of domicile (Cyprus) for favorable tax treatment for repatriation of profits out of Russia, bank accounts for getting funds in and out, templates for agreements satisfactory for Russian law etc.

We arrived in Chita again in the overnight flight from Moscow arriving about 3am. After a few hours of freshening up in the town's best hotel (in those days it would not rate one star by our standards) in a shared hotel room, it was on to Baley to see the tailings and the alluvials. This was an interesting experience as we travelled by road, about 20km sealed and the rest potholed gravel. The road traverses broad rolling rounded hills incised by deep valleys, it roughly follows the Trans Siberian Railway for the first 250km to Nerchinsk, and then crosses the broad, fast flowing Shilka river, a major tributary of the Amur river, then over a range south to Baley.

Nerchinsk is famous for its lead-silver mines, but also as the Tsars' major hard-labour prison site in Siberia, where prisoners, particularly political revolutionaries, were exiled in the 19th century to work in the mines. Most were made to walk from St Petersburg or European Russia. It was also the site of a famous orphanage and, according to Youri, it was also a favourite threat by parents to misbehaving children that they would send them to the 'children's prison' at Nerchinsk! These prison camps were also part of the USSR's gulag system, and the road passed the barbed wire fences and watch house of one such site on the outskirts of Nerchinsk.

The drive was our first true experience of a Russian car, in this case, a Volga. The Volga was very similar in looks and engineering to the HD Holden of the mid to late '60s. With five of us in the car on a rough road it was a tiring experience.

The road is the main east-west highway across Siberia. Interestingly, we passed a number of convoys of Japanese 4x4s travelling west. At that time there was a flourishing business in Russia where entrepreneurs were going to Japan, acquiring cheap second-hand vehicles, shipping them to Vladivostok and then having teams of drivers bring them overland to European Russian where they sold them at significant profit.

On arrival in Baley, we went straight to the Artel's compound. Baley is a town of about 15,000 people on the banks of the Unda river. It lies in a broad valley of low hills with wheat fields, and the occasional dairy farm and forest. The town is dominated by the open pit underground and alluvial gold mine workings, and associated waste and tailings heaps. Its modern mining history goes back to the 1920s, when the Taseevskoe epithermal vein deposit was first developed.

At the compound we were met by the Artel's president, a Russian bear of a man, Alexander Guselnikov. Alexander was a classic Russian mining veteran, having spent his early years on the large bucket/ladder dredges on the notorious Kolyma Goldfields in far north-eastern Siberia. This was the major gulag area of Stalin, with the gold production funding much of Russia's war effort in the 1940s.

This first meeting with Guselnikov was notable not for the negotiations and mining data we reviewed but for an incident that happened about 30 minutes after our arrival. He stood up from behind his desk and waved us to follow him. We trooped across the compound to an adjacent building. Inside, the first room was lined with benches with hooks above – a mine change room?! No, a changeroom for the Artel's banya – i.e. the sauna, and Mr Guselnikov was rapidly stripping to his birthday suit and indicated we should do likewise. With Youri's encouragement, we somewhat hesitantly complied. So within about half an hour of meeting a potential business partner, we were all naked in a room together! Talk about an ice breaker. And then for the ultimate ice breaker, the banya.

Now this is something to be experienced. The benches are stacked up the walls, three levels, and as water is tossed on the hot rocks, the temperature and steam on the upper level is extraordinary. And then there are the birch branches; these are thin twigs covered in leaves that have been soaking in water all day. After we had acclimatised on the lower levels, we were invited to lie on the top level. A 'torturer' then took two handfuls of birch and shook it over you – the heat generated by the evaporating water was amazing. He then beat you (gently, not hard as some people like to tell it) up and down your back and legs. He lifts your legs and does the bottom of your feet and then you roll over, he wets the branches again and then places them over your face!

All this birch treatment with the intense heat and steam takes about 1-2 minutes - that's about all you can take. Neither of us had ever experienced anything like it. And then to top it off, it was into the icy plunge pool in the next room, and then back into the banya. We lasted I think only about 20 minutes, and then we retired, after showering, to their dining room.

I do not have much recollection of that night, but I'm sure there was much beer and vodka consumed, and maybe some Russian champagnski, and of course, accompanied by many toasts.

The following day was a time of intense negotiations and technical discussions. However, it did start with a breakfast, which again was notable because we were served large plates of what we thought was yoghurt – which turned out to be thick cream. The Russians have an extraordinary appetite for anything with fat, including the cream but also butter, which they eat like cheese as thick slices on bread, and hard pig fat, which they often eat as a snack.

The technical discussions, which were all undertaken through an interpreter (in this case, Youri), were a very tiring business for both parties. It was followed by commercial discussions, as we both wished to proceed to an agreement. We found the Russians had difficulties with this, and this is probably the result of 75 years under the communist system whereby they had little 'commercial' experience. However, with time, broad terms for a deal on the tailings were agreed between both parties. The Artel was still in the process of securing a license for them, but this was proceeding rapidly.

We departed Baley in the early evening, for the drive back to Chita, where we had two days of meetings organised. As we drove out of Baley, we noticed that our driver was following a van, and when we reached the first timbered area, that van suddenly slowed and turned off, following a narrow track through the thick forest. Our driver followed; what was going on? Youri did not know, or kept a straight face. John and I looked at each, other quite apprehensively – was this a Russian mafia kidnapping, the likes of which the Western Press was full of in those days...??

The truck continued winding along the rutted track. Suddenly ahead of us we could see fire, and we emerged into a large clearing on the banks of a wide, rocky, fast flowing river, with a large roaring bonfire in the centre. Food was piled up on trestle tables, and of course our hosts were standing there with glasses of vodka for us – a farewell 'Russian barbecue'.

The barbecue was not t-bone and rump steaks. Beef was not common in those days; chicken, pork and freshwater fish were more likely. Barbecues invariable involved shish-kebabs, and that was the case this night. More food, more vodka, toasts and speeches, before finally departing for Chita at about 9.30 pm.

This time, we had Alexander G in the car with us, and he spent much of the drive regaling us with mining stories. One thing that sticks in my mind from this trip was his description of the common occurrence of well-preserved mammoths in the frozen gravels in the Kolyma goldfields. On a later visit to Baley, he showed us black and white photographs of them digging out these frozen creatures. The other extraordinary aside to this mammoth story that night was that there were a lot more discovered than were reported to the archaeologists. In fact, both the gulag prisoners and the mining staff put extra effort into finding them, particularly in the Stalinist years. He said the staff wanted them because the lucrative trade in mammoth ivory supplemented their salaries, while the prisoners sought the frozen carcasses for the meat – 4000 year old steaks! How true this is, I do not know, but he swore by it. Many thousands of prisoners perished in that goldfield from starvation, the cold, and overwork. In fact, history shows that the very first year that the gulags were established, all of the prisoners and all of the guards – 15,000 of them – perished over that first winter, so there could be something to Alexander's story. In a macabre way, it conjures up scenes from the Flintstones, sitting around a fire barbecuing mammoth steak.

The road trip back to Chita was long and slow, as we had to wait for the accompanying van. Alexander, Youri and the driver entertained us with Russian songs, including songs of the revolution, until the vodka took hold and we slept. There was another picnic break about 2am, with vodka, and with us finally arriving at our hotel in Chita at 4am.

That day was spent reviewing more data on the Chara alluvial gold projects in preparation for a planned visit to the project area. Certainly, on paper, the numbers continued to be impressive, with grades unlike any I had seen before in my exploration experience.

Some time was also spent finalising terms on these alluvial projects, plus a diamond project that was thrown in for good measure. This then led to the drafting of the agreements. In anticipation of this, we had taken the bold step of purchasing a portable computer (not sure they were called laptops in 1993) and a small folding printer, before leaving Perth. We also had some 'crash' instruction time (I would not call it a course) on Word Perfect. That computer and printer cost us over $9000, and that's 1993 dollars! However it was a god-send and we had loaded it with numerous agreement templates before leaving Perth which allowed us to rapidly prepare the documents. Unfortunately there were two hiccups; firstly, the need to prepare Russian versions of the agreements, and this fell on Youri's shoulders who had to translate. We had found a typist, Irena, in Chita, and managed to borrow a typewriter. This was a 1960s style machine, no auto-correction and no memory. And in Chita at that time, photocopiers had not arrived, or were extremely rare, so carbon paper it was for copies.

The hiccup that almost stymied the exercise was the lack of paper on which to print the agreements. Russian office paper was light brown and not dissimilar to hand towels or toilet paper in feel and texture and got chewed up by our printer. We had brought some paper with us from Australia, maybe a quarter or half a ream, but that was largely squandered before we were alerted to the dearth of local paper! All of these problems conspired to keep John in Chita the next day while I headed to Chara with Youri.

We headed to the airport about 8.30 am in a convoy of cars and a truck. I was uncertain as to why so many people and why the truck, and started to think that another farewell party was planned at the airport. I was told that a helicopter had been chartered, and in my naivete I imagined that it would be a Russian 4 or 6 seater similar to the Bell 206 JetRanger that we used extensively for exploration in the west. I knew the Russians had similar helicopters as I had seen them flying over Chita (both Chita and Nerchinsk had large helicopter military bases).

On reaching the airport the convoy drove straight out onto the airfield, passing half a dozen large military style choppers, eventually coming to stop beside one of them. This was our charter, a bloody Russian gunship! It had clamshell rear-doors and would probably hold a small truck or tank. My brain went into overdrive thinking that this would be costing us a fortune, particularly as we had a 600km round trip. I had Youri quickly query Guselnikov as this could significantly deplete our meagre funds (at this stage, our Russian adventure was being largely funded out of our own pockets). Mr G was reassuring, saying that they had agreed on a fee of about US$2,500 for the trip! I was flabbergasted, surely he meant US$25,000, but no, that was the number. I subsequently found out it was an Mi-8 helicopter, a success story of the Russian aviation industry, which in many characteristics outperformed the American Blackhawk. There were seven of us travelling that day in a chopper that can carry 30 troops! The truck that accompanied us was carrying supplies as we were staying overnight in the field. This meant we required at least a tonne of food(!), and of course a vodka supply. The latter was interesting; only in Russia would you see alcohol packed in wooden, partitioned crates like old fashioned milk crates. Sixteen bottles of Stoly to a crate, and about three crates were passed up and stashed on board.

The chopper was quite utilitarian, with the seating attached to long range fuel tanks that lined both walls. There were large oval windows that opened inwards and hinged to the roof. They were ideal for photography, but I could just see machine guns being mounted on the sills in Afghanistan. The crew, there were three, pilot, co-pilot and an engineer, came on board and we were away. We needed that extra fuel as it was a long 3.5 hour flight.

No sooner were we away when the food came out and the ubiquitous vodka shots were passed around.

Vodka gets mentioned a lot in this story, but interestingly, in my six years in and out of Russia, I came across very few Russians that I would identify as alcoholics. They enjoyed their vodka and their beer, but they had a way of drinking vodka that became quickly apparent after spending time with them. Vodka was drunk in shot glasses and largely by toasting. Dinners were a continuous round of toasts; first the host would welcome the visitors in a speech followed by a toast. All the attendees would rise and drink with the toaster who, with a flourish, would skol his shot. However, we quickly observed that most of the others would only sip their vodka. And the glasses would be topped up, and the guest would reply, finishing with a toast and a shot. By the time the night was finished everyone present would have been called on to contribute: there would have been toasts to Russia, to Australia, to geologists, engineers, several to the women and wives, their beauty and their children. If the table was large and the night long, often the toasts towards the end of the evening degenerated; I can remember toasts to 'surfing in Siberia' and to 'shooting polar bears in Australia'.

Back to the chopper flight, the route took us over hundreds of kilometres of green forest which started maybe 50km out of Chita. Towards the end, the topography changed, first to being quite hilly, then to mountains and fast-flowing rivers. A most extraordinary sight struck us as we came close to Chara. The town was in an east-west forested valley that was flanked to the north by a spectacular, glaciated, jagged, snow-covered alpine range. We could have been in the Swiss or Austrian Alps. The peaks were over 3000m, and we could see small ice fields and glaciers. Never did I expect such a range in the middle of Siberia.

The other spectacle was an area of white sand dunes. Now we had a desert in the middle of Siberia! It was not large, maybe 10 km long, and

I assume the sand is glacial, a product of the last Ice Age.

I subsequently learned that the Chara River valley follows the Trans-Baikal Rift, and is an active seismic zone. Lake Baikal, just to the west, is a major active continental rift and, to the east and north-east, the fault systems follow the Chara valley with the ranges a product of movement along these structures. We landed in Chara to refuel and refresh, and after a brief meeting with the deputy administrator of the region, it was back to the helicopter and we headed south-east to the Djemku River.

We landed on an elevated river terrace adjacent to the river, which was gravelly, fast-flowing, and shallow. The country is hilly with larch forests. The terrace was flat and covered in low shrubs, and was notable for the coverage of showy, purple wildflowers. On closer examination, these turned out to be rhododendrons which I later discovered were native to this area. The other feature of the flat was the lack of tall trees. But as we moved around the area, we later discovered many tree stumps. The trees had been cut down many years ago.

Also, as we walked through the low, metre high scrub, we came across another unusual sight; what looked like the rusted, iron bowl of an old washing machine, which it obviously was not. It was a drum about 700mm in diameter and about the same length, made of heavy, riveted, steel plate, which on close examination had pipes inside. It was a boiler! And hence the lack of trees. They had been cut down and used to fire the boilers. But why a boiler?

Our hosts quickly enlightened us; the gold-bearing gravels under the terrace were frozen. We were in permafrost country! In fact, it was the very southern limit of permafrost. This area had been the site of a gulag in the 1930s and 40s, and the gold had been worked by the miners sinking small shafts through the unfrozen surface overburden and then working the basal gravels for the gold using hot water! As we looked around, we saw many "turkey's nests" of shaft-spoil as well as other boilers.

In the 1980s the local Expedition tested these river terraces and flats along 6 km of the river valley with lines of shafts about every 250 metres. The basal gravels averaged a bit over a metre thick and graded 1.2 gm Au per cubic metre. There was another 8 km of the valley still to be explored, plus some tributaries.

The Artel had sent in an advanced party and they had opened a couple of the old shafts. We visited a number of these and were shown samples from the basal gravels. A Russian "pan" was produced for dishing the material. This was not a dish as we know it, but a rectangular tray with a deep "valley" in the centre, all carved from a single piece of soft wood. It worked very well, with my samples washed and processed, and sure enough, there across the valley was a beautiful trail of "colours".

However, I was concerned that the area may have been salted prior to our arrival so, knowing that the advanced party had obtained the frozen basal gravel samples by blasting the bottom of the pits, I asked to be taken to another shaft that I had spotted. There, instead of taking a sample from the material stockpiled for us, Youri and I searched in the low scrub around the pit for ejecta from the blast. Sure enough, we found a number of lumps of clayey, gravelly material partly concealed in the undergrowth. These I dished myself and, Eureka!.... a beautiful trail of coarse, flattened gold grains.

I was convinced; which I should have been after I had seen the relics of the gulag times – those poor buggers would not have gone to the trouble they did, if there was not good gold to be had.

After another long, tiring day, this was an excellent finale. I was happy, and our Russian hosts were also happy, and...... another celebration was required!

That night we were camping on site. The advanced party had set-up a number of small tents. I, being the guest, had a tent to myself with a cot, a sleeping bag and a bucket of water. The floor was the spongy, low scrub of the tundra.

After a quick freshen-up, we retired to the dining room!

This consisted of several trestle tables lined-up under the stars and fires were lit adjacent to them. It was late June, just past the Summer solstice, moderately warm and, being about 56°N latitude, the evening was long and softly lighted. We probably started late, and I am sure, finished beyond midnight. The food was basic although I recollect that it did include wild venison that had been shot before we arrived.

We were a "bunch of happy campers" – excuse the pun. The Artel was getting a Western partner who could bring the finance and new technical skills while I liked the properties and the partners that came with them. Guselnikov stood out as be brought strong mining experience, good contacts and was respected by the authorities.

The toasting that night reflected these sentiments, and the vodka flowed. Everyone was very relaxed, and friendships were sealed.

Towards the end of the evening we were all a bit tipsy and an incident occurred, instigated by me, and for which I became a bit infamous, particularly amongst my new Russian friends. I never lived it down. Let me share it.

Firstly, the background; the Artel being a gold producer from alluvial operations around Baley, had appointed a head of security. And in Russia where is a good place to go seeking someone to fill that role? The KGB of course! Well, Guselnikov had recruited the ex-deputy head of the Chita Oblast KGB. (After the fall of communism, the KGB had been largely disbanded, to replaced by a new organisation, the FSB). I cannot remember this chap's name but let us call him "the Colonel". The Colonel, being a senior member of the Artel, had accompanied us on this visit.

He was a tall, somewhat rotund gentleman who talked a lot and was loud. He was very pleasant and friendly with us and showed none of the KGB-stereotype characteristics as depicted in Western movies. He did like his vodka!

As the night wore on and the vodka took hold, his garrulousness took over and he was dominating the tables. Youri, who also enjoyed a tipple or two, was constantly flat-out trying to translate for me the flow of conversation around the table, and he was getting exasperated with the Colonel.

Finally, my vodka started to speak. I stood up and turned to the Colonel who was at the other end of the tables and said loudly to him to SHUT UP and to SIT DOWN. This was not said in English, not in Russian, but it was in Polish. It did not require translation. Polish is another Slavic language with many similarities to the Russian. He knew exactly what I had said, as did the rest of the attendees.

Now there is a story behind my use of Polish; my wife is Polish Canadian and being fluent in Polish had taught our dog to only respond to commands in Polish, on the supposition that there would not be many Polish-speaking burglars around Perth. Anyway, I could tell our dog in Polish to Stop Barking, that is Shut Up. Similarly Sit was another command I knew.

Back to the Colonel: he turned and looked at me astonished, burst out laughing, collapsed onto his seat, took another swig of vodka and then slowly rotated backwards, falling into the soft bushes, where he slept for the night!

The rest of the crowd were crying with laughter which only degenerated further when Youri relayed to them, how I knew the commands. They had never known anyone to have ever told a KGB colonel to shut up.

I am sure there was more vodka and toasts, but it was anticlimactic after my incident. That special night in the summer twilight ended and we retired to our tents and the morning hangover.

The next day we returned to Chita after visiting one of the other alluvial gold prospects. John, in the meantime, was ready with the numerous Heads of Agreements and MOUs, in both English and Russian. All were executed the next day, Russia-style; wax seals with strings through them.

This visit and the agreements executed, formed the foundation of our next five years in Russia. We ended up owning the great Taseevskoe Gold Mine, and the tailings, and listed a company on the Toronto Stock Exchange based on it. We had offices in both Chita and Baley and had expatriates based there for a number of years.

The Russia of the "provinces" was very different to the Russia of Moscow, and when we left in 1998 after the Russian stock market collapsed and the rouble plummeted, it was with great regret. We left many good friends but retain great and fond memories.

ROUBLES AND TROUBLES

ANDREW DRUMMOND
RUSSIA 1993 – 1997

Andrew is 69 years old and graduated as a geologist from Adelaide University in 1972. He has been continuously involved in the mining industry since then in the fields of mineral and energy exploration, feasibility and development, gold mining, and company foundation and management. He was primarily responsible for the ASX listing of Westonia Mines Ltd and Minemakers Ltd, was involved in other floats and has been a director of numerous companies. Minemakers was credited with being the top performing company on any listed stock exchange in the world in the FC year of 2008.

His overseas experiences are in New Zealand, Russia, the Philippines, China, Namibia, and several other African countries.

Saint Petersburg

In 2015, my wife and I spent a few days holidaying in Saint Petersburg, my first visit to that city. This was my first return to Russia since 1997. My expectation was that that leg of our holiday would be relatively grim, based on my experiences from seven visits to that country in the early to mid 90s. I expected a grubby airport, highly suspicious and unfriendly security staff at immigration counters, to be ripped off at almost every opportunity, few facilities for tourists, and a general difficulty in trying to get anything done.

I was shocked at the advances in the country over that intervening 20 years. Street signs and road instructions were in both Cyrillic and English, people were generally friendly and helpful, shops were very well stocked and we felt quite safe when walking around, even at 10 PM in the mid-summer dusk.

On my previous trips, when a driver parked his Lada or Volga car in the street, he removed the windscreen wipers so that they, then in short supply, could not be stolen and on-sold. This time, good cars were plentiful and we even saw Lamborghinis dragging each other up the main street, the Nevsky Prospekt.

We could easily buy tickets to attractions and events, bars were welcoming rather than menacing and plenty of guides, who spoke good English, were available at a reasonable price. The ability to use a credit card and an ATM was most welcome and the consequent need not to carry copious cash was a great relief.

On this holiday trip, we did not get to Moscow or Siberia, nor did we fly Aeroflot, so some direct comparisons could not be made over that twenty year interval. But the differences in outlook and attitude and the general optimism of the people we met and dealt with as tourists were palpable. I began to realise that, at a personal level, I was extremely fortunate to have experienced a unique time in Russian history during my previous visits. The USSR had collapsed and disintegrated, communism had been repudiated and Russia was engaging in momentous social and economic upheaval. At that time, the country was broke and floundering, and trying to establish new systems starting from scratch, with management who knew of no other way of managing than under a state-controlled system. Workers began to realise that a future pay packet would, at least in some degree, be dependent upon their willingness to adapt to new employment conditions which were not necessarily guaranteed for life, and to put in a decent and efficient day's work.

It was hard going for me then at a time when it was hard going for them also but, in hindsight, it was a most rewarding personal experience. The rest of my contribution describes what my company and I were trying to do in Russia and to give an introductory understanding of some of the difficult conditions under which we were operating.

I begin it by offering the well-known Winston Churchill quote from 1939: *"Russia is a riddle wrapped in a mystery inside an enigma"*. When I was chasing up the correct wording of that quote, I found a less well known second part to it which is: *"but perhaps there is a key. That key is Russian national interest."* Having spent significant chunks of 1993 to 1995, followed by another trip in 1997, I can but concur, perhaps only wanting to include the words "and personal" after national.

Background

The story is about the experiences of a junior listed company which started with a couple of small gold projects in Siberia and which aimed to acquire larger ones: all them hopefully to be developed and made profitable using Australian capital, mining and management expertise.

In the second half of 1993, I joined Stirling Resources NL, chaired by the late Ted Ellyard, as its General Manager – Minerals. The company wanted to focus on its energy assets and intended to spin off its mineral projects into a new ASX - listed company called Zephyr Minerals NL. We achieved that in early 1994 and I became its managing director upon its listing.

The countries which had emerged from the breakdown of the old USSR had an acute lack of available investment capital, but plenty of opportunities on offer at that time. There was also an abundance of pitfalls and sharks. One of the better corporate examples, or should I say worse, was ASX-listed Star Mining. The company, whose first board included Malcolm Turnbull and Neville Wran, acquired the right to deal into a joint venture over the giant undeveloped Sukhoi Log gold deposit in Irkutsk Oblast (equivalent to our State). Approximately $25 or $50 million of the IPO funding was dispatched to Russia to fund the evaluation and feasibility effort. The funds were immediately snaffled by their JV partner who then advised that the mining company owed it to the government and other contractors for sundry previously unpaid invoices, back taxes and other due monies. Embarrassed, Star management returned to the market for another large dollop of funds. Little information remains on Star on the web these days, including the exact magnitude of that first loss, but I found a report that Star wrote off the $127 million it had pumped into the failed venture after Russian courts overturned previous decisions by Russian politicians. Little Zephyr needed to be circumspect and try to learn from others' mistakes in these emerging economies and legal systems.

After the expenses of the IPO offering to get Zephyr listed on the ASX, and repayment of debts to Stirling, Zephyr only held a little over $3 million to pursue its Russian interests, as well as others within Australia and New Zealand, so we had to be frugal. We had to convince potential

Russian partners that, if they dealt with us, although we did not have too much money yet, we could raise more on the basis of solid results and performance.

Amongst its mineral properties were a couple of assets in Chita Oblast, of eastern Siberia. Originally acquired by Western Australian mineral entrepreneurs Derek Fisher and John Hopkins, they subsequently dealt into a much bigger and better project and were happy to move their earlier ones into Zephyr. Refer to Derek's story in this book for background. They had previously formed a joint venture with a Russian mining company operating from the city of Chita, which is on the Trans-Siberian Railway to the east of Lake Baikal and roughly 75% of the distance from Moscow eastwards to the Pacific Ocean. Zephyr formed a separate JV with the same Russian company. Our projects were near Chara, which is situated on the northern Trans-Siberian, or BAM (Baikal to Amur) Railway about 600 km north of Chita.

I accompanied Derek and John on my first few trips to Russia before then going solo, or accompanied by Vlad Kroupnik, a Russian – Australian geologist we employed after the Zephyr listing.

Zephyr had a 49% interest in the JV with that local Russian company which held granted tenements over two alluvial gold projects on the Jemkoo (or Djemku) River several hours' drive from Chara. One of them was actually being prepared for operation, and the JV had a separate hard-rock vein gold prospect called Bahtanakh.

Jemkoo Alluvials

The alluvial gold operation was not making a profit or otherwise they would not have needed us to farm in and provide working capital. The partners had to purchase all the equipment, fuel and supplies needed for the first year of operation, transport them to site and get established prior to the actual mining season. With no money to do so and unable to raise it from the embryonic private Russian banking system, enter Zephyr to invest that working capital. However, we were also aiming to make the mining and recovery operation much more efficient by introducing modern Western technology and techniques. If all went according to plan, we could turn the current operation to profitability for a relatively small investment and then use our share of that profit to invest in bigger or better things in Russia – either with those partners or some others. Zephyr's overall strategy was to use our then newly derived credibility as a profitable Russian miner and good business partner to nail those better deals in due course. There was obviously going to be an element of becoming familiar with the mining, business and social cultures during that first operation. That was the theory anyway!

We were under constant pressure to fund the coming year's operation according to the Russian conventional way. They presented budgets and expected us to simply pass over the money. On one early visit, I was proudly shown two whopping D9 equivalent bulldozers which they purchased for the mining operation, without formal approval from the JV, on the expectation that I would stump up the funds for them and everything else without question. In fairness to them, if a mining operation was planned in the forthcoming summer, then all supplies, including consumables and 600 tonnes of diesel fuel, required transport in by winter roads, which the frozen rivers are called, in short order. What it all really needed was a year's break so that we could undertake the required planning and organisation for a new and streamlined operation. This was unacceptable to the JV partner as that meant retrenching all the employees on site for that period, and probably also the partners losing considerable face with the government which granted the licences in the light of their promise to attract foreign funding.

With each visit to JV headquarters in Chita, or to site, new problems arose. In no particular order, some of these included the following.

Our partners helped themselves to some early investment funds, initially saying it was to cover administration costs. When no proper accounting of that expenditure was given, we basically accused them of theft and demanded the money back. They did not bother about doing that and their response was that if we were such a chickenfeed outfit that we were concerned about US $50,000, we obviously were not a suitable partner for their grand designs. Our response that it was not so much the quantum, but the principle, with which we were concerned fell upon deaf ears.

They also had no idea whatsoever about the realities of accounting and returning a real profit. They would produce cash flow models where, say, if Zephyr were to invest 10 million roubles in the mining operation now, we would produce 11 million roubles worth of gold by the end of that first year. However, with inflation still running at many hundreds of percent at the time, that 11 million may have been worth one or two million in constant value at year's end, of which we would only get back 49%. My observations made no impact on them as they kept pointing to the bottom line saying the JV was going to make a million roubles profit, and hence I should transfer the cash immediately. When I replied that we were also obliged to pay a multiple of taxes along the way for matters which would be the equivalent of superannuation, workers compensation, annual leave entitlements, a resource depletion allowance direct to the government based on the amount of gold we recovered, and then the usual company profits tax, and more, they would look me in the eye and tell me that I should not worry about such things because they would fix it up with the government. Our joint company would not pay anything at all – because Russians never did! When one considered that governments, at all levels, in Russia were totally broke I argued to no avail that there was no way that they would allow an Australian company to repatriate any untaxed profit - should it ever be attained.

At a technical level, the operation was very basic, and Zephyr could have done much to improve it had we been allowed. As is usual in alluvial operations, almost all the gold is hosted in the coarser gravels and boulders at the base of the infill of the river valley. Overlying or sandy material is generally barren and, in a sensible operation, would be separately removed rather than put through the processing plant. The Russian mode of operation was to simply process the lot.

The Russians used a Magadan plant, which consisted of a sloping steel channel or chute with riffles to catch the gold. D9 equivalent bulldozers were used inefficiently to push the river alluvium all the way to the base of the front end of the plant and then further push it up a ramp of previously mined material, and through a grid of heavy steel bars which snagged and removed useless, large boulders. High pressure water piped from the river was then used to flush the material down the chute, to disaggregate any clay and to separate the gold from the other gravelly and sandy alluvial material. Riffles were then supposed to retain the gold as it washed down the chute. A simple technique, that we were advised was the standard Russian one in eastern Siberia.

I asked the foreman what the gold recovery was, and he basically thumped the table and emphatically stated: "We get 92% recovery, like all Magadan plants". As we could pan plentiful fine gold from the tail end of the water race, where material was simply being returned to the river system, that figure was a joke. As the mining technique was so crude, even if the initial resource estimate were correct, we still had no idea of the actual gold grades going into the plant. With no attempt to recover the fine gold, overall mining recovery was unknown, but my guess was that it would have been closer to 50%.

Derek and John brought some of the Russian management to Australia to show them some of our gold mining and recovery technology, and to demonstrate our access to capital. I accompanied them and we toured Dominion's open-cut operation at Meekatharra. At one stage the grade control geologist produced a computer-generated plot of gold assays from the latest mining bench using state-of-the-art mining and plotting software. That technology was an impossible dream for the visitors, but they did want to take home one of the plastic strips with punched holes which are glued to the edge of plans allowing them to be hung in a vertical map cabinet: they could see a little business sideline in producing them.

Zephyr thought about taking them on to New Zealand to show them how much more efficiently gold-bearing material could be mined and processed if bulldozers were not used. There the larger alluvial operations use a first excavator to remove and stockpile the overburden, then a second one to feed the gold-bearing gravel directly into a plant which is also set up to recover fine as well as coarse gold. Only a fraction of the manpower is needed compared to the Siberian situation. We could see that there was little interest in the introduction of that mining and recovering technology. Just give us the money!

We kept pressing for efficiencies but to no avail, and eventually the truth leaked out. Our partners were granted the tenements by the new post-USSR government based on promises that they would attract very large foreign investments and also employ hundreds of locals - the two last things we had in mind.

On several occasions, I was asked when the 500 rifles would arrive. At first, I dismissed this as some sort of local joke, but it was eventually explained that it was the equivalent of payments to traditional owners. They had been bought off by the promise of these rifles. In exasperation, I asked the price of Afghanistan war surplus Kalashnikovs, thinking they may have been $10 or so, as old WWII Lee-Enfield .303s had been when I was a teenager in Adelaide in the 60s. But I was told that the promise was for Austrian hunting rifles, which cost about $500 each. In my mind, Zephyr had better things to do with its money than arming the locals, even if we were to get approval from the government to import that arsenal.

The operational base was built over a summer with supplies previously brought in along what is termed a winter road, which is actually a deeply frozen river. It was several hours' drive from Chara on gravel or mud tracks in summer, or on those winter roads. Our skilful driver on those ice roads was an ex-army colonel. As an Aussie more used to overheating radiators it was fascinating to watch him preparing to start the Russian jeep equivalent on a sub-zero morning. Canvas skirts were placed around the vehicle draping to the ground, thus making a wind-protected screen or tent and he crawled underneath with a blowtorch which he systematically used to warm the sump, engine block and transmission: this took him about half an hour each morning before he was ready to hit the ignition.

When one looked around to see snow and ice in every direction, it was odd when I was advised that in summer this same chap was the district fire control officer. Apparently, bushfires are a serious problem in the district: they can be started by lightning and travel laterally quite quickly burning dried moss, even though the ground underneath is damp or even covered by shallow water. We started a fire one summer when a build-up of dry moss was ignited by the muffler. Sloshing around in ankle deep water and scooping it up to douse the surface fire before it spread beyond our control was also rather odd. The supply trucks were ex-army all-wheel drive and quite brutal things. The tyres were bullet-proof and/or self-sealing and I wondered whether these might have application to stop staking tyres during exploration in mulga country in WA.

On the track from Chara to the Jemkoo operation, we passed through a camp which I was subsequently advised was for the exploration and feasibility team for the massive, but then as yet undeveloped, Udokan copper deposit. Our driver kindly advised that last winter bears had killed and eaten five geologists on that project. One of the reasons why there was always a gun in any vehicle, another the possibility of bagging a deer for the camp cookhouse. The driver usually had his eyes open looking for the latter.

Actual mining and processing were carried out in summer once the ground had thawed. It was surprising how warm the river waters became, but with a black slimy bottom they were exposed to about 20 hours of summer sunlight per day. It was strange for an Australian to swim in a river which was a frozen road a few months earlier.

We failed to start any actual work on our second alluvial project before I pulled Zephyr out. However, we travelled to inspect it one summer's day, and found the marker posts where previous Soviet survey teams had laboriously hand-dug pits to test for the thickness and grade of the gold-bearing gravels. On the banks of the river nearby was a trapper's log cabin, which was unoccupied at the time. We used some of his stored dried wood inside the cabin for a brew-up and, in accordance with sensible custom, collected dead wood from the nearby forest and replaced that which we used so that it would be dry when needed by the owner in due course. A small sled tucked in under the roof testified that he had a child, and so presumably a wife, and it must have been a very lonely existence for them. Although I am not a smoker, I always carried cigarettes as a social lubricant in these sorts of places. I left an unopened packet of imported American Marlboro fags on the table of the cabin and sometimes wonder what the trapper thought when he arrived at his home in the middle of nowhere the next time.

The Jemkoo operation itself was based at an old Stalinist gulag, a depressing place in consequence. It was very remote and presumably the prisoners had to walk to from the nearest railhead at the time - probably

Chita some 600 km away - one could not help but feel profoundly sorry for them. Several old guard towers and a very poignant cemetery remained. Some grave sites were marked by Orthodox crucifixes; and some by wooden obelisks, looking rather like a small version of Cleopatra's Needle; and others signified the resting places of committed communists. For the last group, metal discs, which may have originally been the tops of 20 litre oil drums, were attached to posts and the hammer and sickle emblem punched through them in trails of holes: true believers in Red ideals, but look where it got them! Death in a place far from home.

And to finish off my Jemkoo story, I was told how in the gulag days, intrepid Chinese - those master entrepreneurs - used to smuggle liquor on foot from China to the camp and swap it for gold. They did this in winter by dragging sleds of booze along the frozen river. However, they were tempting targets for bandits on each leg of the trip, either for the grog when going north, or for the gold when returning south, and apparently many were killed over the years.

On another occasion, we travelled past another old gulag, this time the site of an historic uranium mining operation. Guard towers and barbed wire were still present at the opening end of a very deep dead-end canyon. I was told that the female geologist who found the mineralisation was awarded a commemorative metal for her contribution to the Cold War effort but died of radiation poisoning while mining the very rich ore. Very sadly, once prisoners arrived, they apparently never experienced direct sunlight on their bodies for the rest of their lives unless they were released. They could see sunlit parts of the uppermost levels of the canyon but worked in perpetual shadow themselves. What can one say?

I was advised that Zephyr could enter into a JV over the project if I wanted, but given that we were struggling on a simple alluvial gold project let alone one involving underground mining of a very strategic metal, I had no interest. And what chance would I have of raising the necessary capital for it in Australia anyway?

Bahtanakh

Based on official mines department records, our gold in quartz veins project, Bahtanakh, seemed a very promising project. Discovered by Russian prospecting parties in 1978, Bahtanakh only received some surface exploration, which resulted in discovery of two vein systems about two kilometres apart. The first was traced on surface for a kilometre and five channel samples returned an average of almost 14 g per tonne. In the second, quartz veins within a pyritized zone returned channel samples of up to 3m at 64 gm per tonne. These were considered attractive numbers.

I hired an ex-military, or maybe still military as it had the appropriate insignia, twin-rotor helicopter and we flew for about three hours to site from Chita to undertake a field review. The noise was horrendous and, with no earplugs or earmuffs available, I sat with my hands over my ears for the whole time. They rang for days afterwards. The helicopter hire cost was US $4000 cash which I handed to the pilots and did not receive a receipt in return. In those Russian economic sink or swim days, I suspect that the pilots, who were in military clothing, simply held the keys or codes to start the machine, knew a couple of mates who could organise the fuel, and then split the cash between all involved. Maybe a couple of senior personnel above them either handed over the keys in the first place or looked the other way for a portion of the money. Anyway, it beat walking to site.

I had spent a fair proportion of the '80s assessing old gold mines and prospects in Australia and New Zealand with an eye to whether they could be brought back into production in the modern era, preferably as an open cut operation. Usually gold is associated with quartz which is generally white and resistant to weathering. The original quartz veins or reefs cut through surrounding rocks, which are often more susceptible to weathering. The result is a significant quartz presence at surface. This is enhanced by lateral dispersal of the quartz by rain run-off, soil creep on slopes, or by animals - all resulting in a much broader white gravel at surface than the original veining at depth. An original 20 cm quartz vein may result in white gravel spread laterally over several metres, and this dispersion often gives initially a totally false and exaggerated impression of the vein or reef thickness than is the reality.

When we arrived at Bahtanakh, I walked around for a while until it became evident that this had also been the case here. The mineralised system, based on a surface inspection and in a few trenches, was observed to be generally less than 1 m thick and near vertical. It was very unlikely that there would be sufficient tonnage of mineralized rock to support an open cut operation and little chance of an underground one unless the gold grades were exceptionally high.

Once I satisfied myself on this we flew back to Chita. When we landed, a lone cow strolled along the strip and, to my discomfort, every other person on board, probably about six including the crew, pulled a rifle out of various storage areas. They took sights on the cow and all laughed at the surprise they could give it – it was all in jest – as the cow wandered on. I had just travelled to the middle of nowhere with US$8000 cash in my pockets and rucksack, so I then decided we did have trusty partners, at least at the armed robbery and murder level!

Word of my disdain for the potential of Bahtanakh preceded my return to the JV office and I immediately faced a hostile reception from my partners. Who was I to so peremptorily dismiss a project on which the Fifth Prospecting Brigade, or whatever it was called, had spent several years of strong State-funded efforts resulting in the reporting of a considerable contained gold resource all the way up the chain to Moscow?

The chief geologist of Chita Oblast was summoned to set me straight, but I stuck to my conclusions. He returned to his office to bring back copies of the official team reports to further the argument. We started out by examining data for the trench which had reported 3m of 64 gm per tonne gold. This was on a government plan of which we had a copy and on which we based our write-up of the project in the Zephyr prospectus. I advised that I struggled to see any evidence in the field of mineralisation over that width in the trenches. He scratched his head, saying he could not recall that reported grade and width, despite what was on the map. He then checked the lists of original assays and concluded that there seemed to be a drafting error. Perhaps 64 g/tonne over 1m in a three-metre trench, should have been shown as 21g/t over

3m. There went two-thirds of the grade in the reported widest part of the mineralisation, and that width shrank to the same third which I estimated in the field! This implied that we may have inadvertently misled investors in the prospectus: no slight matter!

The next aspect for discussion was just what had the prospecting brigade done over the three field seasons, bearing in mind that a contained gold resource of quite a few tonnes was reported, under the Russian resources and reserves system. Admittedly this figure was of low confidence at that early stage and we could not put enough reliance on it to include it in our prospectus – thank goodness. Under the old Soviet system, teams could not report failure and they always had to meet State targets. So of course, they did, covering themselves to a degree by the low confidence ranking of the contained resource. As reporting went up the line towards Moscow, there was an ever-increasing emphasis on discovered tonnes of gold rather than the confidence level attained in the field. Chita Oblast was able to report that Moscow funding provided for X exploration brigades, which found Y gold prospects and which, so far, possibly discovered Z tonnes of gold in accordance with the original Moscow diktat. Everyone is happy until they either try to develop mines on the prospects themselves, or try to sell it to a third party such as Zephyr, which needed to be rather more hard-nosed about it all.

Eventually that particular brigade was revealed to me as not having been a very competent one with the field history being along the lines of: year one, planning; year two, road and camp construction; year three, a few trenches and assays. Oh, and somewhat embarrassingly, the brigade leader was not a particularly good one and was an alcoholic - although that was far from unusual in that part of Siberia.

Bahtanakh was shot down in a few hours in the field and in a heavy duty and heated technical meeting afterwards. On my return to Australia, Zephyr was able to announce that, upon field inspection, further work on the project could not be justified which, thankfully, sidestepped the prospectus problem too.

Once the dust had settled in the meeting room, the Russian partners were somewhat embarrassed, but also more despondent, as they saw their vision of Australian or US dollars for project advancement and development vaporising, along with employment opportunities and the ability to help themselves to potential "management fees".

With the extinguishment of the prospectivity of Bahtanakh, and with Jemkoo showing all the signs of being a potential bottomless money trap for Zephyr, it was time to look somewhere else for a new project which would hopefully prove to be bigger and better.

Krasnoyarsk

I previously mentioned that I was assisted at Chita by Vlad Kroupnik. He was a geologist who had recently emigrated from Russia to Australia. His role was to translate; to assist me in my dealings with my Chara partners, pointing out if I was being blindsided, or if the partners were not acting in the best interests of the entire JV; and to identify new opportunities for the company. He was less than successful in his first two roles because the Russians wanted to conduct all the interpreting for me, rather than through someone on my side. They gave him a most difficult time, basically refusing to acknowledge the role that I wanted for him, and tried sidelining him.

However, interacting with some of his old colleagues from his student and field days, Vlad introduced to the company a very appealing situation elsewhere in Siberia. There was potential to enter a deal on several large hard rock gold deposits north of Krasnoyarsk, which is again on the Trans-Siberian Railway, but this time west of Lake Baikal. Krasnoyarsk had been a closed city in Soviet times as it hosted significant military and nuclear facilities and to access it, even for citizens from elsewhere in Russia, had been difficult to impossible. By the mid-90s, it was glasnost and beyond: Westerners were allowed entry and I was able to access the area without any bureaucratic drama.

Krasnoyarsk is about one third of the way back towards Moscow from Chita. An hour's flight north east from Krasnoyarsk lies the Angara River, a tributary of the mighty Yenesei River. One of Russia's longest rivers, the Yenesei sources from a little north of the Chinese border and then flows in a general northerly direction to the Kara Sea, part of the Arctic Ocean. The huge Norilsk nickel mining operation and city lie adjacent and not far inland from where the river enters the ocean - by

Siberian standards!

The Angara alluvial goldfield hosted major gold production over the decades, and alluvial production was still continuing at that time. It was generally reckoned that several million ounces were recovered but the actual amount was a state secret. Current production was again using Magadan alluvial plants, as described earlier for the Jemkoo operations.

Over the previous few decades, Soviet field parties identified several major primary hard-rock gold deposits. Huge amounts of drilling and underground development and other scientific studies were undertaken but none of the deposits were brought into production due to the vagaries of politics driving the old Red investment and development system. I was advised that economic returns were only one factor for the ever-political communists. For instance, depending on the way the international wind was blowing at the time, development of a major new mine near the Chinese border could be viewed positively as the USSR would then have to build an associated small city, airport, road or rail access and this would enable it to inject greater power into that region. Read military capability. Conversely, if not developed fully or at all, it might trigger potential cross-border incursion and acquisition by Mao's hordes.

First thing for me was to travel to Krasnoyarsk from Chita to meet up with Vlad and his colleagues. I will give some detail on this adventurous trip later. The latter were now entrepreneurs and seeking to set up some sort of company which could gain title to the three known major hard rock deposits and then do some sort of deal with a foreign company such as ours.

The geological, mining engineering and metallurgical reports were all available through the local mines department, whose chief officers were most obliging and seemed to look upon the opportunity for foreign investment to develop those deposits quite favourably. There was also an opportunity to become involved in alluvial operations: in theory at least, using our previous Chargold model of investing a little, learning the ropes and hopefully turning a profit if we could mine it our way, not theirs.

We flew to the nearest town and made site visits from a genuine imported Jeep Cherokee - the first non-Russian vehicle in which I had been transported in all my travels. Our potential joint venture partners gave every indication of being on the ball and receptive to new technology.

The countryside seemed relatively benign from an engineering and social viewpoint, and I was quite technically encouraged. Angara was at about the same latitude as was Chara, but greater tree cover indicated a generally more benign climate, hence potentially easier operating conditions, than we experienced at Jemkoo. I was also advised that there was no permafrost to deal with as there was at Chara.

The chief geologist on site was an extraordinary man. He was about 60 years old and worked all his professional life at Krasnoyarsk. Rather than succumbing to vodka as so many of his compatriots had, he embarked upon self-improvement and was proficient in six verbal and written languages - all self-taught. Being a typical Australian, I could only judge his language proficiency by his English, which was genuinely excellent. I did not expect to meet a Russian geologist in East Siberia giving me his opinion of Henry Lawson's poetry – complete with quotes from his memory.

We became quite friendly and, amongst other questions, he asked my opinion of the standard of development of the district. I replied that not a great deal had been done considering the wealth won over the last century, including to the present day. I asked him where he thought the mined wealth had gone and his response was: "outer space, Cuba and Angola". He was regretful that he had devoted his entire professional career for what he thought was for the betterment of the Soviet Union and its citizens, including having to "live in this hole". But "I was sold a lie, and now it is too late for me".

In order to demonstrate that there was a better way to mine and recover the alluvial gold, we took a technical party to New Zealand to visit some mining operations. It included the state mines department's chief geologist, who had never been out of Russia before, never seen a clean airport and probably never a clean public toilet. He was subject to a series of culture shocks, even before he viewed the advanced mining technology. The first was flying from Siberia into the beautiful and orchid-filled Singapore airport, followed by laid-back Perth then lovely Sydney Harbour. Reluctantly he was taken to a hotel for a meal and drink, thinking that we must be looking for a fight because, from his home experience, why else would one go to a bar?

The last straw for him occurred when he disembarked in Hokitika, on the West Coast of the South Island of New Zealand. Our chief geologist, Greg Bielby, calmly walked up to the Avis counter and flashed his driver's licence and credit card. He was given keys to an almost brand-new station wagon, which did not need its own driver, and had been pre-booked by phone from Australia, and the pick-up was accomplished without shouting, arguing or delay. The old boy just about collapsed in shock and told Greg that he could not have imagined such an event. He was not even going to tell anybody about it in Russia because everybody would believe that he was simply telling lies. Not intended as an anti-Russian story, but an attempt to exemplify the magnitude of the problems we had trying to introduce better mining and business management into our dealings. So many things we took for granted in the West simply had not penetrated into Russia at that time.

In one of the Krasnoyarsk government offices I was treated to an amusing incident. I was talking to one of the senior administrators and somehow, he had found out that the mining and exploration industry often encountered difficulties in Australia because of native title issues. I gave him an overview of some of the problems that have bedevilled the industry and, in often unappreciated consequence, retarded the welfare of Australia. He turned to a shelf and picked up a very large tome which he opened at a random page. Think of something like a full Oxford Dictionary. He advised that Siberia also has indigenous tribes and that they were all fully and faithfully represented in his copy of a census book. He deliberately and slowly ran his finger from name to name on the page, then quickly slammed the book closed with a resounding thud and, in mime form, indicated that he had just squashed a problem as though they were silverfish or worse.

The return to Krasnoyarsk proved the turning point for me and Zephyr in Russia. It was a Russian habit of the time, that if one were lucky enough to own a colour television set, that it be turned on at a fairly high volume irrespective of what other meetings or meals might be happening in the same or adjacent room. I was advised that one could actually watch that good old Australian program, Neighbours, in Siberia, with appropriate dubbing but, oh darn, I missed the experience of a lifetime by not seeing it!

It happened that I sat on a lounge alone with the TV blaring while all the Russians were talking loudly to each other in the next room. Vlad wandered into my room while the local news was on and I asked him to explain the item currently on the screen. A man in a suit was lying on a footpath in a pool of blood. Vlad listened for a while and then said: "This guy is the seventh banker killed in Krasnoyarsk this week".

That was the final straw for me and we headed home. I recommended to the board that Zephyr not throw good money after bad in Russia and that we simply quit our endeavours. The board agreed and decided we would look elsewhere for less corporately, financially, and personally dangerous opportunities.

Post-Soviet Russia in the 90s

To access the vast expanses of Russia, there was generally little alternative but to fly to the nearest airport and then resort to generally execrable roads where necessary and available. In the absence of roads, one resorted to local tracks or, seasonally, driving along the frozen rivers of the winter roads.

While it is easy to be critical of the standard of roads, Siberia is huge and largely undeveloped with inherent climatic problems which are quite alien to Australians. The new part of the Trans-Siberian Railway, from Lake Baikal to Amur (the BAM Railway) involved the building of 3000 bridges along its 3000 km extent, thus exemplifying the difficulties that need to be overcome. Freezing of groundwater below roads occurs which in turn lifts and can crazily contort the bitumen. Once on a highway we ground to a halt on a hump, caused by this, and our driving wheels were lifted off the bitumen. We all had to get out and push the car back onto flat blacktop to then proceed.

To travel to the far-flung parts of this huge country, there was no alternative but to undertake a lot of flying. In the early to mid-90s, my experience was that Russian civil aviation was difficulty and drama on steroids. The country was in the grip of hyperinflation and little hard currency (read dollars) was available to purchase the few goods and products which were on sale anyway. As an example of the latter, in one shop we entered in Chara only three separate lots of items were for sale: one was shampoo, another was Lada clutch plates and I don't remember the third, but it was not related to the others.

Aeroflot was State-owned, and the State was broke and it showed. Times were tough, and shortages were the norm for all. Did this explain why sometimes the hot drink genuinely tasted of a mixture of tea and instant coffee? Once, the food I was given consisted solely of a plastic bubble of berry jam - maybe designed to flavour the drink. A fellow passenger could pull a stick of salami from the inside pocket of his suit jacket and a big carving knife from another pocket - this was before 9/11.

Often whether a plane flew a scheduled flight was determined by whether fuel was available. If fuel was at the airport, someone had to pay somebody or something to have the appropriate amount released. There were not just delays, but also wholesale cancellations for indefinite periods. Were spare parts available and, if so, could the pilots and passengers pay hard currency to acquire them? Were there sufficient foreigners buying tickets to help pay for everything?

On one flight from Chara to Chita, I happened to look down at the plane tyres as I climbed the gangway stairs and noticed fabric exposed through the rubber across the whole width of the tread. As the alternative was a 600 km hike across the frozen tundra, I consoled myself by thinking that the tyre should be good for at least one more landing!

Aeroflot was always considered by the Soviet government as simply being the civilian arm of the air force. A few of the planes I travelled in were designed to serve dual roles with the aim that a civilian cabin layout could be changed across to a military configuration quite quickly if needed. One plane I travelled on was remarkably like a Fokker Friendship, except that a ramp was found at the back of the fuselage under the tail so that jeeps and equipment could be driven on board. Bench seats ran lengthways along the cabin with a static line running under the roof for paratroopers. On that flight, a canvas-covered oblong about 70 cm square and 40cm high was firmly lashed to the floor and guarded by several militia carrying Kalashnikovs. They gave us a pretty thorough visual check as we boarded and sat down on our benches. I asked Vlad what he thought the canvas was concealing and he replied that it was highly likely to be banknotes, given the hyperinflation then operating in Russia. I replied that if we made a move towards stealing the money and they let loose with the guns, the plane would likely be sawn in half. Neither we, nor the other passengers, were very tempted.

It is probably fair to say that nothing Aeroflot did was aimed at providing customer satisfaction, particularly if one were a foreigner. All foreigners paid double the price for any ticket compared to that charged Russian citizens. It was quickly explained that we were not charged double, but Russian citizens always get a 50% discount! When one considers that almost all planes carried very few foreigners on board, this purported generosity to the locals seemed a sure-fire way to bankrupt an airline.

Bag theft was endemic and, at the larger airports, a lively trade of serious cling wrapping of suitcases made it more difficult for staff to open bags and help themselves to the contents. Readers will recall when baggage tags were not sticky but made of thin cardboard tags attached to a bag handle by a loop of elastic. If one were in the habit of just pulling off the tags, as I was, elastic loops accumulated. Vlad noticed this build-up of elastics on my rucksack and recommended that I remove them as otherwise it alerted potential thieves that this bag must belong to a foreigner. Russians do not travel that often, and my rucksack would constitute a priority target for them. Alternatively, Russians carried an inordinate amount of hand luggage, to keep it away from the baggage handlers, and many bags were simply placed in the aisles of the plane, illegal in the West of course. Pragmatically, with little in the way of cabin service, trolleys were not pushed along the aisles, and if the plane were to come down, we would likely all die on impact anyway.

However, to be fair, it was not all a one-way street. Once I inadvertently left my passport on a counter at a regional airport. Sometime later I was apprehensive when tapped on the shoulder by a uniformed policeman, who beckoned me to follow him. We went to the airport manager's office, where a uniformed man sat behind a desk, and said nothing but just stared at my face for a while. Satisfied that my face looked like the photo in the passport, he reached into his desk and handed it to me. I was shocked that I did not have it, and its loss could have been disastrous in the middle of Siberia. I said "spasiba" (thank you) very sincerely and offered him a US banknote as a reward which, to my great surprise, he refused. It would have bought a lot of roubles or access to some otherwise unobtainable consumer item. It was a fine effort from him, particularly under the circumstances in Russia at the time.

Amazingly to me, Aeroflot had no centralised booking system available at all in Russia. This readily led to two obvious problems. The first was that, assuming a plane was available to make a scheduled flight, the booking people may have had little idea how many seats had already been purchased apart from the particular ones that their own office individual may have sold. So if, say, you were trying to buy a ticket at the airport, they did not know how many tickets had been sold elsewhere in the city and, if one's intended boarding airport was an intermediate one for a plane's longer journey, the agents did not know how many seats were occupied on the incoming plane and which would still be used by those already seated passengers as the plane went to its next destination. And how many government bigshots might demand a last-minute seat? Sometimes it was apparently just easier, so as not to cause offence or inconvenience or whatever, simply to say that the plane was probably full – if it came and departed anyway.

On several occasions from Chita, although in fact we were able to get a seat on the plane each time, we went to the airport for the scheduled very early departure on the clear understanding that we might be still overnighting in that city if we could not get on that flight for whatever reason.

The second major hassle was that one could only buy a ticket to the next destination, unless that airport was just an intermediate one on a longer flight, say to Moscow. So, if one started at A, wanting to get to C, but having to change planes at B, one could only initially buy a ticket from A to B. One then took one's chances on acquiring another ticket from B to C. This happened when I wanted to fly from Chita to Krasnoyarsk to view the new opportunity described previously, and necessitated first going to Omsk, where I then faced the challenge of getting that second ticket.

Siberia has developed in linear fashion along the Trans-Siberian Railway and all major cities and airports were strung along it. Omsk was about halfway back to Moscow from Chita, and then one had to do a U-turn and fly about halfway back to the latter to arrive at Krasnoyarsk.

My Chargold partners in Chita were genuinely stressed for me repeating that I knew no Russian and may be mugged or worse for my cash when travelling solo. One could not use credit cards in Russia at that time so being left stranded without money in Siberia may have been interesting. I also knew that a degree of their apparent concern was that I might find better projects than the ones I currently had with them and hence would want to shift Zephyr's focus of investment elsewhere. If I could be discouraged from going to Krasnoyarsk in the first place, that fear may evaporate. They asked how I would organise a ticket at Omsk and my reply was that, if I could not find a ticketing office in which some someone spoke English and could guarantee me a booking, then I would stand in the middle of the airport waving a US$50 note above my head, repeating loudly and slowly in English "I want to go to Krasnoyarsk". Someone was then bound to come to my aid for that attractive consideration. Foreigners were still quite rare in the Siberian part of Russia and, for that matter, I had not seen very many in Moscow either. I was often the object of curiosity, stares or conversation between the locals.

As usual, nothing in Russia proved easy. I duly arrived in Omsk and walked around the terminal looking for my bags. I found the appropriate baggage collection sign hanging from the ceiling and a left-hand arrow pointed to a door leading to outside the terminal. I searched everywhere around the car park, trying to see a carousel or people exiting from somewhere with their luggage, but to no avail. I headed back into the terminal, checked that I read the sign correctly, and then just stood there, the last person left in the now empty terminal. Eventually a woman, from the beriozhka, or hard currency, bar did approach me, and she kindly asked if I had a problem. I replied that I could not find my luggage and she motioned me to follow. We went to that same baggage sign, with its arrow pointing left to outside the terminal, but instead she turned right through another door leading to further into the terminal and there were my lonely bags slowly revolving on a rickety carousel. Perhaps the state signage factory only produced left-hand arrows or, most likely, when the sign with the wrong direction arrow arrived, not one employee at the terminal cared enough to change the direction of the arrowhead with a bit of paint or a marker pen.

Maybe the signage was just to foil spies! Not a totally silly surmise as Aeroflot passenger tickets clearly stated in print that no photographs were allowed to be taken out the windows of the planes while travelling over Russia. This was 40 years after American U-2 spy planes, and the subsequent advent of satellite imagery. For an Australian flying over Siberia in daylight many remarkable natural sites, both geographic and geological, are certainly worth photographing had I been allowed. When not snow covered, incredible detail is seen in the landscape compared to Australia because most of our continent is covered by windblown sand or deep soil, whereas Siberia has been largely scraped clean by glaciation. There is so much more relief and precipitation, resulting in active river systems and mountains. Sub-Arctic sand dunes resulting from grain redistribution by strong winds which sorted material left behind by retreating glaciation was a particular feature.

Anyway, I went with my saviour to her cafeteria/bar and she asked me what I was doing there, and I advised that I was chasing a ticket to Krasnoyarsk. After a while, she went off and brought back another woman who was the booking agent, who advised that she did not know when the next Krasnoyarsk flight was due, but it was certainly not due for a few days. Then she asked me what I intended to do about the situation. I replied that I would hang around the bar having another coffee or two for a while hoping that something would turn up and, if it didn't, I guessed I would seek some accommodation preferably within walking distance of the airport. She then left me with my coffee and making notes. An hour or so later she reappeared and beckoned me to her ticket office. She sold me a ticket which had no departure details and told me to sit on a bench outside her office. After a similar waiting time passed, she again appeared and told me to follow her. We left the terminal, walked across the tarmac to a plane and she indicated I should board it. The plane was a YAK 40 jet, about a 40-seater, which looked a lot like the Fokker Fellowships which used to fly regional WA at the time.

I was the only passenger on board, and the plane took off and hopefully for Krasnoyarsk as there were no announcements, in English anyway. Presumably, when she first told me that there were no foreseeable planes, she either did not know of this one or, if she did, she had no idea of whether any seats were available. Perhaps after my first discussion with her, it was subsequently ordered to fly to Krasnoyarsk for some new job and I just got lucky. One way or the other, it wasn't the way that Allan Joyce would run the airline.

A few hours later we landed, I disembarked and started to walk towards the terminal. Someone yelled at me from the plane to come back and it turned out we simply landed to refuel on the way. Luckily, they noticed me on the tarmac.

I eventually arrived at Krasnoyarsk and, to my relief, Vlad was waiting for me, as we had arranged some time beforehand. Fortuitously, he had only came as a precaution because the scheduled flight from Omsk was not due until the same time the next day. Krasnoyarsk had that scheduled flight from Omsk the next day on its books, but not this day, while I was advised in Omsk that a flight was not scheduled for several days, but then one turned up a couple of hours later: a rather neat example of problems faced when travelling in Russia at the time. At least for the locals, they could speak the language and get a better understanding of what was going on, although that may have been why they always seemed to be shouting at each other!

Luckily for me, we caught up as, otherwise, I would not have known what to do or where to go as the people we planned to meet, and their contact details were unknown to me. With no mobile phones and, even if there were and I knew some contact numbers, would the person who answered the phone be able to speak English? Thinking back on it, I do not know what I would have done. I probably would have assumed that Vlad was not able to make it in time for some reason, found a hotel and hung around for a couple days in case he was just delayed. Then I would have given up and tried to get back to Moscow and on to home.

A few years later in 1997, I returned to Krasnoyarsk and those same gold properties in my post-Zephyr life as a consultant. There was private competition then in the skies and I flew on the Russian equivalent of a very big Airbus. We carried our baggage to the plane and boarded from the underside of the fuselage leaving our luggage on the deck below where we sat, thus avoiding terminal theft problems. We then ascended a spiral staircase to the seating area. A badly dubbed Hollywood movie was showing and starred one of the Baldwin brothers as a drug runner, who was caught by police in Mexico. His girlfriend essentially used her body to get him sprung by the police and all was shown in graphic detail as it was uncensored. Even more oddly, every so often a message in English scrolled across the screen to the effect that we were watching a stolen movie and would we kindly please report the matter to this certain phone number. The hostess was wearing a thick overcoat with boots coming up over her knees and had a moustache that Burt Reynolds would have been proud of. Perhaps the airline company was ahead of its time in its employment of transgender people.

On all Aeroflot flights, standard operating procedure before take-off was at first rather unnerving. The pilot kept a firm foot on the brakes and ran the engines up to full power. When he released the brakes, the planes would shoot forward. Presumably, there was some distrust of the instruments and it was considered better to find out if the engines could not develop full power before the plane moved forward rather than three - quarters of the way down the runway. In John Hammond's story in this book, I note he concludes that it was driven by fuel shortages.

I have previously mentioned that Aeroflot did not have a centralised booking system, but I was even more amazed when I found out that they did not fly any air cargo. So, for instance, if one's bulldozer broke down in Siberia and the spare part was able to be located in Moscow - and this was never a certainty anyway - somebody had to fly all the way there to get it and then fly back. Allowing for the vagaries of whether the planes were flying on schedule anyway, the trusted employee left us carrying our US dollars, would usually extend his stay in Moscow with his mates for some excuse for another couple of nights and then return on a plane or planes back to the nearest airport to our operation. In summary, allow a week or so to get a part to site. This was an example of how I was amazed that Russia managed to stay in the space and arms race with the West for as long as it did and, with the great benefit of hindsight, how the USSR maintained its cohesion for so long. There were so many inefficiencies.

All societies have different social norms and, if one wants to operate there successfully, one must be prepared to adjust one's mode of behaviour and attitudes. As an example, in Australia we would hold introductory meetings with potential business partners in the office over a coffee, or perhaps over lunch, or even a dinner. In Russia, we would go to the sauna, get naked, and hit each other with small leafy birch branches. I really was not comfortable with this, even with a few vodkas under my belt – had I still been wearing one. Another aspect I found difficult, as an Aussie and perhaps due to our general British heritage, was maintaining some space and distancing ourselves from the person with whom we are speaking, whereas the Russian custom is to be much more in each other's face - literally, with faces only 50 cm or so apart when talking but this still does not stop many Russians from shouting at each other.

On my 1997 Krasnoyarsk visit, I was consulting to a little hopeful company owned by Paul Kopetjka and Mike Kitney. As I had done three years previously, they were trying to deal on those same hard-rock properties which were some of the sources of the Angara River alluvial gold. I guess that by that time, the surviving Krasnoyarsk bankers had come to some sort of truce! On this occasion, the sauna was accompanied by a massage. While I am inclusively politically correct of course, I confess to finding it disconcerting being surrounded by a group of naked men while I received a massage from a bloke who was as camp as a row of tents and wearing nothing but a beanie with a pom-pom. Worse was that I was lying face down on a table with my head to the side and in front of my eyes and about 30 cm away was the wedding tackle of the interpreter. I dreaded that I was to get an erection!

The things we do to earn a crust or to advance our employer's interests!

The town of Chara was of some interest and I went for an exercise stroll around some of the streets before breakfast one morning. I was never told why, but this action caused great anxiety for our partners until I returned. Were they worried about my safety, or were there things I should not see? One of my observations was that the office buildings and apartment blocks were of two distinctly different qualities of construction. In one of them the work was very sloppy, and courses of bricks were only vaguely and rarely horizontal. In the other, the courses were entirely level and evenly spaced and holes for the windows were truly rectangular. I enquired about this and was advised that the latter were built by German prisoners of war during the 50s, and they obviously had an entirely different construction ethic compared to the Russians. Trying to be helpful, I offered that in Australia we used a string stretched tautly and horizontally from one built up end of the wall to the other such that the intervening course of bricks could then be laid evenly up to that string line. They assured me that Russians did that too. I can only presume that they had guys holding the string at either end and both were getting solidly stuck into vodka at the same time.

Many Russians I spoke to and came to know had faced up to a difficult mental readjustment post the breakup of the USSR and the dictates of the state. Raised and educated under a Communist system, they had believed that it was the ultimate in societal development and so all other systems, and their adherents, were inherently inferior. The State provided jobs and food, and people were not encouraged to think or choose. Realising that they had been lied to all their lives was not easy to stomach.

By way of a simple example, the wooden window frames of traditional houses were painted a particular mid blue colour. It was that same colour everywhere without exception. Out of curiosity I asked why this was the case and they found my question perplexing. The answer was circuitous but basically the State decided that windows were to be painted that blue and hence that was the only colour that was manufactured and made available for windows. No discussion needed, and no one thought that, say, red windows would look nice as a sometime alternative, and why couldn't they buy that colour paint anyway? I wondered that if, for some reason they were able to source paint otherwise destined to paint a tractor red, what would happen to them if they dared to paint their windows red without authority?

And they seemed rather xenophobic. They had a particular disdain for the Chinese, and for people of dark skin colour. The Chechens of the previous USSR were often referred to as blacks. But they did seem to like Australians – or our money anyway.

Another thing I noticed was that there was no end of specialists in Russia. Once somebody was trained in a particular job, that was their lot for life, and they were then deemed a specialist. A pleasure to associate with, they were often very highly trained. However, many found it difficult to move beyond their comfort zone if we requested that they try something for which they held no formal qualification or general acknowledgement that they were trained for, or had been assigned, that task previously. Multitasking was not seemingly encouraged by the Soviet state. Rather, a personification of the old joke that "an expert is someone who knows more and more about less and less until he finally knows everything about nothing."

To finish my story, by analogy with trying to catch fish, Zephyr spent what it could reasonably afford on our expeditions but did not have the equipment, stamina and financial resources to lure and land the really exciting ones. Some companies spent a large amount of dollars on management and personnel time chasing big fish, but ultimately to no avail either. When I decided that no further investment could be justified, all Zephyr had to show for about a quarter million dollars were photos of some recovered gold nuggets, none of which were returned to Australia. The consolation was that our failed enterprises provided incredibly cheaper Russian lessons for the board, management and our shareholders than they did for those of Star Mining!

I would like to return to Siberia again one day, but purely as a tourist, to gain an appreciation of how much it must have changed since the mid-90s. Vast distances, the relative lack of infrastructure, and ever-present overriding factor of the harsh climate will retard progress relative to the European western portion of Russia. A few more degrees of global warming, should it ever occur, would not be a bad thing in Siberia. The last thing it needs is a return to global cooling.

JOURNEY TO
THE KIZILCUM DESERT

JOHN HAMMOND
UZBEKISTAN EARLY 1990s

John Hammond graduated as a geologist from the University of Western Australia in 1969. His career took him to more than 25 countries, primarily as a grassroots explorer, but he also had significant involvement with feasibility studies and corporate acquisitions. He began working with AMAX exploring for nickel in the Eastern Goldfields of Western Australia and then spent nearly eight years on the Namosi porphyry copper project in Fiji. In 1988 he transferred with Australian Consolidated Minerals to Perth and from that base he managed exploration in Western Australia, the Northern Territory and the Mediterranean region. In the latter area he led further exploration success in Turkey and Greece. Before retirement in 2013, John spent the previous 20 years working as an internal geological consultant, firstly for Normandy Mining and, later, in a similar role with Newmont.

In the early 1990's, soon after the breakup of the Former Soviet Union (FSU), I accompanied Roger Craddock, then General Manager for Eurogold Madencilik in Turkey, on a visit to the newly independent Republic of Uzbekistan. Our visit was by invitation through official channels, via our parent company Normandy Mining, and our belief was that we were going to assess mineral exploration opportunities. Normandy was at that time the major shareholder in Eurogold which was developing the Dikili gold mine in Turkey.

The trip turned out to be well beyond my past experience, starting with obtaining a visa. Uzbekistan had just opened a consulate in Istanbul and we needed a visa at short notice. We were advised to attend the consulate on a Sunday morning and we should have our visas in time for a flight scheduled for the following evening. Roger and I flew from

Izmir to Istanbul and proceeded to the consulate. After being admitted, we were offered coffee and had a long friendly chat with the consul about the virtues of his homeland. Finally he asked for our passports and took them to his desk in the corner. He pulled out a large official stamp, stamped the passports and, after a few more pleasantries, wished us a happy and productive trip to his country.

The next evening we boarded a Turkish Airlines flight to Tashkent, the capital of Uzbekistan. At that time there were only one or two of these direct flights per week and all foreign flights seemed to arrive in the middle of the night. My first impression on landing was how rough the runway was as the plane vibrated towards the terminal. As we disembarked it was evident that there were large expansion cracks throughout the concrete surface, apparently the cause of the bumpy ride. We entered a very basic customs hall, cleared our bags after considerable waiting and then, in the early hours of the morning, went to look for the driver who was to pick us up.

The car park was dimly lit and most people had cleared out of the airport ahead of us. It was very cold. We couldn't find anyone waiting for us and could not have adopted a Plan B even if we had one. There were no taxis or buses there at that time in the morning. We went back into the terminal to seek help. There was hardly anyone about but Roger managed to find a lady who spoke a little English. It took a while but she finally understood our predicament. She made a phone call and then directed us back to the car park. Within about 15 minutes a car showed up and the driver explained in limited English that he would take us to our accommodation. My impression driving along was how poor the street lighting was at 2 or 3 am. It was eerie. After about 20 minutes we arrived at our guest house. A watchman unlocked big iron gates and we were driven to the front door. They expected us and we were shown to our rooms. I should mention that we later found out there were no hotels where foreigners could stay in Tashkent at that time.

Next morning after a basic breakfast of bread, honey and coffee, we were told to wait and a driver would come and take us to a meeting at the Ministry for Mines. We had no choice as the guest house had a high perimeter fence and a locked gate with a guard. The place had apparently been a Politburo meeting house in the days of the Soviet Union. The interior of the building had extensive fancy wood paneling on walls and ceilings which looked like a good place to conceal listening devices. Anyway, a car arrived in due course and the driver told us that he would transport us around while we were in Tashkent.

The meeting at the Ministry of Mines was very formal with the Vice-President of the country, the department chief and his entourage all present. This high level reception apparently came about because our Chairman, Robert de Crespigny, had met the President of the Republic of Uzbekistan at the World Economic Forum in Davos a few weeks earlier and they had discussed possible involvement by Normandy in Uzbekistan's gold business. We were extolled at length on the benefits for western companies doing mining business in Uzbekistan. Newmont Mining was cited as the prime example. That company had just completed a deal and was mining the low grade Muruntau rock dumps (which still had plenty of grade for us Westerners), then crushing, fine crushing/grinding and heap leaching. At that time Muruntau was the largest open pit gold mine in the world and had been a major source of income for the Russian military prior to break up of the FSU.

After lengthy discussions about doing business in Uzbekistan, we raised the issue of exploration possibilities for foreign companies. We were told approval had been given for us to visit to some satellite deposits around Muruntau and that we would need to travel to Navoi which is the administrative centre for the region that hosts these mines. We asked if we could also see Murantau and were told that would have to be arranged locally. A key question in our minds was how to get to Navoi and, on asking, we were told we would have to make our own arrangements to fly there.

Prior to the meeting we had been assigned an interpreter called Gozel who spoke perfect English with an American accent. This was amazing as she had never been outside her country. She had also had her two upper front teeth removed and replaced with gold fillings which were supposed to be a mark of beauty and status. As the trip went on we noticed that this was quite common, and not only with women.

Following our meetings with the Ministry we were taken back to the guest house and the driver informed us he would be back to pick us up at 10am next morning. It was late afternoon and we were effectively locked up for the night. We had an early dinner (which was neither sustaining nor appetizing) in the guest house. With no entertainment, no bar and not even television, the only option was an early night. Next day, after another sparse breakfast, our driver showed up as arranged and the big iron gates were opened.

More meetings were scheduled and so, accompanied by our interpreter, we met representatives of the national geological department and were given a general overview of the geology of the country. The meeting went until mid-afternoon and our driver returned us to the guest house where we followed a similar routine to that of the previous evening.

The following day the driver took us to an Aeroflot office, this time accompanied by a young interpreter who worked for Gozel. He was to assist us to buy our plane tickets. There were many people milling around, seemingly waiting for something that didn't happen. Now you would think buying tickets should have been a simple matter but it took several hours. First we handed over some US dollars and the interpreter went away to change them to local currency. Then the cash with our passports were presented to the Aeroflot desk and finally late in the day we had tickets to fly to Navoi the next afternoon. Our interpreters notified the relevant people there that we would be coming.

After another night at the guest house, we were driven to the airport for the mid-afternoon flight. The young 'minder' who had helped with purchasing tickets was at the airport and he came with us on the plane. Now, at that time, getting onto a domestic aircraft in Uzbekistan bore no resemblance to anything we were used to. The flight was announced, terminal doors opened and there was a mad scramble across the tarmac to the old Russian plane which was about the size of a Fokker Friendship. There were no allocated seats and we took the last ones available. Roger sat in a window seat with a coat on against the plane's side which was ice-cold. This plane had no heating or insulation. From Tashkent we flew to Bukara, a famous Silk Road city, where the plane refueled while we had to get off for about half an hour. Finally we were airborne again and it was only a short hop to our destination. We remember looking down as we came in to land and being told by our young minder that we were very lucky to get into Navoi as it had been a 'secret town' and up until about two years before it hadn't even been marked on maps. It was a new town of 120,000 residents, purpose-built in the desert, with water piped in and its own power station, all to facilitate uranium mining and processing.

Unlike on our arrival in Tashkent, here a driver was waiting for us. We were whisked away in a big, shiny, new, black Volvo and taken to accommodation that was somewhat more salubrious than the guest house in Tashkent. The Volvo stood out like a sore thumb on the streets of Navoi where the commonest cars were small Russian-made Ladas. It turned out that this car was the personal vehicle for the Chief Administrator for the whole Navoi Region. The driver indicated we should rest for an hour and then he would return and take us for a meeting with the Administrator. It was late afternoon when we were ushered into his office. We spent an hour or so there hearing about Novoi Region through our interpreter, who was not nearly as fluent in English as was Gozel in Tashkent. From the Administrator's office we moved on to a swish dining hall (which was part of the uranium treatment plant) and had a fancy dinner with traditional Uzbek toasting until late into the night. I must admit that several of the vodka toasts went straight over my left shoulder to land in a pot plant behind. No one noticed in the general merriment. We joked about glowing green in the dark when we left!

Next morning, after a good sleep, we started out fairly early to go to the field. The driver arrived with the Administrator in the black Volvo and we set off through the sparsely vegetated red sands of the Kizilcum Desert, heading for Muruntau. The drive, which took a few hours, was hair-raising as the Volvo sped along, often in excess of 100 kms/hour on a track that might have been safe at half that speed. The driver had obviously made this trip many times before. Along the way in the blurry vision we saw occasional double-humped Bactrian camels and industrial plants off each side of the road that were uranium solution mining operations.

We arrived at Muruntau mid-morning and, of course, a meeting had been arranged in the geological office. The geology and mining operations were explained in very general terms. After the office meeting we were shown the process plant with, I seem to recall, 14 ball mills side by side in a building over one kilometre in length. Then we were taken to a lookout to see the open pit which was truly impressive in scale. Standing on the pit rim we asked the chief geologist some detailed questions about the operation; ore grades, production rates, et cetera, and he said he couldn't tell us, as such things were State secrets! That was as close as we got to the rocks and by then it was getting towards lunch time. We were ushered back into the Volvo and travelled at break-neck speed on a good sealed road to Zarafshan, a small city which is the administrative headquarters for Navoi Mining and Metallurgical Combine (NMMC), the operators at Muruntau. The company buses its workforce from this base daily to the mine.

In Zarafshan we had a splendid lunch courtesy of NMMC and then set off to the Amantaitau – Daughyztau mine/deposit area where the British company, Oxus Gold Plc, had a joint venture with local entities. Some of the joint venture local representatives accompanied us. It was evident that there had been attempts to mine and process the ore but there were apparently metallurgical problems and also difficulties with disposal of large volumes of highly arsenical tailings. When we asked about the "exploration opportunity" we received blank stares. The reason we were there in our host's eyes was to provide insights on solutions to the arsenic problems. It turned out that the Normandy involvement may have come about because of its perceived expertise in dealing with the arsenic-rich gold deposit the company was mining at Wiluna in Western Australia. After spending a few hours around these deposits we retraced our steps, again like a bat out of hell, through the desert and arrived back in Navoi just after dark.

The next morning we flew back to Tashkent and our driver was waiting at the airport. Our minder/interpreter took us for 'lunch' at what he described as a very new kind of entrepreneurial restaurant run by a local, in what could only be described as a shack, where we ate delicious meat cooked on charcoal and some soup. He seemed very proud that this was the beginnings of entrepreneurship in Tashkent. We spent the afternoon seeing the sights of the city under the watchful eyes of our driver and the young interpreter. At the time Tashkent had big wide boulevards and no high rise buildings. In one area we saw sheep being herded along a fairly main thoroughfare. Needless to say traffic was not too busy. We asked the interpreter about hotels. We wanted a drink and eventually were taken to what he described as the main hotel in the city. It was probably four storeys high and he explained that only locals could stay there. We went in but were dismayed to find that we could only order a soft drink. Westerners were quite uncommon at that time in Uzbekistan and I felt like everyone was watching us. Later in the afternoon we were taken back to the guest house.

The driver came next morning and we went to a short debriefing meeting at the Ministry and then to a geological museum which had an impressive 3D overview model of the country's geology. Later we returned to the guest house to wait for our late night flight. Our driver arrived about 10pm and took us to the airport and, after nearly a week in Uzbekistan, we were on our way back to Turkey. The trip was an eye-opener and good preparation for later journeys into Kazakhstan and Russia. However, this time there were no deals to be done.

GETTING THE COLD SHOULDER
– AND MORE

DAVID MILTON
KAZAKHSTAN EARLY 1990s

David Milton graduated in 1972 with an Honours degree in geology from the University of Adelaide. He commenced work with CRA and was sent to Hamersley Iron, working at Mount Tom Price and Paraburdoo as a mine geologist. A transfer to Broken Hill resulted in a wider involvement in mining areas and with CRA in specialised ore resources work.

This was the start of travelling to many parts of the world to evaluate the technical and economic merits of numerous diverse mineral and mining projects and operations. A range of corporate and management roles of increasing seniority spanning 25 years with CRA was followed by nearly 20 years in consulting roles, again in many parts of the world and for many commodities.

He is now semi-retired in Perth and, with a vengeance, has taken up woodturning as a hobby.

The shock news

In the early 1990's, after wrapping up an urgent several week drilling campaign to improve the knowledge of a nickel deposit undergoing a feasibility study, I left the site on Friday for home. Glad to be out of the 45-degree Celsius March day in the north eastern goldfields of Western Australia and knowing my wife had arranged a house inspection for a potential new home, the weekend was going to be busy. The inspected house was quite nice, so an offer was made on the Saturday afternoon, and a deposit paid. With the acceptance on the Sunday morning of the offer, was created a flurry of considerations concerning selling our existing house and financing the new purchase, and the myriad of details that surround such activities kept us busy all day.

The phone call from the Managing Director at 8 pm that Sunday night completely changed this situation. Basically, I was required to depart Wednesday morning to urgently assist in evaluating a gold project in Kazakhstan, a former Soviet Bloc country. The brief was to assist a senior manager, who was already in the country, to gather all possible geological and other relevant technical information for the deposit. This was to allow an assessment to be made back in Perth of its potential value as the basis for a proposed joint venture to develop the project into an operating mine and processing facility. The timeline was that we had four weeks to do all the evaluation; that is, gather the information, and pull it together into a coherent form to support a study and recommendation on joint venture negotiation positions. I was also informed that the intention was, if the joint venture were successfully negotiated, I and my family would be moved to Germany immediately to take a leading role in the project development and initial operation phase. Furthermore, I would be contacted later that night by the in-country manager with more information. His request immediately threw a spanner into the works re the pending new house acquisition. After a brief discussion with my wife, we decided to withdraw our bid, so duly contacted the agent that night explaining the rapidly changing and unforeseen circumstances. The agent's view was not very sympathetic, and she advised me that she would contact the vendor the following day to see if they would cancel the bid.

Late that evening a phone call was received from the in-country manager, Bill, who filled in essential details, such as a visa would be given on arrival at Almaty, and all flights had been booked and there were to be no exiting airports on the way. I was to bring five reams of A4 photocopy paper, as many muesli or energy bars as I could pack (breakfast food was lacking in-country), and some seriously decent Arctic winter clothing. Most importantly, I needed to also bring several "umpire" samples of different gold grades for check assaying of the project's retained samples and of the accuracy of their laboratory. A sleepless night was filled with churning thoughts: what if the house offer was not cancelled, how could I then get all the conveyancing signatures and finance sorted out while overseas? With three children under five, my wife already had her hands full. Were all my vaccinations up to date, who was going to pull together all the urgently needed work on my current project and maintain quality? There were a myriad of loose ends needing to be sorted out, so I tossed and turned all night long.

Monday and Tuesday preparation

An early rise and into the office to start handing over the current project issues and the matters requiring follow-up to anyone I could co-opt, and to brief the project manager on its status. Then I met with the Managing Director for a further briefing on the company's aspirations on the Kazakhstan project and to provide me the information that had been acquired from a visit a couple of months prior by a scouting business development staff member. The day seem to fly by and all the time I was worried about the pending house acquisition issues as there had been no news from the agent. The important item of photocopy paper was sorted out by raiding the company stationery supply. The umpire samples had to be ordered and would be available for pick up on Tuesday afternoon as the owner was away and not back until then (there was only one reliable supplier at this time in Perth).

Finally, late in the afternoon, a call was received from home after the real estate agent had contacted my wife to advise that the vendor had accepted the withdrawal of our offer as, fortunately, a second offer had been in the wings and that had been accepted. Luckily, our deposit had not been cashed, and the cheque was to be returned. Well at least a major problem had been solved, but my wife was not happy as she was very keen on the house which now had slipped from her grasp.

That night a quick look around home to find suitable Arctic cold weather clothing only turned up some woollen socks, jeans, a fleecy jacket and a couple of jumpers. Not unexpected, as I had spent the last five years working at Olympic Dam, near Darwin and around the Meekatharra areas and had not really experienced any cold weather. An appropriate suitcase was found.

On Tuesday morning it was a trip to the supermarket and a big buy up of muesli bars and energy and nut bars along with tissue, soap, shampoo, toothpaste, and toothbrushes. Quick dash into the city to get some warm clothes. But where do you get Arctic clothing in late summer in Perth, hardly a place where shops fill with "snow" gear for winter? A succession of shops provided bits and pieces: Paddy Pallin, fleecy gloves and balaclava; Mountain Designs, thermal underwear, double knit socks; Scout Shop, flannelette shirts and scarf and, at Kathmandu; a Gore-Tex jacket that was windproof and water-resistant and a woollen cap that pulled over the ears. Just enough time to race out to get the umpire samples after a final briefing and handover of the nickel project issues. A quick pack cramming everything into one case which had to be strapped with an old belt to secure it as the contents with the extra paper and food threatened to burst the zips.

Flights and Almaty

A very early start on the Wednesday morning and a relatively uneventful set of flights from Perth to Bahrain, then to Frankfurt and ultimately to Almaty, except for the delays due to carrier changes and the normal issue of trying to stay awake in the terminal and trying to sleep on the plane. The Lufthansa flight arrival in Almaty was about 1am. Over the preceding years I had learnt that it was best to travel in aircraft wearing a selection of warm clothes, including a sports coat over a jumper. However disembarking of the aircraft was via a gangway onto the apron and a 100 metre walk to the terminal in a minus 5 degree Celsius night with a brisk breeze blowing. Result being very cold hands and head by the time I reached the terminal and slipping every step of the way. Fortunately the company had an agent representative who met me at the door and ushered me through the visa and passport areas; my bags were collected, and not inspected, and a very cold me was delivered to a warm greeting by Bill in the airport lobby. We were quickly bundled into a minibus and soon arrived at a the local agent's house where I proceeded to be warmed up by a quick couple of shots of vodka, which I soon became aware was the local "milk".

As I had spent most of my career in relatively moderate to warm climates, I was taking a little time to adjust to the cold when Bill informed me we should get a quick couple of hours sleep as we had to be at the airport again at 6am for a flight to the closest local town to the project site, named Kokshetau. Fortunately, I was able to have a nice hot shower and freshen up after the 30-odd hours of travel and this warmed me up. After about an hour's broken sleep, I was up again. This time I rugged up with at least thermals, flannelette shirt, jumper, and the Gore-Tex jacket, two pairs of socks and my field boots, balaclava, and gloves. A quick bite of some bread and jam and out to the airport.

Flying to site

On the way to the airport Bill informed me he had yet to go to site, so this was his first trip there also. He explained he was aware that the food situation may be a little difficult and hence the request to bring as much of it as I could. He also explained that he had acquired a small portable photocopier which he had in his bag and, if we were asked about it, we were to say it was for the vendor. Its actual purpose, and hence all the paper I had brought (he had not been able to get much of it locally), was that we were going to copy as many of the reports as we could arrange. All his work in the preceding month had been dealing in Almaty with officials of the various government departments and the limited stock company which had the ownership of the target deposit, which was called the Vasilkovskoye Gold Mine. By his account, this was a tenuous process and it was unclear what status had been reached but, importantly, we had been given permission to visit site and carry out a technical "evaluation" of it in a four week window, so to site we went! On arrival at the airport I was surprised to see quite so many people catching flights. With our trusty agent, we were shepherded through a book-in process in which we took no part except for passports to be handed over and then returned to us with a card which was our approval to fly and had seat numbers on it. Then we were moved to the next area, a small departure lounge. Incidentally, our bags were not checked in, so we carried them with us through the whole process. The company agent departed the scene and we were left alone and neither of us could speak any Russian, so it was a watch and act with the herd approach.

By this time, it was about 8 am and, after another two hours of waiting, we were suddenly ushered out onto the tarmac with about 30 other people by some sort of official. We approached a small three-engined jet which I soon learnt was a Soviet-built Yak 40: from its rear there was a set of internal steps for boarding into the aircraft. By this time, we were the last in line and, ascending the narrow steps with our bags, we saw why the others had hastened ahead. The baggage storage area, which is at the rear of the plane on either side of the stairs was full, there was even a cage of chickens. Obviously, no room for our bags so we dragged them forward to our two seats which were unoccupied, squeezing past three people who were standing in the aisle. We then noticed a door at the front left-hand side of the aircraft, but it also had bags piled high in front of it. No other option than to put the bags in front of us and somehow fit our feet in. By this time now I was sweating as I would on a 45-degree Celsius day in the Pilbara and cursing my thermal insulation clothing. A cold feeling soon came over me as I pointed out to Bill that there was no pilot or co-pilot in the aircraft. So we waited for about 15 minutes when a noise at the back of the aircraft signaled the arrival of two men dressed in some type of uniform who climbed up the stairs. One pulled out a big hexagonal-type key which was then inserted into a hole by the cabin rear door. On turning it, the stairs elevated and presumably sealed shut. One of them then took the key and put it into his pants pocket and they then proceeded to the cockpit. We had a pilot and co-pilot. But we also had three persons standing at the rear of the cabin and bags blocking a door which was apparently the emergency exit. Surely, we were not going to take off like this.

Within a minute of their getting in the cockpit, we could hear the whine of starting engines which seemed to me were fired up and brought up to high thrust quite quickly. In an earlier part of my career I had learnt to fly a small single engine aircraft and, after getting to a restricted pilot's licence level, had stopped my training. But I have always taken an interest in flying. So, it came as a completed surprise when the aircraft suddenly started to move, turned around about 180 degrees and full thrust was applied. What, no taxiing to a runway? I gripped my case tightly. Surprisingly, within a few seconds, we were airborne, proving that the Yak 40 is an impressive work horse, capable of taking off in less than 700m and I am still impressed with how well it flew. It generally carried 27 passengers but can take up to 32 so, even though there were 27 seated, the extra three were well within its capacity for this flight which was going via Karagandy - a famous copper mining district.

Despite the initial concerns the flight was uneventful and a spectacular view of some of the open pits was seen on our approach to land. More interesting were the remains of several other aircraft at either end of the runway, which looked more like military jets and which had suffered some catastrophic fates. There were clear signs that this was still a military airfield as there were embrasure earth works and several parked MiG aircraft along the sides of the runway. A brisk taxi from the runway to the terminal was punctuated by the engines being shut down a couple of hundred meters from the apron and we coasted in. As soon as the aircraft came to a halt, the crew jumped out of their seats, walked to the rear, pulled out the hexagonal key, unlocked the door and the stairs unfolded to the ground. The crew then left the plane before the passengers who then ambled out. Those who were departing took their bags, the rest of us went into the terminal and waited. While waiting we observed a small tractor pulling a small tank on a trailer out to the plane and then its wing tanks were refueled but it only seemed to be there for a few minutes before it left, so perhaps they only needed a top-up of fuel. The boarding and take off procedure were the same as at Almaty. Wait an hour or so, then passengers herded out to plane, this time a couple more cages of chickens were loaded, and we now had four people standing at the rear. The take-off was equally short: start the engines, roll out to the runway, turn and depart using only half the length of it; next stop Kokshetau. By this time, we realized we were getting a little hungry.

Arrival and the first night

The arrival at Kokshetau was uneventful and now we were within 70 kilometres of the Russian border and on the edge of the steppes in the very northernmost region of Kazakhstan. Bundle ourselves out of the aircraft and into the terminal, where we were greeted by the project's Manager of Mining, Anatoly, and a translator hired for our stay. The first impression of the former was dominated by the top and bottom rows of gold teeth, eight in all, that complemented his ample size - both of which are rarely seen to that extent in Australia. This hearty, generous and likable man was to become a close friend to Bill and myself for the whole of the project's life for our company. The interpreter was a woman named Frieda, who was in her early fifties, and had been a lecturer in English at a local university, which was now closed since the breakup of the Soviet Union. We were quickly bundled into a minibus: it was about minus 10 and a brisk breeze was blowing so now the thermal mass I was wearing was worth its discomfort. The driver had only gaps instead of teeth and constantly smoked but his ability to miss the deepest potholes in the 12 km drive to the accommodation was quite skillful.

After a quick drive around the town where the highlights were pointed out in the fast dimming daylight, we were taken to our accommodation, the best in the town. This nondescript-looking building was the previous Soviet regime's accommodation for officials. The lobby/reception was no bigger than a small lounge room, with a window behind which the manager took our passports plus the travel and permit documents that Bill was carrying. After about 15 minutes of discussion with Anatoly, who said he would see us at 8am in the morning, he departed with the translator and we were escorted by the manager to two rooms on the second floor via the stairs - as the lift was unserviceable. The rooms were about halfway along the corridor which had at its entrance a room with an open door, beyond which a woman was sitting in a chair observing all comings and goings and tending the adjacent samovar. The rooms were quite small, with just a single bed and a small table, and with a shower and toilet tucked in - not unlike a mining camp Donga. There was a big hot water radiator at the window end of the room which kept the room nice and warm.

Both of us were by this time quite hungry so we decided that we would go out and find a place to eat using a dictionary to help with communication. After about an hour wandering around and finding no place to eat, we found our way back to the accommodation and, through the use of the dictionary, were pointed to an area in the basement of our building where there was a small canteen-like affair with a server and a couple of tables, one of which was free. Again with the trusty dictionary, we were able to find out that they were about to close but had some borscht and bread rolls available for $5 US a serve, a price that by this time we were willing to pay. To drink we were offered, for $2 US, a litre of vodka in a type of bottle not unlike one of our old milk bottles, but with a rip top, as it was assumed you just drank the lot in one sitting. Two soup bowls filled with the last of the borscht and two rock-hard bread rolls were passed over in exchange for the cash and we sat down at the empty table. In my haste to tuck in, I picked up the bread roll and tried to split it apart but only succeeded in dropping the brick into the borscht, splashing most of it out of the bowl and over my front and lap. I am sure that the little sips of borscht and the now slightly softened roll were one of the memorable meals of my career. Well at least I had a reasonable supply of muesli, et cetera, bars in my bag and, once back in my room and cleaned up, Bill and I had a small feast. This was followed by a quick hot shower and into a nice bed in a pleasantly warm room for a good night's sleep.

To work

After a quick breakfast of a muesli bar and a welcome hot cup of tea from the samovar, Bill and I were ready to meet Anatoly at 8 am, who then had us driven around to the project office, some five minutes' away. The office was a complete three-story building which appeared to house several hundred people. After the obligatory meeting with other senior company executives, in discussion the issue of meals came up wherein the problems we had on the previous night were aired. This prompted immediate action: we were taken to the executive dining room (just a separate canteen) and treated to a hearty stew with rice and an initiation into the continual vodka toasts by all around the table. This was the pattern for the rest of our stay when in the main office: we would be taken to lunch for a hearty hot meal and copious amounts of vodka from the ubiquitous rip-top bottle. More importantly, Anatoly showed us where the only place to get a meal in the evening was located. This was a ballroom that had a small canteen that served rice and a chicken curry every night. Some nights we had some cabbage or peas added, all for $5 US cash a serve. And they served rip-top bottles of vodka.

The issue of meals out of the way, we settled down with the key technical team members to gather as much information as possible. Bill was looking after the metallurgical, mining, and studies area and I was focused on the geological, resource, geotechnical and hydrological aspects. As we had only the one translator, it was decided that Bill would mainly use her to assist him and I would have to manage with a secondary use of her services. Fortunately, the chief geologist had some limited English and we were able to get started on the details I needed.

Obviously the first information I sought were the ore resource estimates and supporting data. After a lecture to me on this information being a State Secret, and after some discussions with Anatoly, the chief geologist had a series of hard-bound ledger style books brought to me. These books were bound with a ribbon with a stamped wax seal: this was duly broken by the chief, and the contents revealed. The volumes had the precise information I was after - and more. I was stunned to learn that this deposit, which outcrops, had been found in 1963. Depending on the grade cut- off, it was a 10 to 20 million-ounce deposit, able to be mined by an open cut, and with average grades of 1.5 to 2.5 g/t. In the ensuing nearly 30 years since discovery, it had been extensively drilled, many holes being more than 1000m deep; a shaft sunk; three levels of extensive evaluation driving, cross cutting, rising and winzing completed; numerous geological academic studies completed, along with metallurgical and other technical studies.

The only mining activity had been a small heap leach trial of approximately 100,000 t of mined material. This was all reported in the duly sealed Secret volumes before me. What was not comprehensible was why it had not been developed. If it were in Western Australia, it would have been almost completely mined out and processed after 30 years. The reasons for this were later revealed but, in the meantime, the focus was to get this important geological information copied. It was very clear that we could not just roll up to the office and start copying everything, nor could we get the potential partner to photocopy it all. In the western world most of this sort of information was in digital form, but here it was strictly only on paper. The process was that I sat on one side of a desk and the chief geologist the other side while I looked through the reports. As I initially still needed a translator, we found out one of the other geological staff had a reasonable grasp of English, so he was assigned to help. Thus, a reasonable working arrangement was reached.

After a couple of days of tediously hand-transcribing key information and, in some cases, putting information into my laptop computer, it was clear I was going to struggle to accumulate a comprehensive set of data. Just getting to understand the nomenclature and legends on the numerous diagrams was a challenge.

As all the 15 reports were piled high on the desk we were using and left there for the next day, it was decided that I would try and take one of the volumes back to our rooms for clandestine photocopying. So, at the end of the third afternoon, and unnoticed by the others, I was able to put one of the crucial drilling log volumes in my briefcase and left the office. By this time we had become known to the various guards at the entrances to the office and at the accommodation as we were now walking to and from them - a brisk and enjoyable 20 minute walk in the snowbank-lined streets. We passed through the guarded doors with great relief as we would have great difficulty in explaining the Secret report in my briefcase. Back at the accommodation, Bill brought the small portable photocopier into my room and I proceeded to copy the report. This was done under a blanket on the bed to prevent the photocopier light from spilling out and attracting the attention of the woman at the end of the corridor. Next morning the volume was returned without arousing suspicion. Over the next days we managed to copy the most important parts from all 15 volumes of reports, on some days bringing out two volumes at a time, duly copying them - including the fold-out maps and cross sections - and returning the reports unnoticed. I now had a suitcase of copied reports rather than blank paper. Back in Perth, the jigsaw of photocopies was put together and they became the basis of our digital models of the resource and its evaluation.

Conditions for the locals

In the first couple of days we had got to develop a good rapport with the people we were dealing with. We had also quickly found out that if we were going to maintain this, we needed to master the ubiquitous toasts in vodka and cherry brandy at any occasion during the day, particularly at meals. The issue was that we could not offend by refusing to skoll. But we found that, if one of us did not, especially if they had an underlying reason like feeling unwell, but the other did, it was acceptable. So, a pact was drawn up were we took it turn about each day to be the hospitable one while the other played a minor role.

This led to a manageable situation where we maintained a reasonable level of sobriety but also to some memorable occasions. The first of these was when Anatoly invited us to his home one evening for a meal. Now it was my turn to be the hospitable one and it was no great surprise that, after being greeted into a small but well-appointed and warm apartment, the rip-top was pulled off the bottle of vodka and we were into a round of toasts. This was fine as there were only four or five persons present (I cannot really remember) but it soon escalated to "traditions". Now as I have mentioned previously, Anatoly was a well-built man and a large platter appeared which was filled with hard cheese, solid pork backfat, and a very large number of pickles. Anatoly explained that he was of mixed tribal Kazakh and Russian descent and that we would be treated to a traditional meal of horse meat - a very generous compliment - after we had a few toasts to traditions.

He then explained that Peter The Great was highly regarded and that the platter before us was in honor of him. The "tradition" was that he was a great leader who established his reputation by toasting until all the food ran out, in his case reputably the pickles, which are ubiquitous and in abundant supply. So, to the ritual: slice of cheese, shot of vodka, then a slice of pork fat with a pickle. Bill, who was feigning some type of excuse was not taking part, so it was a 78kg moderate drinker versus a probably 120kg well-practiced one. In a short space of time the first bottle had gone, the second appeared, and the hearty gold-toothed laugh got more pronounced. I think the platter was finished, well Bill told me it was and that I was a good happy ambassador for our cause, country et cetera, and did not disgrace myself. But I cannot remember anything after the initial couple of toasts to Peter The Great. Strangely, I woke next morning with no ill effects nor hung over and felt remarkably well.

On the first weekend we worked we had Sunday afternoon off and wandered around the town. As it was still heavily snowbound and probably about minus 10 Celsius, we were surprised to see a couple of shops open, so we wandered in for a look. In both cases they were grocery stores, but they were almost devoid of any goods despite having lots of space and shelves. The few goods that were to be seen were all preserved foods in large screw-lidded glass jars, being mainly tomatoes, cabbage and pickles. There were a few large cans of tomatoes, some blocks of cheese and some large bags of rice. We found out from our translator that, prior to the break-up of the Soviet Union, these shops were always filled year-round with pretty much any food imaginable - much like a Perth supermarket. The Soviets had a food distribution system that provided, say, oranges from the Black Sea area in the middle of winter to these stores, no doubt helped by the fact that Kokshetau has a major rail network passing by it. During the Soviet regime they wanted for very little but, within a year since independence in December 1991, food supplies had rapidly deteriorated and, in many cases, people were living at subsistence levels and mostly relied on growing food during summer for their survival during winter. During our stay we were to see many examples of this.

One afternoon we had been out to visit the small pilot plant facility on the edge of the town and were late getting back so offered to drop the translator at her apartment. Instead she asked to be dropped at the local "spa" which was basically a heated swimming pool and sauna for the community. The reason for this was that she met her family there and they stayed in this building until it closed because the central heating in her apartment had ceased to work at the beginning of winter and was not yet repaired. The apartment was subject to a constant temperature of below freezing so they kept warm in the spa, dashed home and had a quick meal and into bed. As all the water pipes in the apartment were also frozen, they had to use the spa for bathing and to help prepare food: they either brought hot water home in a small thermos flask or thawed ice out over a small kerosene-fired burner which was also used for cooking.

We found out that it was not actually only her apartment block but several streets of apartments which were thus afflicted. The whole town, which was centrally heated by hot water piped from a coal fired plant, had a network of insulated underground pipes that fed the heating of buildings, but these often burst. In Soviet times they would be repaired immediately as it was an allocated job for someone but now no money was available to pay people to do this - so it just was not done. This led us to discover that in fact there was little to no commercial activity at that time in the nation of Kazakhstan: the country had been born, but it had no financial foundations. In the place of the socialist Soviet system, where everything was provided and everyone had a job to do and had access to social benefits, a commercial vacuum had been created. To a degree it was everyone for themselves as there were as yet no well-established transaction-based activities. This explained the great welcome we were receiving as we had hard cash currency - the mighty US dollar - which was highly coveted.

Geologist Day

On the Wednesday after the beginning of spring we were informed by the Chief Geologist that there would be a special big celebration on Friday to which we were invited as guests of honor. This would be held in the geology group's meeting room and would start at midday and be a big feast. It was for Geologist Day but would be a couple of days earlier. What is Geologist Day?

A quote from the "Myanmar News Agency Date: April 06, 2020"

"This professional holiday was established in the Soviet Union in 1980, but after its collapse the tradition to celebrate Geologist Day wasn't forgotten. Nowadays it's a professional holiday of geologist in Russia, Ukraine, Belarus, Kyrgyzstan and Kazakhstan. Geologist Day was established under the initiative of a group of Soviet prominent geologists, headed by academician Alexander Yanshin. Discovery of deposits of the oil and gas West Siberian province in 1966 became the reason for creation of the holiday. The geologists applied to the government with their request, that was approved by decree of the Presidium of the Supreme Soviet on March 31, 1966. The date for the holiday was not chosen randomly. It marks the end of the winter and beginning of preparation for summer field work. Geologist Day is traditionally celebrated by all geological and mining organizations. Other people, who

are involved in relating to geology fields, also consider this holiday as their professional day."

So, at around midday we were taken to the meeting room which was filled with all the geological staff and their wives and partners. Before us on the meeting table lay a wealth of foods which must have been quite an achievement to accumulate given the austere conditions under which all were living. On a side table ominously was a pile of vodka and brandy bottles. After a series of speeches followed by toasts, an enjoyable feast was had which helped counter the continual flow of vodka and brandy from the never-ending toasts. The afternoon wore on, and there were more increasingly incoherent speeches and toasts with participants gradually passing out at the table or leaving. By about 8pm Bill and I bade farewell to the remaining three or four staff, including the chief geologist who, despite being of slender build, did not seem to be affected by the copious amounts of vodka he had imbibed. Now this was somewhat a hasty departure and I left without my trusty ear-warming hat. Earlier in the week there was a forecast for blizzard conditions from a storm from the Arctic and it descended on us as we made our way back to the hostel. Fortunately, we knew the route well, but the icy conditions underfoot made for slow travel. By the time I got back to the accommodation, I had ice on my beard and my ears were very cold. It took me most of the night under the bedcovers to warm up and now, with no central heating in the room as explained later, the room felt like it was about the same temperature as it was outside.

Project site visit, fishing on the lake and logging core

After several delays, a visit to the site was permitted. The project is about 20 kilometers north of Kokshetau and we set out in a minibus with Anatoly, the chief geologist, our translator, and a driver. On the previous day, the temperature rose above zero and there had been considerable snow melt. It was a nice clear day, but the temperature was around -10 degrees Celsius and the snow melt had partially refrozen and the ground was icy. Over the past few days, we had found out that there had been considerable development and construction on site and much equipment had been delivered. There also had been some trial mining and a several hundred thousand tonne heap leach facility was in place which included a resin extraction circuit. However, we could not find out anything about the grade of the heap leach nor its actual production performance. There seemed to be a wall of silence around the amount of gold recovered and its recovery rate, and around any reconciliation between the grade the geologists thought the heap would be from mining versus that determined from testing of it. Any questions on this were rebuffed by invoking that it was a state secret and we needed to speak to more important people than the staff at site. These people were never identified to us.

Thus, it came as a surprise when we reached the site to see only three large buildings, only one of which was closed in, the other two being partially completed shells. Also nearby was a quite large "trial pit", with three approximately 5m deep benches mined out, and a substantial heap leach pad covered in snow and presumably ice. One of the partially completed buildings had a railway line that ran straight into it. On inquiry we discovered the way the Soviet system had worked with reference to what was laid out before us. Basically, everything revolved around the Five Year Plan (FYP) which was re-cast each year. The basic premise was that every task was planned but no attention was paid to the need for what was planned or whether it made economic sense. For example, the railway line went into the main store building, an eminently sensible strategy given the harsh continental climate of the region, thus allowing supplies to be railed efficiently direct to the store. However, the store building was not closed in nor completed as it was not the priority building in the FYP: the main office was the priority, so it was at least closed in, but was far from complete. The FYP for the locomotive factory (somewhere else) had to produce two locomotives for the project, which had been delivered some years previously and were deteriorating, in reality were probably inoperable, in the partially completed store and were exposed to the snow, ice, rain etc.

A FYP somewhere else had resulted in the manufacture and delivery of the electric motors for the milling circuit, but there was no room in the store so they were outside with miscellaneous other items such as pipes, rolls of cable, liner plates and various steel work. These could not be installed because, firstly, the plant site buildings had not been completed as their FYP priority was lower than that of the office. Secondly, nor had there been any final engineering layout done as the FYP for the engineering team had changed and they were working on another project instead. None of this worried the project people as all they had to do was their part of the FYP and get paid: the situation would sort itself out later. Basically, there was no economic urgency or, more critically, any understanding of effective use of capital and resources. Anatoly explained how, under the Soviet system, the FYP process kept everyone employed who possibly could be, and was the basis for social harmony, but it provided no idea of the inefficiencies or extra cost it caused.

When we looked at the mine area, the mining equipment was all in poor condition and had been parked up for about two years. It was revealed that there was no workshop or repair facilities and spare parts were nearly impossible to get, so as one machine broke down it often became a source of spare parts for other machines. On inspection of the heap leach pads and recovery facility it was clear that if there were any records of production, they would be highly dubious. The pads were literally frozen. Construction and stacking had started in summer time and some irrigation was carried out but winter froze the system and, the next summer, more material was deposited on top and irrigated before the first area thawed - as the FYP dictated the actions. The next winter froze the pad further and there never really had been any significant leaching.

The recovery plant was a very hazardous area; bare electrical wires, significant leaks of pregnant cyanide solutions and virtually no ventilation. The resin columns the Soviets used to recover gold were inoperable and bags of presumably partially loaded resin were stored awaiting stripping at a facility in the town. There was no metering or measurement of flows, so no one had any idea of what quantities of pregnant liquor had passed through the facility. It was little wonder we could get no figures on production. Bill and I concluded that there was great opportunity for our company to introduce world best engineering, mining, milling and technical practices into this almost virgin project.

The last stop for the day was at a diamond drill rig to see the quality of this operation and how it compared to those practices used in Western Australia. The rig, as is needed in any Arctic environment, was housed in a hut which was sealed against the cold winds and heated with a well-tended stove and was basically a copy of an American Longyear 38 drill. The hole was being drilled to about 800m depth and was currently at about 700m. There were four men in the crew and the drill string was conventional, not wireline. This outmoded method means that every time the hole had advanced such that the core barrel was full, or blocked, the entire 700m string of rods had to be pulled out and unthreaded in three-metre sections to recover the barrel which was then emptied of that core. Then the entire rod string had to be progressively re-screwed together, rod by rod, and lowered to the bottom of the hole so drilling could again be progressed.

Wireline drilling has been the standard in the West since at least the early '70s: a barrel recovery contraption is lowered by cable down within the entire rod string, and latches on to the barrel which is brought to the surface to be emptied, before the barrel is dropped back down the center of the rods. A 45-minute operation versus one that could take over half a shift. Thus the rate of drilling in meters per day was not great, not helped by the poor quality of the equipment and the basically 1960's technology. More importantly the core quality was mediocre, again not unexpected due to the relatively narrow drill size; to the rocks being drilled, which were often fractured granite or intrusive; and to the lack of modern drilling fluids (they were using water and soluble oil only).

As there was only a partially filled core box in the hut, for the sake of wanting to see more core, I was directed to a stack of boxes outside. I had some concerns about the wooden boxes which were made from slats and narrow boards and potentially allowed losses of fines through gaps in the floor of them. I also wanted to get a better feel for the consistency of drilling over a larger interval, how well the core had been marked up by drillers, and the actual percentage recovery by measuring some runs against the lengths recorded by the driller. I thought it would be a perfect day to do this - sun shining, no wind - so with gusto I laid out about six core boxes on the ground.

Now the thick gloves that I wore virtually all day had to come off as writing in a note book while measuring core length with a tape and pulling 35mm diameter core out of a tray for a look can only be done by bare fingers. In life there are lessons that are learnt by unpleasant experiences. Now as it was only minus 10 Celsius, I thought I could do about 15 to 20 minutes of work before gloving up. I neglected to understand that the core had been outside for several days, including during the nights which were considerably colder than the days; that there had been a blizzard a few days before where day temperatures had dropped to about -25 Celsius; and that there was the effect of ice in "gluing" core to the bottom of the wooden boxes.

The first piece of core I touched immediately stuck to my fingers, fortunately only lightly, and I was able to pull them off without losing any skin, but the core stayed in the box unmoved. Well, I will try to use my pencil to dislodge the core, but to no avail, it is frozen to the core box. Well then, I will just do some rough measurements, so pulled my steel tape out and proceeded to measure up runs. Lesson number two: steel tapes cool quickly and are good conductors. The tape started to stick to my fingers, so I shook it loose, allowing it to fall into the snow. At least I had one approximate measurement of recovery, so pulled out the notebook to record the pertinent data with now rather chilled fingers. All this after about two or three minutes' effort. I thought I would continue, so picked up the tape to find it gummed up with snow and quite cold. With a bit of effort, as some of the snow was melting from the residual warmth of the tape, I was able to pull out enough of the tape to measure a second run, this time putting the tape down on the core and making the note book entries. On the third attempt the tape was frozen as ice had now formed inside where the previously slightly melted snow had re-frozen completely and seized the tape. After a few minutes of trying brute force, the tape was abandoned as was the attempt to measure up core.

By now my gloves had been off for about 15 minutes and my fingers were feeling very cold. In an attempt to salvage some information, I took a series of quick unscaled photos of the core, put my gloves on, and restacked the core boxes. It took several hours for my hands to warm up and I was lucky not to have suffered a more serious frostbite injury.

After the failed attempt at logging we set off in the minibus back to our accommodation. On the trip out, the road skirted the western side of a large lake, Lake Kopa, which was in part ringed by small cottage-like buildings, stretching over a couple of kilometres and going right to the frozen edge of the lake: each had a small enclosed area, often filled with trees. We were told these were the summer houses or "dachas" for the population and the trees were mainly apple, pear and cherry grown to provide additional food for wintertime. On the way back we asked if we could be driven through the area and we were intrigued when we did to see numerous huts out on the frozen lake along with many people. As neither Bill or I had seen ice fishing before, we asked to if we could go out and have a look.

Now I was still struggling to get feeling back into my cold hands, so I reluctantly went out onto the ice. We slipped and skidded our way out about 200m to a small group of men huddled behind some makeshift shelters of thin plywood. It was still a pleasant sunny day with only a light wind but, to me, it felt bitterly cold and I had my hands deeply thrust into my pockets; but now my feet were taking on a decidedly chilly feel too as the ice sucked the heat from them.

The fishermen, who all seemed to be wizened old men, had a common set of appearances: their faces were leather-like, they all wore what looked like thick military winter-padded clothing and big boots; they were constantly smoking, and the ubiquitous rip-top bottles of vodka were in plentiful supply. The fishing was through a small hole drilled by a hand auger through the ice which was about half a metre thick. The prize were small herring-size fish which were obviously abundant.

These men evidently spent all day out on the ice fishing and were really pleased to see us, offering the bottle of vodka. I declined due to the poor state of my hands, but Bill played the grateful guest and, after several swigs, had a bit of a go at trying to hold the fine fishing line but not catching any, and was offered a recently caught one. I am still not sure to this day whether or not the fish was meant to be eaten raw with bites taken by a first person and then handed to the next, who had their share, and so on, or whether it was the effect of a vodka devilry by the locals having a bit of fun with a "tourist". But Bill was truly caught in having to take a bite of raw uncleaned fish anyway. Much to my amazement, he did not gag or throw up, washed it down with another swig of vodka and made excuses that we had to go. As we walked back to the edge of the lake we noticed that the end of the road that ran down to it was basically the dump for all the cinders and ash from the dachas and possibly all the other rubbish, so it may have affected the quality of the lake water and hence the fish. Bill survived anyway.

The assay lab and core yard

One of our key objectives was to ensure the accuracy of the stated assays. As I had brought over a range of certified standards from Australia, we arranged to go out to the laboratory that prepared and analyzed most of the samples from the project. This facility was a fully functioning laboratory and analyzed samples from many projects and was built along the lines of academic research laboratories in Australian universities when I was a student. In fact, when we went to the facility, it could have been Adelaide University's Physical and Inorganic Chemistry building. The first thing that stuck me was how well maintained it was, how clean and how it exuded an air of a highly organized facility. We were taken to the director's office so we could meet and explain what we wanted done and to get a quote for the job's cost. We were stunned to be introduced to a very slightly built, small woman who was in her seventies at least and to her assistant director, another woman about 45 years old. The director, Zoya, was of German heritage, the assistant of White Russian, and they were very proud of their facility and expressed their willingness to help with our request. We were shown around and given details of how they handled, tracked, prepared, and analyzed samples, which were as good as, if not better than, procedures at Australian commercial laboratories. The analytical technique used by them was gravimetric fire assay with classical parting of silver and gold.

The lab was well equipped with good grinding and crushing equipment, balance rooms, furnaces, wet chemical areas, well organized chemical stores, facilities to make key components like big glass stills for water, and cupel and cupola making areas. All most impressive and run with Germanic efficiency.

In going around the laboratory we got an insight into another aspect of previous Soviet history and current conditions. The reason Zoya was at the laboratory was because, at the end of the Second World War, a very large number of Germans were exiled to Kazakhstan which was a frontier country then. In fact, about 30% of the population of Kazakhstan in the 1990's was of German descent. It was a similar situation with the assistant, where White Russians were another group which were exiled during Stalin's reign to the frontier. People of Russian descent made up about one third of the population of Kazakhstan, and the remaining third were mainly the original nomadic Kazakh peoples who were a mixture of Asian and Middle Eastern heritages. Interestingly, during the Soviet regime, education was universally available so there were few barriers to becoming trained or skilled in areas like analytical chemistry.

The other striking thing in the lab was that perched on most available well-lit warm places like window sills, or above central heaters on benches near windows, were a whole array of vegetable seedlings - tomatoes, capsicum/chili, egg plants, cucumbers et cetera - in various states of growth. These plants, we were told, were raised until quite mature in these areas until the risk of frost was over, then they would be planted out around the dacha, or any other place they could be grown, during the brief continental climate summer with its long hours of daylight. Often flowering and bearing fruit in as little as six weeks, these plants provided very important food for the long winter period. So, maintaining and nurturing them was a vital task for the workers. This was not only confined to the laboratory; in the main project office the same situation was present, there were potted vegetable seedlings everywhere.

Several days after handing over the samples we were given the results. On plotting up the assays against the recommended values we were very pleased to see that they all fell with the 95% confidence lines and were mostly in fact remarkably close (within 1%) of the recommended values for both gold and silver. After such a good outcome and prompt service we arranged to pay the laboratory. Payment was to be made in cash from the US dollars which we were carrying and, as we had to visit the core yard facilities, we decided to pay in person. On hearing of this, Zoya insisted we have lunch at the laboratory, which suited us fine as it was always difficult fitting work in around the lunch at the project office block.

So, we duly arrived at the lab and were shown into the directors' meeting room. Before us was a table laden with all manner of preserved meats and vegetables, blini (pancakes) and a small bottle of cherry brandy - but no vodka in rip-top bottles. We were welcomed warmly by Zoya and her assistant director. We handed the money over and it came tumbling out that this was the first income the lab had received for seven months. They had not even been paid their normal salary (they were a government facility) and were so pleased. Through our translator, we found out this was a major issue at the time as the transition to self-rule was not going well from a financial perspective. We settled down to lunch. Now it was Bill's day to play the host role and he thought this lunch thankfully would not degenerate into a drinking contest. In celebration of payment Zoya, who I doubt weighed 50kg, proposed a toast. She proceeded to reach to the side bench, picked up two large glass-stoppered one litre reagent bottles filled with clear liquids, and placed them on the table. She then picked up a 200 ml and four small 20 ml beakers. Into the 200ml beaker she poured exactly 100ml from one reagent bottle and 100ml from the other. By this time, we both realized what the score was: this was neat ethanol being diluted with distilled water, both products of the lab and its distillation gear.

Now for the toast: Zoya and the assistant stood up, as we did, and together down the hatch the full 20mml beakers in one go. It was an experience and certainly had no taste, but I only had one toast. Now Zoya, before we could sit down, had refilled the beakers and again proposed another toast. As I was sitting on my beaker with some pretense for not drinking, Bill was left with reciprocating toasts with the ladies. The ability of the two women to drink this neat ethanol cut with distilled water was legendary and unsurpassed in my experience. The litre bottles were emptying at a frightening speed. Time and time again the toasts were about the honesty of "westerners" and payments being honored in cash dollars. Within an hour Bill was nearly comatose and we had to help him into the minibus as he was having great difficulty walking. For me, the meal was great, with only a couple of ethanols, a couple of cherry brandies and great food. It appeared that all the drinking had no effect at all on Zoya or her assistant, who invited us to come again for lunch if we could. Unfortunately, this did not occur, and I often wonder, if we had made the time, whether the drinking pace would have been stepped up.

Spring is sprung

The Tuesday of our second week was the first of April, officially the beginning of spring in this part of the world. Looking out of my room window that morning you could be fooled into thinking it was still deep winter weather, as there was a light blanket of fresh snow and small snow flurries in the breezy conditions. A quick normal morning routine of a brief warm shower, a muesli bar and a cup of hot tea from the communal samovar, then off to the project office we went. At the office everyone seemed to be in a buoyant mood, most were looking forward to being able to get out to their dachas or small plots of land to prepare them for the growing season and the longer daylight hours. It was a normal day for us, ploughing through more documents, some meetings, lunch at the cafeteria and a nice walk back to the accommodation. By this time we had found a place where we could get a consistent hot meal in the evening and our routine was to drop our gear for the day off in our rooms and head out to the "restaurant" for a meal, return to the accommodation and continue photocopying documents.

All was well until then but, on return to my room, I noticed that it was feeling a little colder than normal. I checked the room radiator at the end of the room, and it was a little cooler than usual. But this had happened a couple of times previously and had returned to a normal warm temperature after a few hours. I put this down to just fluctuations in the "communal" steam supply system. At least we had heating, unlike our interpreter, who had none. After some photocopying under the blankets I went to bed. While travelling I usually slept in a light track suit and, if my feet felt cold, a pair of socks, which had been my attire for here. About one am I woke up feeling quite cold, finding the room now had chilled down considerably and the radiator not giving out any heat at all, in fact it was cold to the touch, as if cold water were flowing through it. Obviously, no one was going to deal with this at this time of night, so I piled on a jumper, another pair of socks and hopped back into bed and went back to sleep.

On waking early in the morning, the room now was quite cold, so a quick dash aiming to have a nice warm shower and to shake off the morning chill. Turn on hot water tap, usually wait 15 seconds to get an enjoyable flow of hot water: but after a minute still only ice-cold water today. Obviously, the shower hot water problem and the radiator one were related. So, a quick wet cloth wash was initiated, followed by a towel off and into some appropriate clothes to warm up. A check with Bill in his room confirmed that it was the same for him. After the obligatory muesli bar and at least a hot cup of tea from the communal samovar, it was off to the office and a warmer environment there. On mentioning the lack of heating and hot water in the accommodation to the staff, we were informed that for certain "public" areas at the beginning of spring the hot water supply was turned off. Unfortunately for us the accommodation was such a facility and so for the rest of our stay we had no heating or hot water in the rooms, so we reverted to cloth bathing and rugging up - with extra jumpers, pants and socks now to be the sleeping attire. Fortunately, the photocopier under the blankets did warm up the bed a bit so this was an extra bonus.

Visit to Burabay

On the second Sunday our hosts indicated that they would take us out to see some of the countryside, which included a visit to an area the previous regime had reserved for the elite of its organization, and which was now the Burabay National Park. During the communist rule the general population was excluded from this magnificent area of rolling wooded hills and numerous lakes which had been both the winter and summer playground for the party elite. The drive out was through what appeared to be limitless very flat steppe-like countryside where the leafless birch trees, which lined the roads were half buried under snow drifts. After a couple of hours of slow-paced driving because of the potholed and often ice-covered road, a series of small hills appeared on the horizon. Entering this area via a winding road, we were soon into a steadily and increasingly more rugged terrain. At the end of winter as it was, the area we visited was stunning, with many substantial stone-built dachas and buildings surrounding now frozen lakes which backed onto heavily wooded hills - their pine trees draped in a blanket of pristine white snow and ice. It was not hard to visualize the summer scene when all the trees were free of snow and the lush forest floor was covered with mushrooms, wild strawberries, and other berries; the forests bordering the blue-colored lakes; and with a large number of people frolicking in the long summer daylight.

Now that there had been a change in the "ownership" these areas were now accessible to all, you just needed to be able to get there which, for most people, was virtually impossible as the numbers of vehicles and the availability of fuel were still restricted. Therefore, there was absolutely no one else here, the silence was almost eerie but in keeping with the landscape. A fine picnic was had by us on the edge of a small lake (now an ice pond). It was memorable for the cold meats and breads and for being warmed by fine cherry brandy rather than vodka, for snowball throwing, freezing cold feet from walking on the ice and for good companionship. Out of earshot of others we were able to talk about many things that had previously been off limits.

As both Bill and I were not used to outdoor cold weather, it was decided we would head back to town early and there we would be taken to the market and spring fair. We were dropped off there and wandered through the mixture of fairground-type attractions and simple craft and food stalls. There were quite a few people there, mainly families with children clamouring to have a ride on the small ferris wheel and clustered around the food stalls which offered a paltry selection of root vegetables and brassicas. Around one stall we noticed a larger crowd and went over to investigate. Peering into it, Bill spotted some large apples which people were mainly buying, so he pushed his way in and managed to flash a one dollar note for two apples. With glee the stall holder snatched the note and passed over the apples, immediately replacing them with two more from a large straw-filled box next to the stall table.

This was the first fruit we had seen in two weeks; Bill passed one apple to me and immediately we sunk our teeth into them, both imagining a nice crisp juicy fruit. With horror on our faces we both looked at each other as the apples were almost inedible, soft and mushy with a pale brown flesh. But, despite this, we ate them with some effort. The next day we found out that fruit such as apples were kept from previous seasons in those straw-filled boxes and used for cooking, not for eating. They were now a rarity since the Soviet supply lines had been severed and virtually no foods were imported. It just reinforced the difficulty the Kazakhs were having transitioning from the Soviet system to being an independent nation.

Departure

After nearly three weeks we had collected as much information as we needed and conducted meetings with the project staff and local Oblast officials. The latter meetings had been very difficult as there seemed to be a large element of previous Soviet regime senior office holders still running the Oblast and there was a hint of bribery being important in progressing business matters. Our company had very clear views on not entertaining any bribery situation, in fact Bill had wide experience in dealing with this matter having work extensively in the Middle East and southern Asia: the position was clear from us that we would walk away from the deal if bribery had to be involved. This led to several meetings being aborted, but we persisted in trying to understand what the local "tax" conditions were likely to be. Mostly they were outrageous conditions based on either tonnage mined or ounces produced and not related to fundamental economics and always seemed to involve "middle" men. We were, in the end, not able to reach any agreement and so decided to leave.

Duly we packed our bags, mine containing several reams of photocopied reports, and headed to the airport. At the counter there were only two persons, who appeared to be a pilot and co-pilot, greeting us and about twenty other people, most of whom seemed to have little or no luggage. After some discussion with our interpreter it was revealed that we had to pay cash for the flight, in American dollars. It became apparent that the pilot, not the airline, had to pay for the fuel to fly back to Almaty, and it would only be loaded into the aircraft after the cash was received. As I had now been booked on a connecting flight out of Almaty later than night it was important that we got back there this day. After some haggling, Bill handed over two hundred dollars to the pilot who then disappeared.

About half an hour later we noticed another one of those small tractors pulling a small tank out the aircraft, again a trusty Yak 40. Presumably, fuel was pumped fuel into the plane as the tractor returned about 10 minutes later. The pilot reappeared and we were once more, as in our flight up to Kokshetau, herded out to the aircraft lugging our bags, which were now piled into the rear of the aircraft. Again, after a short delay the two crew came out from the terminal and entered the plane; one locked the rear door with the large key, and both climbed into their respective seats. The engines started, the plane wheeled around and did a quick taxi out onto the runway and we were airborne in less than a minute. It now was apparent why the take-offs were so rapid; they were saving fuel and hopefully we had enough to get to Karaganda. We did, and there the process was repeated, and we finally arrived in Almaty around midday. This gave me just enough time to have a quick walk around the city and buy a gift for my wife before heading out for a decent meal, the first in almost three weeks.

At the time Almaty was experiencing a surge of interest from international businesses, particularly the oil and gas industries and from mainly the USA. There was some lesser interest in minerals, but oil and gas representatives were thick on the ground trying to do deals with the government ministers, all of whom were based in Almaty. Of course, they had to stay somewhere while negotiating and some quite opulent hotels had opened and were doing a roaring trade with well patronized restaurants and bars. We thus booked into a restaurant, duly arrived at the required hour, had a cold beer and ordered from the mainly western style menu, but which had an American slant. A nice steak and salad with fresh fruit platter was ordered. The salad arrived first and was soon dispatched, followed by the steak and the fruit salad, all washed down with a couple more beers.

As I was flying out at midnight, we did not stay on and I wanted to have a quick shower and freshen up before my very first, First Class flight (it was the only seat available), something I was looking forward to. I had just got out of the shower when the first bout of nausea hit, and I just made the toilet in time to empty most of the evening's meal. Cleaning up and just getting ready to dress when the second wave hit, and the remaining portion of the meal was emptied. Cleaned up once again and managed to get dressed when the diarrhoea set in. This was one of the worst cases of food poisoning I ever had in all my travels and made the next 24 hours a misery of dry retching and girding my loins until I could find a toilet. Getting out to the airport and through customs was a bit of a blur, as was being the only passenger in First Class with several very attentive staff trying to feed me all sorts of food and drink with me in a fetal position wanting for it all to end. It did eventually and I duly arrived back to Perth with luggage in tow, now mostly recovered from the bout of food poisoning, but required by quarantine/customs to open my bags and have them go through the reams of paper copies in Russian. After some questioning, I guess I was not considered to be carrying anything illegal, so I could go home to a nice balmy Perth autumn day.

Post Script

After a week's effort in the office I and a team of helpers put all the diagrams, sections and reports together and had created a digital data base and hence model of the resource which allowed us to form a fairly good view on its potential. Negotiations continued in Kazakhstan for a couple of months and major advances were made on key economic issues. As a result, a selected group of the Kazakh technical team were invited out to Perth and spent 10 days inspecting the company's operating mines and processing plants. There were many dinners and numerous toasts and a memorable trip on the weekend out to Rottnest Island. Here the water, despite it being late winter, was too inviting for Anatoly who stripped off to his underpants and went for a swim. Now it was one of those nice sunny winter's days and the whiter than white skinned Anatoly just wanted to stay in the water. In the half hour or so he was frolicking in the shallow water he managed to change to a pale pink and next day was well and truly sunburnt but seemed to be happy about it.

Unfortunately, despite assurances that we had exclusivity after increasingly more investment of time and effort and paying for various services and fees, the company was not able to make any progress in doing a deal. When we found out that there were other northern hemisphere mining companies competing in a "Dutch Auction" type situation with the consent of the government, we withdrew from the project and region.

And, for the non-geological readers, just to indicate that mineral exploration and evaluation are science-based....

The Vasilkovskoye intrusion-related gold deposit is located 17km north of the city of Kokshetau (Kokchetav) in Akmola Oblast, northern Kazakhstan. The deposit was discovered in 1963, and a pilot open pit mining project undertaken from 1980 to 1986.

The deposit is characterized by concentric metasomatic, mineralogical, and geochemical zoning. The mineralogical zoning is expressed by distinct paragenetic assemblages and characteristic minerals. The gold grade is irregular and varies from 1.5 to 3.6 g/t (cut-off 0.8 g/t). Native gold is fine grained (up to 0.12 mm) and associated with pyrite-arsenopyrite-quartz and bismuthinite-pyrite-arsenopyrite-quartz assemblages.

Gold mineralisation is spatially associated with a stockwork of hydrothermal quartz and quartz-arsenopyrite veins and veinlets that forms a zone that flattens towards the surface, and steepens with depth, dipping SW, persisting to 1000 m below the surface, before pinching out.

CHINA AND SOUTH EAST ASIA

IT'S ONLY GOLD

PAUL ASKINS
THE PHILIPPINES 1975

Paul Askins is a retired geologist currently living in Perth, where he has been for nearly 40 years, although he was born and partly educated in Sydney. After graduating with a BSc in 1965 he has worked in mining geology and engineering geology, but mainly mineral exploration, in every Australian State, and with overseas stints in The Netherlands, The Philippines and Papua. He has also completed a BSc Honours at Adelaide University, and an MSc in Townsville.

Under the regime of GREAT LEADER WHITLAM and his Cabinet circus of C names: Cameron, Crean, Connor, Cairns, Cavanagh, and Cass, especial mention should be made of Reginald 'Rex' Connor - Minister for Minerals and Energy, who had a thought bubble in his Marxist-segmented brain and so decreed he wanted to nationalize the Australian mining industry.

This, quite obviously to everyone, except THOSE IN COMMAND, resulted in an exodus of Australian investment in mining and exploration and a transfer of activities to overseas alternatives. And retrenchments to many.

I was lucky enough to get a small stint in The Philippines, at the lovely age of 32, working for Combari, a subsidiary of Comalco, and our brief was to do things to support Comalco's bauxite and alumina and aluminium business, such as evaluating that country's coal deposits as an energy source for an alumina refinery. Given the quality and disrupted nature of Philippine coals, it did not take long to write them off and so we drifted into looking at more interesting things like alluvial and epithermal gold. So off to the great Filipino south of Surigao del Norte where it had a permit to explore an area on the coast not far from

Surigao City. We had a small apartment rented for a while in that city and I seemed to spend my nights virtuously trying to write up my thesis for the part-time MSc I was doing at James Cook University, whilst the Filipino geologists and staff were out enjoying the night life. This was/is a Christian part of Surigao, well away from the southern Moslem parts where there was a bit of insurgency, so I could assume the night life was safe in terms of terrorist acts but I could not vouch for the safety of the other activities.

At the prospect there was gold in every dish in all the small creeks draining onto a large alluvial plain, so we wanted to dig some test pits to get decent samples, but the local farmer did not want us to disturb his bananas and his coconuts. Well ... they were the days of Ferdinand Marcos and martial law and we were legally entitled to explore, so we complained to the authorities. In short time I was despatched in a pump boat to the prospect with a humourless member of the Philippine constabulary sitting *behind* me with his fully loaded automatic Armalite-type rifle, with which he was fiddling constantly. I tried to be very friendly lest he be tempted to use me for target practice.

We found the owner and he had an enforced conversation with the rifle man in a dialect I did not understand, but the result was clear. The farmer was terrified and our work went ahead. Luckily for his bananas, and unlucky for us, our sample results were not encouraging and we moved on.

Down the road from Surigao City was a place called San Francisco, which in some respects had only its name in common with the US one. It was a tiny town, where I am sure everyone toted a gun when the constabulary were not around so, apart from its wet tropical climate, it was much like the old lawless Wild West.

There was a lot of illegal artisanal gold production: accompanied by two Filipino geologists I went into a valley to have a look. One prospector showed me his alluvial gold. Stunningly beautiful cubic gold crystals on wires of gold attested to an epithermal vein source only metres away. As a once in a lifetime opportunity I wanted to buy some of this fabulous stuff as specimens, but one of the Filipino geologists quietly tapped me on the shoulder, pulled me aside, and explained bluntly that I stood out like the proverbials amongst those locals: I was clearly under curious and macabre scrutiny and, if I bought any gold, everyone would know I had money or gold or both, and all of us may never make it out of the valley alive. I can tell you the hairs on my back involuntarily rose. I quickly thanked the prospector for showing us his wonderful gold, and pretended to be a poor eccentric, and we left the valley without delay. That WAS a place geologically HIGHLY prospective, but alas not a recommended place to go. To this day I do not know whether exploration and company mining has ever taken place there, but I am curious to find out.

On another trip looking at potential gold areas we were shown a prospect in Luzon, north-east of Manila. The owner proudly showed off his alluvial gold concentrates and explained he did not get a full price for his gold from the roving gold buyer because it was contaminated with other heavy silvery stuff. Well, that silvery stuff was platinum, so the gold buyer was ripping off the poor guy by getting the gold at a discount *and* the platinum for nothing. I didn't have the courage to upset the owner and let him know, so at the time there continued to be a very happy buyer and an ignorant but reasonably happy seller. I have subsequently also wondered whatever happened in that prospective region.

Well the Whitlam era ended. At THE DISMISSAL we all learned instantly over the expat grape vine *before* the evening news in Australia! The local Embassy staff were horrified – we geologists were delighted.

It was not long before I got shipped out to Tasmania, just for fun, straight after seven years in the tropics. I remember when we arrived in Burnie it was a beautiful day of 18 degrees, all the locals sweating in summer gear, and us in leather great-coats because it was so cold. And there began another story.

STREAMING GREED

ANTHONY WILLIAMSON
THE PHILIPPINES
MID 80s – EARLY 90s

Anthony Williamson has over 40 years' experience as a geologist in Australia, Asia and PNG, in both the government and private sectors. Taxi driving wasn't his most challenging or rewarding experience, but being the Director of Mining for the PNG Government was. Helping to grow their multi-billion dollar industry by developing a globally competitive fiscal regime, by maintaining investor confidence and promoting the country, was stressful but satisfying. He has been credited with pivotal roles in numerous successful private ventures in PNG and South East Asia, none of which returned him a windfall profit or even early retirement. In summary, honest and equitable dealings with beneficial outcomes are important to Anthony, and his career and life reflect these values.

Introduction

This little yarn is set in the turbulent mid-80's to early 90's post-Marcos, near-anarchy period of the Philippines under President Aquino where the newly "liberated" people had to reinforce old allegiances, forge new alliances without fully abandoning cronyism, and create their new identity. Some people couldn't make the change and were viewed as agitators against the "new" Philippines. It was a time when the *Americano* was no longer welcome, the NPA (New People's Army) communist rebels slowly dissolved into less effective cells of bandits, and many members of the AFP (Armed Forces of the Philippines) were stuck in a time warp mounting coup after coup. In contrast, the country declared itself open for business seeking the capital to explore many of the closed copper-gold mines and prospects. However, the Philippine situation was not new to the world: it had endemic corruption at all levels complemented by a complete breakdown of military, judicial and law enforcement to the extent that it couldn't have been much different from the 1880's gold rush days of almost anywhere in the world. It was a time when weapons ruled, and the powder was greed.

An Indonesian Prelude

Jakarta, Indonesia early 1986 was the start to my Philippine adventure. Indonesia was welcoming foreign investment and continually amending their mineral licensing system to meet the security of tenure needed by listed companies and their investors. Geologist Peter Macnab of Brisbane-based explorer Pacific Arc Exploration NL had recently arrived in Indonesia and worked with entrepreneur John Benson to package a copper-gold tenement portfolio. Peter was fresh from the Lihir discovery in PNG and, realising there were many geological opportunities in Indonesia, applied for several Generation IV Contracts of Work (CoW's). Seven of those CoW's were granted in December 1986 and, together with a Joint Venture over a Mining Lease in Lampung, made up the final Indonesian portfolio. Management wasn't Peter's strong point and he was the first to admit it, so he recruited me from the PNG Geological Survey to set up and manage the upcoming exploration programs. I enjoyed Jakarta, the people, and the challenge of both the job and learning the language. Unfortunately, I didn't last long as resident manager after I seriously cramped the financial benefits flowing into the local partners' bank accounts from our monthly cash calls.

It was a frustrating time because most of our domestic companies that were required to sponsor internal travel papers and work permits were not yet in place, so our exploration could only bumble along on a limited scale to stay under the radar of the Indonesian authorities. This was not easy in 1986 as geologist Mark Gillam continually reminded me. It was a busy time in Jakarta and our competitors were getting CoW's granted and were also recruiting staff. Many of the Jakarta "old hands" were there or soon to return. Those exciting times are a contrast to the absence of the *bule* (white foreigner) in Jakarta these days. With more of our CoW's being granted, the time was right to pursue some serious financial backing, so John & Peter invited several potential Australian investors, including Alan Bond and Kerry Packer, to come to Jakarta. Packer displayed the most interest and, in mid-'86, he set about acquiring all Pacific Arc's tenements in PNG, Indonesia, and the as yet non-existent Philippine properties.

Dave Shatwell, and later Graeme Fleming, did a lot of the early mapping on our Indonesian epithermal and alluvial gold property in Lampung and, later, Mark Bouffler and Fiona Fenton were hired to do the exploration drilling on the Bukit Jambi epithermal property. The numbers on the alluvial gold resource potential were looking interesting so I sought advice on alluvial gold mining equipment from Philippine-based metallurgist/chemical engineer Robert Hisshion (Bob). Bob suggested I consider purchasing trommels, jigs and tables which led us on a quick trip to Ipoh in Malaysia. Before I committed to purchase anything in Ipoh, he suggested we continue on to the Philippines to compare manufacturing costs. Costs were indeed cheaper, but the expertise wasn't as good as Malaysia. Anyway, Bob and I got on quite well and I was impressed with his lifestyle in the Philippines. While I was on my way back from the Philippines, I rang Jakarta and was told by Peter it was probably best that I don't try to return to Indonesia as one of our partners would make sure I wasn't allowed back in. Apparently, I had stemmed the money flow to one of our local partners a little too well. I was a bit miffed that the investors I represented would allow the local partner to get away with diverting funds, but I was gently informed that I had failed to see the bigger picture. I thought I took my first brush with corruption fairly well and, being a geologist, the potential of the Philippines was obvious to me, so I nicked back to Jakarta anyway, grabbed my gear, and flew back out the next day to set up shop in the Philippines.

Wide Eyed In Manila

Metro Manila was a congested, polluted, hot and humid amalgamation of sixteen individual cities that in late 1986 was home to a bit over seven million people. It felt more than that after having lived the previous five years in Port Moresby with its population of about 160,000. Manila superficially presented as Hollywood action movies, sodas, Jollibee burgers, skyscrapers, malls, Madonna and Michael Jackson in an unusual mix with Spanish influences, such as horse drawn carriages, the names of people and places, old churches, Catholic morals and the other Madonna. This juxtaposition was evident in the large number of dual carriage driveway, short time motels, with their music videos and vibrating waterbeds, that catered for the randy Romeos and the perpetually virgin female flock, with all morality supposedly decreed by the appropriately named Cardinal Sin. Fortunately for me, Bob had a thriving metal refining business and, more importantly, he knew the people and the place, so I stayed with him for the first few weeks at his house in Quezon City. I must say though, we were hardly ever home as there were many rounds of breakfast meetings, lunches, dinners, San Miguel beers and bars involved in introducing me to many of his Filipino friends and colleagues. After a few weeks of discussions with some of his selected friends, Bob and I set out early one morning to Makati law firm Guerrero & Torres to incorporate a company. I had yet to get my Alien Certificate of Registration (ACR) and Immigrant Certificate of Residence (ICR) so couldn't be a founding director. But, unlike the drawn-out affair in Jakarta, it all went so smoothly with attorney Natividad Kwan, that by mid-afternoon we were having a cold one in a Quezon City poolside bar celebrating the soon to be approved (October 1986) Banahaw Mining & Development Corp. Progress!

The drive from the CBD of Makati to Quezon City was an endless stop-start along EDSA (the main ring road of Metro Manila formally known as Epifanio de los Santos Ave) and, although a journey of only a few kilometres, it was the endless meandering of trucks, buses, jeepneys and cars that slowed any trip down. Fortunately, I had a driver, Ray Pejo, who'd look in the rear-view mirror occasionally and offer a sympathetic "very traffic, sir". Comforting to hear, but it still felt like a long way home with all the traffic lights being a haven for street vendors roaming between the vehicles selling the Manila Bulletin, strings of fragrant sampaguita, candies, a peso for a "stick n stork, sir?" (eucalyptus candy and a menthol cigarette) and free light! Amazing too, that along the same stretch of road after midnight, I could travel home at 100km/hr, and hardly see a soul or car.

The initial funding for the Philippine venture came from myself, with a pledge from Pacific Arc Exploration NL to repay my outgoings in return for a commitment from them to give me a 15% free carried interest of the proposed venture to do with as I saw fit. As an incentive, I granted 7% to be divided between our Filipino partners Mr Pio Caccam, Mr Douglas Villanueva & Mr Balgamel B Domingo (Lito). Of the foreign equity entitlement, I retained 5% and I gave Bob 3%. Subsequently, Muswellbrook Energy & Minerals (MEM) took over of all Pacific Arc's tenement holdings in Indonesia, Papua New Guinea and the Philippines. The final shareholding in Banahaw was 60% Filipino and 32% MEM or, using the grandfather rule, MEM had rights to 56% of the total Banahaw stock.

Unlike Jakarta, there didn't seem to be many foreign geologists resident in the Philippines in the mid to late 1980's, even though there was a fair bit of exploration activity. I knew they were there somewhere, so assumed that most were fly in fly out. Paul McKibben, E Max Baker, and Mike Spadafora were some of the resident geologists that I frequently had a beer with. Maybe I just moved in different circles, probably because I was often in the company of Bob and/or Filipinos.

From the time of Banahaw's incorporation, the recruitment, exploration, and development all moved very quickly. Well respected and seasoned Filipino geologist Lito was hired to identify areas that were likely to yield a good prospect. He knew of many, and so began a series of rapid field trips that extended from Appari at the northern tip of Luzon to Zamboanga in the southernmost part of the Philippines bordering Malaysia. I continued in this acquisition phase for the next three years, not just for Banahaw and Greenwich Resources in the Philippines, but also in Taiwan, Japan, Thailand and Malaysia. In 1987, I acquired and managed several properties in Sabah, Sarawak, Peninsular Malaysia and Thailand that were soon after vended to Pacific Arc Exploration NL in exchange for three million fully paid shares in the company, but that's another story.

Travel in the Philippines probably hasn't changed much since the late 80's. They had all the usual forms of commercial transportation; scheduled flights, buses "The Philippine Rabbit", and ferries, all of which either crashed or sank with sickening regularity. I was ticketed on Manila - Baguio flight PR206 on 26th June 1987, but missed the flight as I was delayed with legal meetings in Makati. I decided to drive to Baguio that afternoon to meet our visiting Thai geologist Khun Metha Amornsirinukroh who was waiting for me. When I arrived, he was sitting in the lobby of a small hotel in tears reading a newspaper about flight PR206 that had crashed and all 50 people on board had been killed. He had no idea I wasn't on the plane, and I had no idea it had crashed. Needless to say, after a few wai's and hugs, we walked up the road for some burgers and beers.

Metha always said I was a lucky man. A few weeks later in a typically dark Bangkok cocktail lounge, sipping Mae Khong whisky, I made light of his "lucky" comment because he was Buddhist and luck shouldn't come into it. He didn't want to hear any of the nonsense I was talking and, as if to prove it, his cigarette lighter that we didn't notice had slid into his ashtray set off a metre high shooting display like a firework. We both wasted our drinks trying to put out the flaming ashtray, serviettes and burning peanuts. Anyway, I have since retracted my comments to him after I missed a chopper run in Aceh and the chopper crashed and killed most on board. The "Lucky" confirmation came again when I was caught under my arms by the co-pilot as I slid out of a chopper circling about 400m above the ground while I was taking photos in Mongolia. That one frequently wakes me up at night. I should have listened to Doug Kirwin who told me the night before over a Chinggis Khan voddie at the Matisse nightclub in Ulaanbaatar, that, if I was "going to hang out of the chopper, don't forget to put a safety harness on!" A slow learner I may be, but also lucky.

The Philippines has many transport charter companies providing fixed wing and rotor aircraft, cars, boats and, for local transport, jeepneys, motorcycles, tricycles, horse 'n buggy, outriggers, canoes and the rarer "skates" and "skylab". Skates are car-sized, wheeled, wooden, raft-like platforms that run on disused railway lines. They have central seating for about 20 people and are powered by a person who scoots the contraption along with one foot, just like a scooter. Arguments invariably occur when two skates meet travelling in opposite directions.

The Mindanao Skylab is a motorcycle with a plank centred across the pillion seat. Passengers sit about 2-3 wide on the plank each side of the motorcycle. It's interesting rounding a corner to see the eyes and faces of the passengers on one of these coming at you. Especially when they are on the wrong side of the road, which they usually are on corners.

Mindanao Focus

In the early days, the eastern part of Mindanao island in the southern Philippines rapidly became Banahaw's focus. Having spent five years mapping in PNG, I was comfortable working in the partly forested, humid, near sea level climate of Mindanao. The local people, or Manobos, were generally helpful and we got them involved whenever we could. Unfortunately, rampant logging throughout the Philippines meant that very little remained of its tropical rainforest and a lot of the rivers were clogged with landslide debris so I didn't trust regional drainage sampling. I reasoned that the obvious way to a discovery was to visit areas which were currently being mined, or had been mined by illegal miners. The downside was that these mining areas were unsafe, and not small affairs as one may think: Diwalwal had about 100,000 miners, all working in unregulated, lawless and unhygienic conditions. These sites had their own economies, food stores, bars and, of course discos, all reliant on gold. One drizzly morning I was walking along the Co-O ridgeline with some field assistants and pointed down at a fellow lying on the edge of the road outside one of these bamboo discos. "Late night, still sleeping" I asked pointing at him, "no sir, he is still dead" was the reply. Hmm, let's keep moving.

Many of these mining areas were within mineral tenements held by substantial Filipino exploration and/or mining companies who tended to overestimate themselves and the value of their prospects. A no-win competition that I never wanted to get into. However, Co-O was one mining area that was held by a small local syndicate named Central Mindanao Mining and Development Corporation Inc. The area met our criteria, and Lito visited the owners at their houses in the nearby barrio and they indicated that they may be interested in doing a deal. Lito and I met up in Surigao a couple of hundred kilometres north and it was off to visit the area. I wandered around for a while amongst old pits and scratchings in altered rocks wearing my standard geo gear of the day; non hi-viz shirt, stubbies (shorts), and running shoes. It was clear that the main mining activity at Co-O was on three separate gold veins. I grabbed a few likely-looking surface samples and then went underground. While I was underground sampling one of the main veins, a large piece of pointed quartz broke off the roof and skinned down my shin and stuck in the mud. Blood everywhere and, to my surprise, hearty cheers from the lease holders who accompanied me underground. Apparently, a human "blooding" the area was far better than any chicken or goat blood sacrifice. They were convinced, there and then, that we should do business together, so we all headed off the hill and Lito got down to negotiating with them in the barrio over a couple of bottles of Fundador Rum.

Later, in 1987, CMMCI entered into a Mines Operating Agreement with Banahaw whereby the latter agreed to act as Mine Operator for the exploration, development, and eventual commercial operation of CMMCI's 18 mining claims located in Agusan del Sur. Yay Lito!

Meanwhile, I don't like rum, so I drove south to Davao City as fast as I could to call in an order for a drill rig. I didn't even wait for the assays; why bother, the miners were digging and recovering gold right there. I may have looked like a muddy, blooded pagan convert to our driver but, more so, I was a geologist and an excited one at that. I knew I had just been on a future mine; three gold lodes in altered and mineralised agglomerate that were running around 6 - 8g/t gold, and each about 600m long, which I believed would coalesce at depth in the underlying andesite. No formal ore resource investigation needed, just a few drill holes and we were off to the races!

Exploration

Within a year of starting Banahaw the staff level grew to over 100. Construction was overseen by director Pio Caccam. Exploration at Co-O was in full swing and there were other leases in the portfolio demanding exploration attention. At one stage there were a little over 40 geologists, all Filipino except Australians Graeme Fleming, who moved across from Indonesia, and, for a short time, Chris Green. Even though labour and the Peso were cheap, it was difficult to keep a lid on staff levels. There were staff for every conceivable duty; drivers, cooks, maids, cleaners, gardeners, clerks, clerks checking clerks, and more clerks. It took a couple of months, but the space for our head office turned out to be relatively easy to select: Manila City was too busy, Makati City too expensive, Quezon City was too old and too far north, and Pasay City had no class and was too far south. But the developing area of Pasig City, just east of it all, fitted the bill. A real estate broker had part of the 5th floor and all of the 4th floor available in the Hong Kong Bank building, so we agreed to meet so I could check it out. The empty 4th floor looked great and the price was right. There's not much to see when there are no fittings or fixtures, but the view was great to the southeast over Laguna de Bay. I'd seen enough and suggested we go down to look at the outside of the building and parking availability. On the way down, the lift doors opened on the 3rd Floor and the sign on the wall in front of me read RGC Exploration. It was all the confirmation that I needed, we were in the Right Place, so off we went back to the broker's office to sign a lease agreement.

Exploration amongst the 25,000 miners and operators at Co-O wasn't an easy exercise. Fortunately, I didn't have to do it, but did manage to visit regularly. Graeme was invaluable and managed all the initial mapping, drilling and sampling at Co-O. My visits were two-way, I was able to walk around and be updated on the program, but also to listen and discuss what problems he faced and the best way forward. Most Filipinos speak good English, but working on foreign soil can often lead to a feeling of isolation, so it's good to talk with a fellow countryman, especially one that shares similar values. Food is often a hot topic in many field camps, but fortunately not in this case because Graeme was not too fussy and ate what the locals ate. But he did have his goodies, *e.g.* chocolate was one of his special treats that got him through the normal roster of 6 to 10 weeks of work, followed by a welcome 10-14 days off. On one occasion, Graeme radioed me and said he needed to get off site for a short overnight break and would meet me in Davao City.

I encouraged these short field breaks and they were all about recharging the spirit, and contributing to the local economy, shopping for goodies to take back to site, beer, dinner, a good hotel room with shower, and lots of talk. Returning to site the next day was always a bit depressing and anxiety abounds. Did we get everything, etc, etc? We had a late breakfast and then off we went, with his cargo in the vehicle that set out ahead of us. After about an hour drive up the road at the turnoff to the exploration campsite he asked the people standing around where his cargo was. "The fackers took it" was the reply. "What fuckers?" asked Graeme. The bloke looked at him and said "the usual fackers, sir". Graeme wasn't having a bar of any of these fuckers taking his goodies and by this stage stood there all tensed up and I could see he was wondering what to do next. He slowly turned around to look at me for help, but I'm of no use because I'm just standing there laughing my head off. The Filipinos with their unique American accent and ability to interchange "f" and "p" were actually trying to tell him the packers had taken the gear to site. All was well, he turned to go up there but, as I was about to drive off north towards Butuan, to look at another property on offer, I leaned out of the car and said that he was lucky it wasn't the "howlers" that took his gear because I hadn't met them yet. He rolled his eyes, he'd been working there a while, he knew the haulers were no problem.

Exploration management in Banahaw changed after the MEM acquisition of Pacific Arc's interest in the company. Lito remained at the helm and I kept my Board position, was retained as a consultant, and handed over the technical manager role to Jakarta-based John Carlisle, who was supported by Professor Andrew Mitchell. Ex Pacific Arc Exploration company secretary Mr Phil Higgins took control of the finances and Mining Engineer Alistair Grant, the MEM managing director, flew in from Jakarta to oversee everything. All geologists remained on staff and Banahaw continued its exploration through and beyond the construction phase.

In addition, as part of the working arrangement with the leaseholders, Banahaw constructed a tailings treatment facility and were busy buying and treating tailings. Many of these early technical employees who started at the tailings treatment plant, and later the mine processing plant, were former Benguet Corp employees who had been laid off because of their involvement in scams with Benguet's own tailings treatment operations. Benguet Corp was one of the most reputable and largest mining companies in the Philippines. This meant that, due to their misdeeds, these professionals were virtually unemployable. Their situation appealed to Banahaw because it desperately needed qualified staff but wasn't willing to get into a salaries war to obtain them. So, with a simple promise that they would behave, they were taken on by Banahaw. I'm glad it wasn't my area of responsibility because, not surprisingly, we were never able to get an accurate monthly reconciliation of metals. Many of these same staff didn't stay too long and left Banahaw to become names directly involved with the infamous Indonesian Bre-X affair.

Co-O Early Construction Phase

Establishing the Co-O operations wasn't a routine matter of construction. No, this was the Philippines, let's make it difficult. Construction was simplified in a heartbeat by signing a 550tpd processing plant turnkey agreement with BHP Engineering. Commissioning of the mill took place in 1989. The difficult part was achieving what is now known as a social licence to operate from the stakeholders; namely, the existing miners, the lease holders, the rebels and the military. For the former, we agreed to let them mine down to 25m, which at Co-O was just about the depth limit of their mining capability. To ensure compliance, we drove an adit [a horizontal mining tunnel] the length of the main vein at that 25m depth. Later on, armed guards were positioned along the tunnel and escorted out the miners that occasionally fell through the roof. The lease holders remained supportive as long as they could continue to buy tailings from the miners. To ensure the deal to secure tenure over Co-O, we ensured they received a substantial share of the profits from the tailings treatment operations. That took care of the miners, transporters, gold processors and, most importantly, the leaseholders.

Next, we needed to placate the military and the rebels. At that stage of our development we were without substantial security. This led the military to accuse us of collaborating with the rebels, and they insisted we get proper security. Yes, the AFP controlled the approval and sale of weapons. The AFP insisted the side arms carried by our security personnel were insufficient and we should have shotguns and more guards. We duly followed instructions and then word came from the rebels that we had become a pawn of the military and, in a twisted logic, the rebels relayed to us that if we were truly running an independent operation, we surely needed more guns and people. On it went, backwards and forwards, always escalating, until we had about 100 guards. Of course, all guards had armalites, side-arms, and most carried grenades.

Hmm, over the top? Well, the rebels thought so too and, in February 1989, they mounted an attack on the fledgling operation which was still under construction. I had left the compound the day before for Davao and, on arrival in Manila, I received word from Alistair that the plant had been attacked and five people had lost their lives due to gunfire. A sad day and we made the news. But it didn't stop there. The AFP insisted we beef up security to 365 guards, all with M-16's, most with grenade launchers and, later, an M2 tripod-mounted machine gun was positioned at the plant entrance gate. The military also insisted on positioning a 105mm Howitzer on the mine access road. With an 11km range this was not a self-defence item and this proved to be the case when they shelled the locals at the Anoling mining area, about 8km to the north of Co-O. As a consequence of the shelling, the directors of Banahaw, which included myself, were placed on an NPA death list. Not a good list to be on in the late 80's in Mindanao. Nevertheless, Banahaw carried on construction and site security became more vigilant. Though this didn't offer much peace of mind to the regional exploration teams.

Banahaw was still looking for acquisitions in Mindanao, and several months later I went with Professor Mitchell to look at an old mining area. We drove to Davao del Norte with a few Banahaw field assistants and trekked a short distance to a village so we could hire some locals to guide us to the old workings. After a bit more walking, we arrived at the partially overgrown old mining area, got our gear organised, and went in an old tunnel to do some sampling. The tunnels are only about 1.5m high by 1m wide, sometimes are about 100m long and can meander around following the gold veins. Sampling in cramped, muddy, hot and damp holes in the ground does nothing for appearances. In this case, even though the veins weren't very big and they didn't extend very far, it was enough to get us filthy.

When we had collected enough samples we turned around and headed out of there. About 20m in front of me I could see the Filipinos exiting the mouth of the tunnel and wondered why they were still walking hunched over and what looked like their hands protecting their heads. When I came out, I saw that we were surrounded by a seven person NPA cell who were circled around the area immediately in front of the mouth of the tunnel. We were ordered to kneel down and to put our hands on our heads. The bad turn of events just got worse when we were informed that, in revenge for the deaths of their colleagues in the Anoling shelling incident, they were looking for the directors of Banahaw, in particular, one Australian named Anthony Williamson. Ooh shit! Andrew and I immediately declared we weren't directors but English consultants, and pointed to the hole we came out of, after all, look how muddy we were, real directors don't do this! The rebels seemed to be going along with this notion but the fact that they were taking our Filipino field assistants away one by one to a hut for "interviews" had me worried. The first fieldie got taken away but turned around and said "Tony, don't move, you stay here with Andrew". I was still on my knees next to Andrew with the business end of a Garand rifle at my temple. I don't know where he thought I might go but then I got it, by calling me Tony I wasn't an Anthony.

The rebel spokesman asked us trifling questions; where we came from, what's in our bags, what we were doing here and so on. Then, the rebel holding the gun on me spoke in a language I couldn't understand, but it was translated by one of the most stunning women I have ever seen. Standing a few meters in front and above me, feet shoulder-width apart, nursing her armalite, dressed in camo pants, khaki T-shirt and crossed bandoliers emphasising her generous upper figure, this beautiful Asian-sized Amazon softly said: "he asked do you I believe in Communism?". Well fuck me, what was I going to say? What would one say? Well you probably guessed wrong, this kneeling idiot with a muzzle resting on his right temple, turned his head slowly clockwise to look directly up the gun barrel at the rebel who had his finger on the trigger, and without blinking an eye, replied that: "No, I didn't." I was shattered that I had said that, WTF was I thinking? It took me a couple of milliseconds to simultaneously realise that I read of many people who had died in these situations, that the questions were possibly over, and that the rebel could instantly react to my comment. Quickly, I added "your country does not allow people to vote for communism, but my country, England, does, so what's missing here is the people's freedom of choice". It was the best I could do on the spot. No gunshot, no words from them, I finally broke eye contact with him and looked down and forward again. No sound, then she asked me a couple more questions along the same lines, giving me time to explain myself, and then she was drawn into a discussion with the rifleman.

Meanwhile, with the guns still on us, Andrew called some of the rebels over and was wisely handing out our smokes, and anything else as fast as he could. After all, weren't we all just friends in the jungle, pommies and commies? The last of the interviewed Filipinos came back to join us at the tunnel entrance. We waited, surrounded by these rebels and I thought: "what now"? Finally, he moved the gun away from me and then demanded "*tinapay*" or bread for his team. I thought, there goes our lunch, and maybe, just maybe, we're going to get out of this. Having donated as much as they wanted, they told us to say nothing, move out quickly, and remember their cause. Exit for us, back to the car and off we drove, all of us a bit giddy as the moment sunk in. This was the third time I'd had a gun pointed at my temple, close range, and wondered if it would be the last.

After construction, production started facing the same problems as most new mines; not enough ore, grade too low, metallurgical problems, purchasing and logistics issues, and staff problems - so it was not a place an exploration geologist would naturally gravitate to. However, the site was there, and it did have food and accommodation for the passing Banahaw exploration geo.

Well before Co-O came into production, Manila was my home base, and it felt like home. I was in my early 30's and had been away from Australia for about eight years and I now thought Leaders were Presidents and VP's, not Prime Ministers or Managing Directors. My house was in a gated estate, in fact, not too far away from Max Baker. It was also used as an expatriate staff house and came with a cook, maids, and gardener. Like all developing countries, Manila had its problems with the basic utilities resulting in frequent power brownouts, insufficient water supply, and intermittent phone connections. This was pre-internet days and we used a radio to contact site. Government services varied, though there was a good emphasis on education and health, but most agencies required bribes, especially those of law enforcement. The reality of life for most Filipinos was no luxury, and it was there to be seen, in town and rural areas. I faced the conundrum that both living and life were cheap.

There were six coup attempts during 1986-7, mainly in Manila but also in Mindanao, and like many other Manilenos I had become used to the early morning wake-up call of tank shells pounding the walls of nearby Camp Aguinaldo or Fort Bonifacio, or from the explosions following aerial strafing by T-28 Trojan planes or rockets from helicopter gunships. Life went on for most, but sometimes these skirmishes came a bit too close. While walking back from yet another Mindanao prospect through scattered tall stands of wild elephant grass with geologist Mike Spadafora, he suddenly pushed me to the ground and I heard him say: "keep your head down, we're in crossfire". I hadn't heard any crack of gunfire, but I did hear whizzing sounds overhead. We lay there for a while until Mike, a Texan and an ex-Vietnam veteran, merely said: "that's it, c'mon, I gat us some rum back at the car". The proliferation of guns was extreme and I'm convinced that all males would have carried one if they could afford it. This of course is not true, but that's the feeling one got. Weapons were in view everywhere; all security guards around shops, in parking lots, and outside many houses wore weapons, ditto with the military and police on the street.

In stark contrast, Bob and I had a meeting in Sydney sometime in 1990. We had to report to Trevor Kennedy at Kerry Packer's Park St office. We flew into Sydney that morning from Manila, cabbed it to town, found the place and walked up to the security guard who was sitting on a stool in the narrow lobby reading a paper. He kept reading and we just stood there, fumbling around trying to find our ID or at least the letter stating the reasons why he wanted us to be there. Finally, the guard looked at us as if we were lost and asked what we wanted. "We're here to see Mr Trevor" I replied. He dropped his head to carry on reading and pointed behind him at the elevator and said "over there". This was not what we were used to at all. Anyway, we shuffled slowly to the elevator, frequently glancing back at the security guard to make sure he wasn't about to make any tricky moves and pull out his non-existent gun.

Not all my exploration was in steamy Mindanao, I often went to the high Cordilleras that formed the spine up the centre of Luzon. Clear blue skies, fresh air, pine and grass-covered rugged topography, cold fast flowing streams descending to substantial rivers. I recall, late one afternoon after a day of wandering around sampling with locals (Igorots) in Nueva Vizcaya, we were all tired and had gathered round to make a small fire and drink some Ginebra (local gin).

Just on dusk, while we were sitting and talking around the fire, a well fed, sandy coloured, long haired dog about the size of a Labrador came up and sat next to me and rested its head on my leg so I naturally petted it for a while and then it just got up and wandered off. It was a few more drinks later, and the sky a bit darker, when there was a "bumpff" sound and a waft of smoke and spray of embers from the firepit as dead Rover was thrown on the fire right in front of me. Very timely, because I had been thinking the few biscuits that I had brought with me wouldn't be enough to get me through the night, and was wondering about dinner. Dog is a Filipino specialty but, no thanks, not Rover, especially just after patting it. I had another sip of gin and sat a bit longer. Rover by this stage was well singed and while he was being turned over, another guy leaned across and offered me a plate with narrow strips of some type of food on it. I had just taken a small slice and was told: "it's good, you'll like it, its pickled dogs' ear". Not only had Rover been rolled over such that his dead eye now looked up at me, I also had a slice of his ear at my lips in one hand, and my gin in the other, with everyone looking at me. Do I, or don't I, ponders Tony trying not to think of getting my own ear chewed off by the locals if I didn't eat it. In it went. I would recommend anyone else wishing to try this Igorot dish to at least allow the ear to soak for a couple of days to break down the cartilage a bit. After a few simulated munches, I stood up, wished all a good night, blamed the gin and retreated beyond the edge of the fire for a freezing night of intermittent sleep. Fortunately, I had a driver because the next day I was dog-tired all the way back to Manila.

Almost all of our activities still required funding from Australia because we were only getting a modest return from the gold treatment plant. Unbeknownst to all of us geologists busy collecting samples from the bush, compiling maps and diagrams for presentations to high net worth individuals and/or the likes of slick brokers in Cannacord, the 1987 market crash was looming. We were cushioned a bit from the events of Black Monday 1987 because MEM was bankrolling everything and it took a couple of years before the fan was hit. By 1990 Banahaw had almost a thousand employees, but was really struggling to make a quid. Even though I was long out of Banahaw management by that stage, it was a sad realisation that people I thought I knew well were fucking up. Ah well, nothing new here, Tone, move on.

The Sting In The Tale

By 1990 my usefulness to Banahaw was well and truly at an end. Also by 1990, the '87 stock market crash had finally taken its toll on many junior exploration companies. John Gaskell of Greenwich Resources told me that the funding was finally turned off to Pacific Phoenix (another company I had set up in The Philippines) and, similarly, Pacific Arc Exploration was also too low on funds to continue exploration on its Malaysian and Thai properties.

With no likelihood of employment, consulting, or business opportunities in the Philippines, a relocation out of Manila was needed. A simple flight back to Australia was not a reality due to visa compliance issues. It was back to Indonesia, back to Jaksel (south Jakarta), in the hope of gaining some income in Indonesia. Most people have been through the humbling process of seeking income, and it's even harder when you're an expatriate. Times were grim and there was no work in Indonesia either, my personal funds were dwindling, but fortunately I still had my 5% share of Banahaw. Within a few weeks of settling in to Jakarta, I got a phonecall from a person with an Australian accent, saying he was in Manila, and that he wanted to discuss the purchase of my share in Banahaw. It was my last resort, so we struck a sale/ purchase agreement. I had the share certificate with me, so I signed the transfer on the back as requested and couriered it to Manila, and waited for the funds from the sale. That was the last I heard of the matter, the funds never came, the Philippine phone number went unanswered, so I assumed the phone line was dead. I tried calling some of my friends but to no avail. In addition, Lorna, a colleague and former employee of mine in Pacific Phoenix Corporation, who held a significant Filipino percentage of Banahaw stock that was previously in the hands of MEM, but assigned to her when MEM pulled out, was also relieved of her holdings. It amazes me to this day that I walked a careful, watchful line, wary of locals wanting to rip me off but, in reality, it may have been a fellow countryman who did the deed. Anyway, I can feel proud to have been a significant part of the economic development of the eastern Mindanao region, and employment of hundreds of people through the establishment of the Co-O operation.

It's a bittersweet end to this tale for me. Because the Co-O mine is still operating, I can take great satisfaction in my initial geological and economic judgement, but I am disappointed that subsequent due diligence in the period well before the establishment of what is now ASX-listed Medusa Mining Limited, did not pick up any anomalies in Anthony Williamson's share certificate transfer, or what happened to Lorna Pangco's Filipino holdings in Banahaw. I'd be curious to know what entity, or person, was the transferee!

HUNDRED YEAR OLD EGGS
AND WARM COKE

LYNDA FREWER
CHINA 1989

Lynda Frewer graduated in 1985 from Curtin University in Western Australia. Before and during her studies she worked for a junior diamond exploration company in South Australia, and in the Western Australian Kimberley and Goldfields regions. After graduation, she worked for Freeport of Australia and then Normandy Exploration. In 1988 she opened a diamond sample processing laboratory in Perth, Diatech Heavy Mineral Services, to cater for the rapid growth in Australian diamond exploration: it still operates today.

"It's a survival of the quickest, the rudest. Wimps won't last. China is one big crowded brawl."

I thought about this snippet that I had read in the airline magazine just 48 hours prior. Now I was sitting on an antiquated diesel train, dehydrated and with a pounding headache to match, headed for Changsha, in Hunan province. I was already trashed by the first day of action as I stared at the snow from the train window. Hang on. It wasn't snow. It was a mountain of discarded polystyrene dinner boxes spread for kilometres along each side of the tracks. Welcome to China.

Peter Ferrick, diesel mechanic, diamond mine supervisor, and currently an angora goat farmer, was an old mate from the Bow River Diamond Mine situated in the Kimberley area, Western Australia. He had called me from his farm in Gidgegannup a few weeks beforehand. "I have been invited to go to China to look at a diamond prospect. No one on the team knows anything about diamonds. Can you come along?"

"Really? Great! Who's the team?" I can't deny I was a little excited by the thought of this adventure into a country that had been closed to the western world for decades. My grandfather had travelled by ship to China from England in 1921 and 1922. His adventure took two years. I was the keeper of his 100 year old ink-penned diary and had read it with great interest several times.

"A local builder and his friend" said Peter. "My bank manager in Midland introduced us. The builder has some Chinese contacts, and has been offered an area to explore for diamonds and gold in China. But he's mainly interested in searching for business opportunities for the supply of building materials like granite and marble, and he's also interested in opening fast food outlets. The idea is to keep the diamonds, and exchange the gold for the building materials."

Hmmm. This sounded disturbingly like a Boys Own adventure. After working as a dutiful diamond exploration geologist for a large company, I had started up my own diamond sample processing laboratory in Perth less than a year before Peter's call. Not being connected to any one particular company suddenly brought all sorts of opportunities. I should really have stayed in Perth tending my fledgling business, but this was too good an opportunity.

What the hell. Peter was a friend and I trusted his judgement on this. "Yes, when do we go?"

Diamond exploration might sound very exotic, but it is not exactly the glam job it sounds. It revolves around digging gravel samples from riverbeds, screening off the larger lumps, and putting the samples into bags. It's all about getting wet and dirty. The bags are eventually emptied, and each sample is winnowed down in the same manner one would use in gold panning, separating the lighter minerals from the heavier more dense minerals, but using slightly different equipment. Strangely, the exercise is not aimed primarily at recovering diamonds (although finding one is obviously a good sign), but to find any "indicator" minerals that are more numerous in a deposit than the diamonds themselves and can act as pathfinders to lead the explorer to a diamond mine.

So, Peter and I packaged up the practical items we knew might be handy for the not so glam job of diamond sampling – a small motorised screening device, a suction pump to lift gravel from the river bottom, a small portable jig to concentrate the screened samples (which uses the same principles as a gold pan), and numerous other handy items such as bags, shovels, and wooden framed sieves. These were dispatched to Hong Kong on a pallet to be sent into China when we arrived.

The builder, Gary, and his friend, Greg, were very excited about their own diamond exploration equipment.

Wet suits and diving tanks.

It seemed that Greg had one skill to share on this trip – he knew how to dive, and he would be able to get to "the big diamonds in the middle of the river". He knew nothing about building materials, fast food outlets or mineral exploration.

Despite our best efforts to talk them out of it, they thought they would be able to don their diving gear and plunge into murky Chinese rivers and pluck the diamonds off the bottom.

In Hong Kong, Peter, Gary, Greg and I met with the first of our own travelling entourage. Mr Wo, a well-connected Chinese businessman from Indonesia, and the very elderly Mr Wong who lived in Los Angeles but hailed from the province we are heading to – Hunan. Mr Wo would act as interpreter, but later he told me his job was to also look out for any unusual increases in the number of local funerals or shootings. "There are some hideous plague-like diseases that can race through the small towns due to no running water, poor sewage, poor hygiene and general filth. And the AIDS epidemic is a new problem in China, and too difficult to handle so HIV positive people are just shot". "Great", I thought. Maybe I should have added more vaccinations to my list. I never did work out what Mr Wong's role was, but I sensed that he was greatly revered as a successful Chinese man living in the United States and was returning to his home province, and of course he commanded respect because of his age. We called him "Old Mr Wong" because every second person we later met was called "Mr Wong".

The pallet of gear had arrived in Hong Kong and was forwarded to our destination, Changsha, the capital of Hunan. We travelled for 2½ hours by mini-bus from Hong Kong to Guangzhou. Chinese Border Control would not let us in – they thought the compressor and diving tanks might be bombs or rockets. We were shuffled from one exit door to another, then called back, and apart from a dozen Border Control officials talking, smoking and arguing at once, no-one was making a clear decision. It took 1½ hours with much animated discussion between Mr Wo and Border Control to be allowed to walk out into the square. And straight into communist China.

We walked over to the train station. These Aussie guys did not travel light. Peter and I had one small cabin bag each, just as we were told to do. Gary and Greg had full size suitcases, cabin bags, brief cases – and a fridge trolley stacked with two enormous cardboard boxes containing the diving gear, the compressor, *and* the bright yellow diving tank strapped onto the top.

It was 12 noon. The heat standing in the square outside the Guangzhou train station was oppressive. Peter, Gary, Greg and Mr Wo had gone into the station to work out firstly where to buy train tickets for the 4pm train, then where to dispatch the 'checked' luggage. I was left to stand guard over the hand luggage outside the check-in building. I was surrounded by what seemed like thousands of people all talking at once and all trying to check in at once. The noise was deafening. The air was thick with cigarette smoke. Hour after hour ticked by. I was extremely thirsty and at the same time learnt a very useful tool in China – extreme bladder control. With white knuckles I stood trying to hold onto all of the hand luggage at once and to keep watch all around me, terrified someone would run from the crowd, grab a bag and disappear. People continuously jostled and argued with each other; they poked and prodded, they yelled, they waved their arms. Luggage tags were put on; then torn off. There were hundreds of people with TV sets in huge cardboard boxes. Some were put on a luggage belt to go through an X-Ray inspection, then the box and owner would be sent to the back of queue again. There was no order, and there was constant queue jumping. And the jostling, poking and prodding would start all over again as disagreements broke out. Few people seemed to actually move forward from the square into the station to catch a train. I stood in that square for nearly four hours cursing the 'team'. I had been stared at from every direction by thousands of eyes over that time. The train was leaving at 4pm. At 3.45pm the team appeared with tickets. I could have hit them.

We had a four berth "First Class" sleeper. Mr Wo had taken a separate cabin. He had fallen ill. Our sleeper was dirty, stuffy and hot. No air conditioning. It was decorated with scruffy satiny embroidered frilled cushions and embroidered throws in dirty blues and peach colours to hide the dilapidated seats underneath. And none of it was necessarily cleaned after the last occupants. The only drinks available were purchased through the window from the jostling vendors on the platform. The drinks were warm, flat and very sweet. The train finally pulled out of the station. It would average 55km/h, and take us 16 hours to arrive in Changsha.

Of course it was possible just to fly to Changsha in a couple of hours. However, before flying on the aged Russian-made planes that made up the China Airlines fleet, it was necessary to sign a statutory declaration written in Mandarin characters to say you wouldn't take action if the plane crashed, or something to that effect. A beneficiary had to be named. I still have a copy of this declaration. So an early decision was made to play it safe, and with the yellow bomb, catch the slow train to Changsha.

As the sun set over the very picturesque paddy fields, I took in the pleasant green scenery through the grimy window, trying to rehydrate myself with the warm soft drink, and trying to ease the pounding headache. I jumped when suddenly something white flashed past the window. And then another. And then there were more and more airborne white flashes which quickly gained momentum, flying past on both sides of the train through the slipstream. White polystyrene boxes by the hundreds were being thrown out of the windows of the train. The ground below was now covered in metres and meters of white boxes, tossed out of the train around this point daily. The food boxes had been purchased through the train windows by the Sitting Class passengers (no dining car for them) at the previous station, and the fastest eaters threw theirs out first approximately 15 minutes out of the station, and the rest followed. This continued for several kilometres until the slowest eaters had finished and ditched their boxes too.

We ventured to the dining car although we had caught a glimpse of the filthy kitchen car. The Sitting Class cars were packed with people holding chickens and children, now with full stomachs. The smell coming from those cars was something to remember and the toilets are worthy of mention. At the front of each carriage is the "starting block" room, and at the rear is the "western toilet" room, complete with shoe prints on the rim, and no seat or lid. The optimum time to visit these small rooms is at the start of the journey. The Chinese are not good shots, and it doesn't take long before the toilets are completely fouled, and the thick suffocating stench of urine permeates from each end of the carriages towards the centre and into the cabins. First Class carriages are not immune. A trip to the toilet means taking a deep breath, covering the mouth and nostrils with a moist towelette to filter the smell, then running into the small room not knowing what would greet you, loosening clothing as fast as you can with one hand, trying not to fall over in the rocking train, and trying not to look down or gag. And when you did look down because there was no way to stop yourself, there was the realisation that any shoes worn in China would not be coming home. We were lucky – our cabin is mid-carriage so the toilet smell was diluted. Maybe that's why it was considered First Class.

Mr Wo, who was feeling better after his exhausting train station and Border Control experience, ordered a number of food dishes and warm beer to wash it down. We met "Colonel Chee" who was very curious about us, and who rolled about laughing when we "conversed" with him using our Chinese/English phrase book.

Back in the stifling hot cabin, we tried to sleep with no fresh air coming in. No one has had a shower. There aren't any. We all smelt quite bad. The door is closed, and the window was also closed tightly because the train stops constantly during the night at small villages, and it seems that thousands of Chinese are bustling and banging outside of the windows trying to sell a warm soft drink at 3am.

At 6am sharp, the speaker right next to my ear exploded at full crackling volume with the Chinese version of Mozart's "Rondo Alla Turka" (the theme music from the 1980's Jif Cleanser advertisement known well in Australia). One minute later, the door flew open and a woman barked orders in Mandarin. She demanded our sheets. The four of us stared at her from our bunks in shock, and she glared back before she finally gave up and went to the next cabin to harass them.

At 8am we arrived in Changsha, and went straight to our hotel, named the Xiangjiang, for breakfast. It was art deco style and hadn't been cleaned since. Our rooms are on the top floor, and the view is of endless grey multistorey concrete buildings disappearing into the thick grey smog. There is a strong musty smell, partly coming from the carpets that are wet mopped, not vacuumed. Spitting is rampant, and the carpets are not immune. A spittoon is stationed outside every door and in the bathrooms although the carpet is used in preference. More mind games trying not to look down. On every floor there is a desk with two immaculately dressed young Chinese ladies amid the grime. They are dressed in stiff white shirts, black skirts, black stockings and shiny black shoes. Their jobs are very clearly defined:

1. Keep a close eye on the lift door. For guests arriving, hand them their key with a slow stiff bow and say "Your keys", and gesture with a slow sweeping arm towards the hallway with one hand. For guests leaving, say "Your keys", take them, then press the lift button. When the lift door opens, slowly gesture with a sweeping arm to the open lift door and say "Your elevator is waiting".

2. As soon as a guest departs the floor, go straight to their room and fold the end of the toilet paper into a triangle.

After breakfast we went to collect the luggage. The compressor was gone. "Yay", I thought quietly to myself. After two excruciating hours waiting in the square outside of the railway station while Mr Wo went in to do battle for the compressor, it finally turned up. But it was enough time for us to check out the local streetscape – almost everyone was wearing navy blue or khaki Mao-style shirts (men and women), bicycles were everywhere, butchered meat sat on wooden tables, vendors were selling food cooked in giant woks filled with pig fat which sat on fires contained in 44 gallon drums, and all sorts of food I couldn't recognise was laid out on trestle tables, including a table piled high with rats tails and the vendor enthusiastically spruiking his goods with a loud hailer. Peter, being a farmer, butchered his own meat. But there were cuts he just didn't recognise. The cuts look like they had been smashed open with a sledge hammer, with bone splinters everywhere. Breaking the language barrier by waving his arms, he asked the vendor what the meat was. The vendor smiled and barked loudly. We had been in China for just 24 hours.

Today would be the start of endless meals and meeting people who all wanted to eat as a first priority. Nothing it seemed was possible without eating. On this day we met with numerous government officials in the hotel conference room which doubled as a disco. We were joined by a Mr Ye, Mr Wang, Juan the interpreter, and also Mr Ye's daughter, Li, and Mr Wang's son, Yin, who were hoping to be granted education sponsorships if they tagged along with our Western team for long enough. Plates and plates of food and copious bottles of wine emerged from the kitchen. I was seated next to Li who spoke good English. She was very demurely dressed in a tartan skirt and sensible shoes. As we chatted, mid-sentence, she leant slowly over the gap between us, and carefully dribbled out the contents of her mouth onto the floor between us, and then continued chatting. Clearly there were going to some eye-opening cultural differences, but now I knew how the bone splinters in the meat were dealt with.

It was almost impossible to work out who was who, who was important to the exploration cause, who was a government official, and who was just along for the free feed. Old Mr Wong had materialised out of nowhere just in time for dinner, as he often did, and sat with the officials, and Mr Wo was kept busy interpreting between tables. We noticed that at the end of an important meal such as this, the other party would stand up in unison immediately after the last mouthful and march promptly out of the door.

"Gary, how did it go? Do we have a plan? Any maps?" I asked tentatively. I would need something to plan out a basic exploration program. This is Exploration Rule #1: There's no point taking a sample if you don't know where you are.

He was looking a little drawn. Maybe it was the agate-like grey coloured 'Hundred Year Old' preserved eggs and the wine consumed for dinner that were causing the problem. "It went OK. I have been offered a 50km stretch of the Yuanling River with four main gravel deposits to assess. A Chinese geologist will meet us but that's not for two days, and we can pick up some local labour. We leave tomorrow on a minibus for

Chang De, then Chenxi, our base. There are some places to visit along the way. No maps though – they are classified".

Any map would do, and a geological map would be a bonus. A flesh and blood geologist was actually fantastic, even if we didn't speak the same language. We could use diagrams to chat, and Mr Wo to assist. And we had our sampling equipment in transit to achieve something practical on site. It had arrived in Changsha, and was dispatched on a small truck to meet us in Chenxi in a few days. The only map I had been given so far was a map of the entire province on an A4 piece of paper showing rivers and railways, and the text was in tiny Mandarin characters. I had tried to locate myself using this map on the train trip into Changsha. As the train passed through a station, I would try to memorise the complex characters I had glimpsed on the platform sign, and then searched in vain on the map for something that looked the same. I never made a match but I provided Peter with endless entertainment. He had sat with me through 10 gruelling weeks of evening Mandarin lessons at the University of WA before we left Australia. We managed to learn just three useful phrases:

你好

Nì hǎo ("Hello")

谢谢

Xièxiè ("Thank you")

And my favourite:

这是我第一次访问中国。我很高兴来到这里

Zhè shì wǒ dì yī cì fǎngwèn zhōngguó. Wǒ hěn gāoxìng lái dào zhèlǐ

("This is my first visit to China. I am very happy to be here")

At 6am, the Chinese 'good morning music' blasted though speakers at the hotel. Wake up comrades! This was to be a daily occurrence on our trip. The minibus awaited, but not before a long breakfast with the growing Chinese entourage. Early on, we started a new mealtime trend. The western cutlery was often wet or fatty, and not exactly clean of debris. So as Gary, Greg, Peter and I chatted, we would polish it furiously with our paper serviettes before the food arrived. Our entourage must have noted this 'western custom' with curiosity and adopted it themselves before every meal. We would laugh and nod at each other across the tables as we all polished away.

The bus was bristling with excitement and chit chat as the swelling entourage loaded on, which included young Li, Yin and Juan who all spoke very good English. An hour later, the bus broke down in the small village of Bei Lo and we spent two hours waiting for a replacement vehicle.

The Bei Lo locals poured out of buildings to check us out. Had they seen any Westerners in the flesh before? I didn't know, but the atmosphere became almost festive and they welcomed us warmly. We watched as handmade red-brown mud bricks were carried in woven baskets hanging either side of a bamboo pole balanced on people's shoulders, while they walked up wooden ramps onto building sites smothered in bamboo scaffolding. Chinese music blared loudly through crackling speakers to entertain the labourers. Men and women both laboured equally. Carts and barrows made entirely from wood (including solid wooden wheels) were used by the industrious workers in the nearby paddies. There were just a few shops on the single main street, and when we went into the government store, the smiling and curious hordes followed. Every item for sale was behind a glass cabinet, and each cabinet was filled with an enormous variety of unrelated items next to each other. All brand new, but eerily from a bygone era. There were mantle clocks that my mother would have had in the 1950's in England, and black 'Singer' sewing machines with manual handles that were almost identical copies to the one I had inherited from my grandmother and that dated from the 1920's.

Items were weighed by placing them on a hand-held scale with a dish on each side; one for the item and the other for a selection of weights. A wooden abacus was at the checkout, and the operators' fingers raced over the beads as they added up our purchases.

Next stop was the pharmacy. It was a small room with a counter, and behind the counter were numerous wooden drawers, porcelain jars and brown bottles. This was Chinese medicine, and I was lucky enough to see the pharmacist fill a duplicate prescription while the customer waited. Two brown sheets of paper were laid out on the counter, and the dried up bugs (huge cicada bodies I thought?) and roots were weighed and piled equally onto them. They was folded neatly, and the abacus beads flew about to calculate the cost.

The road into Chang De was like the Gibb River Road in Western Australia. Full of potholes, and we progressed on average at maybe 50km/hr. And to make matters worse the driver spends his entire time blasting the horn. Another hour passed, and another compulsory meal stop was made. We groaned as the entourage again ordered what looked like a full Chinese banquet, and for us too. Cutlery and chopsticks were duly polished. Everything was hoovered up from their revolving 'Lazy Susan', and the food that we just couldn't eat was passed over to the entourage, and that was also hoovered up. Our party can't seem to do much about this and it's still not possible to work out who is who. No one is showing leadership – this is communist China. Was it Mr Ye in the scruffy shoes with no laces? Over the next few days, it became clear that the pattern would be three meals everyday no matter what, no matter where, and that breakfast, lunch and dinner would consist of the same enormous volume of food.

Chinese food in China is far from the westernised version we have in Australia. We ate rehydrated sea slugs (a delicacy), chilli-spiced cucumber (Hunan Province is known for its spicy food, even the cucumbers are not immune), spiced chicken feet, grey-ringed 'hundred year old' preserved eggs and all sorts of things we couldn't recognise. One day I accidently ate some dog liver when the "Lazy Susan" spun so quickly on the western table (because we all liked same single dish of food and were fighting over it like children), that when the plate was in front of me I misjudged and snatched a morsel in my chopsticks from the next plate and tossed it in my mouth. I knew straight away my fatal mistake.

It took us eight hours to travel 170kms from Changsha to Chang De with meal stops and breakdowns. This was going to be a slow field trip.

We left Chang De the next morning having declined breakfast but it was served anyhow. We were finally on our way to our base at Chenxi, but our first stop was at an empty diamond cutting factory, and then we stopped at a near-derelict diamond and gold dredge on the wide but gently flowing Yuanling River, in the Taoyuan area downstream from the alluvial exploration area offered to Gary. So at least we were confident there were diamonds and gold in the vicinity even if they were a hundred or more kilometres downstream. The scenery on our way to Chenxi was stunning. We wound our way through the Xuefeng Mountains with hairpin bends and deeply incised valleys. The rivers below cut steep gorges through slates, sandstone, quartzite and limestone. It was peacefully beautiful when the driver stopped blasting the horn and quietly free-wheeled downhill to save fuel, winding his way through the hairpins. Peter looked like he was going to faint. He mumbled something about engines being turned off and brakes failing, and "Angel descents….."

Suddenly the bus pulled to a halt. We were at a temple. Lunch time. At least it wasn't the full-on banquet again, but noodles from the temple café which was perched across the road from the temple, right on the precipitous edge of a very steep slope.

The slope was perfectly aligned for food flow through the café. There was a line of woks against a bank of adjoining windows, iron-barred but open. Each wok was filled with fat and food, and once the food was scooped out into a noodle bowl, the excess fat was flicked expertly with a huge spatula through the bars of the open window, before broth and noodles were cooked in the same wok, and those excesses also flicked out of the window. The bars were dripping with stalactites of white congealed fat, and the build-up of debris outside the window started under the outer window ledge and headed straight down slope. Years of temple visits and a visit to the noodle café by pilgrims suggested this mass probably had a life of its own and migrated slowly down slope as it was loaded up from the window every day.

Chenxi is a very small town. There are no hotels, so our accommodation is in a military hostel. It is filthy, and I have no running water in my shower. There was no time for showering anyhow – it was 7pm and the compulsory dinner was ready. A 20 minute meeting after dinner finished at 11.30pm, and I finally met the Chinese geologist, Ernie. He exchanged his thoughts on the exploration area through sketches and Mr Wo. Apparently there have been historic pits dug into the gravels which are up to 20m thick. Finally, we could get on site and do some work. Our pallet of gear still hadn't arrived though.

On one of the first evenings as we headed out for our sunset walk, the Mayor of Chenxi appeared from nowhere and joined us. He is terrified our little group will break up and wants us to stick together and take NO photographs. He is very welcoming though and wants to show us his town. There are grubby fire hydrants closely spaced along the streets, but on closer inspection they are actually hydrant-sized sculptures of carp with their tails fixed to the ground and their mouths gaping open, or frogs in the same stance – public urinals for gentlemen. Around a corner, we come across a "clinic", and the Mayor wants us to walk quickly past. I don't think he meant for us to see this. There are very graphic hand painted canvases hung on the walls inside and out, and flapping in the breeze, of genitalia adorned with all sorts of detailed images of the outcome of sexually transmitted diseases and other genital issues. Most of the locals must be illiterate, and this was their way of sharing important community health information with the public. After that gem, he led us to a bicycle shop, and a washing machine shop.

Walking the streets is a hazard. If you don't watch where you are stepping, you will wade through all sorts of nasties. And if you do look down, because the eyes are traitors to the stomach's better interests and just can't resist, you will probably gag. I was very glad of the gumboots that I had purchased to work in, but also wore at most other times for trouble free walking. They were a good pair of boots but they were also never going to leave China.

At sun up, the sound of people spitting reverberates up and down the hostel corridors. The corridors are open terrazzo verandahs with spittoons outside each door, and 'No Spitting' signs ironically above them. The military hostel has a large compound filled to capacity with khaki-covered army trucks. We only see the occasional soldier, and the hostel seems largely empty. It's raining, and after breakfast and an eternity of waiting, the minibus is again filled with people. The entourage are not so keen after seeing the rain. Interest is waning at last. The bus will take us to the river to visit the first of four gravel deposits, and after that we have a 60t barge to move us up and down the 50km stretch of river to explore the other deposits during the next couple of weeks. As we come over the first hill, the river comes in to view. We get out and stand in the rain staring in disbelief down to the river.

Greg groans and I try not to laugh. Peter is actually laughing out loud. The mighty Yuanling River is in full flood. It is a tributary to the Yangtze River, and over 800km long. It's roaring like a jet engine as the turbulent brown water, complete with massive chunks of buoyant vegetation, churns through the valley. The first gravel deposit is completely submerged under dangerously fast flowing brown water. Greg's dreams of donning his scuba gear and diving into the river to pluck out the big diamonds are completely dashed.

Walking though the rain across flooded and muddy paddy fields, and shivering from being wet and cold, we manage to get to the barge which was 2km away, to take us to the next gravel site. The barge is immaculately clean. Ernie enthusiastically shows us the river and the beautiful geology as we head upstream. The rain is easing and things are looking up. However, the entourage are looking weary after the 2km walk. I am starting to think we can wear them out and shake most of them off, and actually get some work done.

The next morning our pallet of gear arrives from Changsha. The pallet was broken and the engine from the screener was stolen. In fact the whole outer box was missing and the contents had been piled onto the back of the small truck to bounce around freely on the potholed roads. It was Greg's turn to laugh.

Ernie was trying to be helpful when we told him about the stolen engine. "We can make a copy?" he offered. "I bet you can" I thought. "It will take too long Ernie. But we will have plenty of labour, right? Let's do the screening by hand" I replied.

A record number of freeloaders joined us that day for the digging of the first sample. I stood looking at the gravel bank with Ernie. Holy shit. This was a massive gravel layer in a massive river that had just flooded. Ernie was very agitated because the diamonds were found at the bottom of the gravel layer, which I was aware of. There had been exploration carried out in the 1960's. But practically – there was also a 16-20m thickness of gravel and boulders in the way, and in the absence of an excavator there was no way to reach the basal layer. So I was hoping to at least take samples to find the elusive indicator minerals. Ernie insisted on digging a pit by hand into the 20m of gravel, so I run with it to see what happens. This is going to be good.

Five expert labourers had been brought in from Changsha. They start to dig a hole, very very slowly. By 1pm they have made little progress and are fired on the spot by Gary. They then demand three days wages, which they say is 1000 yuan ($300) each. The going rate is actually 10 yuan per day, or $3. Gary, being an Aussie builder and no shrinking violet, "sorts them out" and sends them home with a few new English words to share with their family and friends. A handful of locals appear to fill the void, and start digging. They do a great job, and are paid 10 yuan for the remainder of the day. They are asked to come back the next day, but they demand 400 yuan ($120) each per day. Gary sorts them out too, offering a few more words from the English vocabulary, and they march off as a group. Was this a union strike in communist China? A couple of them seem unhappy with the march-off and want to stay and work, but the others drag the scabs away.

We arrived back in Chenxi at 7.15pm. The officials are upset because they don't know what we are aiming to do. And we are upset at the late starts, endless standing around, time wasting, long periods sitting around eating, broken equipment, no maps and union style demands from the labourers. Discussion clarifies almost everything. They seem to think that recovering a diamond is paramount, hence the pit digging anxiety. However, our plan is two-fold; to gather information that will assist in building a plant to recover the diamonds in the basal layer at a later date, and to sample for indicator minerals to check if a primary source is in the area and feeding the alluvials.

Gary gives them the hard word about maps and lack of helpfulness and threatens not to come back. Peter, Greg and I nod in solemn agreement. We didn't want to tell Gary that it was actually Greg who broke the suction pump. The tone among the officials changes, and they are suddenly more open and friendly.

To celebrate, Gary sets off firecrackers in the courtyard of the hostel. He is outdone by the firecrackers going off in the town as a funeral procession walks past. We go out to watch. Mr Wo is looking very worried. He has been keeping watch and thinks the number of funerals is increasing.

He also informed us that pro-democracy students were stopping trains in protest around the country.

We gathered together a team of fabulous locals who become our field hands. Popeye turned up with a group of friends one day, very keen to help dig and sieve samples. They really worked hard, and were happy with 10 yuan a day ($3). Keeping in mind that a government official's wages were 220 yuan/month ($75 /month or $2.50/day) this was in line with what would be considered a good deal.

Popeye must have had a terrible accident at some time. He only had one eye, and the other socket was a gnarled mess of scar tissue. He had only a few buck teeth, and these were pointed in every direction, and there didn't seem to be enough lip to cover them. On top of this, his right hand had a small stump for a thumb and no fingers. Yet he was the hardest worker of all. We had two other field hands of note; "Smiley", named for obvious reasons, and "3.5" who would only ever handle the 3.5mm screen. We think it was because that was the most likely diamond size that he could actually see. There was also a Chinese woman who acted as a formidable forewoman. She kept everyone in line by waving her arms and barking orders, and the workforce obeyed obediently. We named her Rose. She insisted I wore a woven conical Chinese hat to protect my fair skin from the sun, which I did, and I handed her my Akubra hat to wear, and then later handed it to Ernie to keep. He was reluctant to accept but quietly pleased I think. I guess receiving a gift can have repercussions in a communist country.

The language barrier wasn't too much of an issue. After the hilarious success with Colonel Chee on the train, I did try to use the phrase book from the UWA course on my field hands, and would point at a question. It was an epic failure. They couldn't read.

A routine was quickly established to remove a sample from the hole, screen it, weigh it, package it, and take the full bags on bamboo poles to a dugout wooden canoe which ferried the samples back to the barge where they were packed into drums to eventually make their way back to Perth.

As we worked further and further away from our base in Chenxi, the travel time on the bus and barge increased, and the only way to arrive on site in good time to have some work actually completed was to depart the hostel earlier and earlier each day. This was a great strategic move that Peter and I had discussed to shake off the freeloaders and time wasters. With a strict 6am or 6.30am bus departure, and not arriving at the barge until 9am, the number of freeloaders finally began to diminish. They wanted the full Chinese breakfast banquet. We packed boiled eggs and warm coke, and stood our ground next to the minibus, and won. The furtherest sampling point was not reached until 11am. The freeloaders were horrified that we didn't down tools at midday for lunch. The novelty was wearing off at last.

By the sixth day of barge work, we had winnowed the freeloaders down to two.

Some days we could even pick up the field hands on the way with the spare seats now on the bus. Things were looking up. The only issue now was to deal with the lazy deckhands and the last two freeloaders who would sit on the barge while we worked and would drink all of our soft drinks. One day was so hot, and I was so thirsty, I broke a cardinal rule. I drank water out of a Chinese River.

One early morning, I looked out over the terrazzo veranda into the compound. In the darkness I could see that every single army truck was gone. The compound was empty. They must have left during the night, and I didn't hear a thing. Nor did the others.

Lunches were usually eggs boiled in the hot water drawn from the pipes around the heat exchanger on the barge, and warm coke. Gary asked for chicken one day to break the monotony. The next morning in the darkness, I was the first to be waiting next to the bus, and my eye caught a small movement. Two live chickens had their legs tied together and were left on the ground. I guess this was lunch. I would have eaten them for lunch if it wasn't for the graphic plucking and gutting that occurred on the deck below me while I was enjoying the beautiful scenery later in the day, and the guts were thrown overboard as the feathers swirled around the barge.

We still had no maps but were managing. And I constantly questioned Ernie about the previous diamond exploration, and he would drip feed me information until I could piece together the story. In the 1960's, several pits were dug into the gravel layers. The program took 10 years, and the pits were hand dug at an average rate of *20cm per day*. The overburden on the higher terraced gravels is 16m in thickness, and this had to be removed before even hitting the top of gravels. Those pits required boarding. I often wondered why Ernie thought we could do this in a few days. I would have been speaking fluent Mandarin by the time it took to dig a couple of pits.

Our last sample for the program was the one that was located under water on the first day. Small pockets of the gravel were now exposed again after the flood. Sampans ferried us and the gear over to the gravel islands. After eggs and coke for lunch again, we gifted Popeye, Rose, Smiley. '3.5' and the other field hands our gumboots, leather boots, gloves, picks and extra plastic sample bags. They made great raincoats apparently. They were just the best labourers. They were really happy and everyone was sad that this was goodbye. It took three hours to get the 44 gallon drums full of samples off the barge and onto the truck for Guangzhou, and then on to Perth. No help from the lazy deckhands again.

Ernie seemed happy enough with the exploration action, and waved goodbye wearing his Akubra. Throughout the trip we had visited places that might have been of business interest to Gary, like a chipboard factory and a slate deposit, and he met with people to talk about fast food opportunities, as well as cigarette vending machines. The fate of the diamond exploration, and gold for that matter, depended on what we would uncover in Perth, but quietly we all knew it was just too difficult.

Mr Wo was anxious and wanted us to move out of town. There had been more funerals in the tiny town of Chenxi, and now students were protesting in Changsha.

The Mayor's car took us to Chenxi Railway station. It was bustling with people. A young man came up to me and said in perfect English "Tell the world about China. We want democracy", and then he quickly faded back in to the crowd. What? I had no idea what was happening in China. What did he want me to tell the world?

When we arrived back in Changsha, students were demonstrating, and trucks were driving around the city loaded with protestors, but they were very peaceful and orderly. We were excited for a meal at a western restaurant, but the chef had gone home so it was a Chinese banquet again.

The Deputy Premier of Hunan Province invited us to dinner. This was an honour, and would be quite formal.

We were led into a private dining room adjacent to the hotel restaurant. There were two magnificent circular tables surrounded by high-backed ornate rosewood chairs. Chinese officials in Mao shirts at one table; Westerners, Mr Wo and Mr Wong at the other. Everything was covered in protective clear plastic. There was the usual 'Lazy Susan' at the centre of each table. We thought we knew the drill. Eat the dishes you liked and hand the rest to the other table. However – this was to be a little different. One dish was brought out at a time, spun around on the 'Lazy Susan' for everyone to see, and then taken to a sideboard to be served into individual bowls. There would be 8-10 courses of food. The next course couldn't be served until every bowl was finished. It would be impolite to leave any, or to hold everyone up by not eating quickly enough.

The first dish – rehydrated sea cucumber. It was spinning around in front of us. Our table was silent. None of us had coped too well with the sea cucumber offerings previously. The chocolate brown rubber stared at me through the gelatinous sauce. There was only way around this. I smiled and I wolfed it down, showing my obvious pleasure. The others inwardly groaned then joined in. This would be the last time. Course after course came out, quite edible if not recognisable. Before the last mouthful was swallowed, the officials jumped up, bowed, and left the room.

Our train arrived in Guangzhou at 7.30am, and the military boarded the train. They were looking for "student troublemakers". There were groups of military and groups of students and locals gathered in the square where we would exit China. My last view of China was of those students, the military, and a real life 'elephant man'. He was sitting in the square, his face and body grotesque and distorted, and his flesh hanging off his frame in rolls and wrinkles. It was raining.

We departed communist China on May 28th and arrived back in Perth on June 2nd.

300 000 troops had been mobilised to Beijing on May 20th, the day our military compound in Chenxi was emptied.

On June 4th 1989, martial law was declared, and troops with assault rifles and tanks fired at pro-democracy demonstrators in Tiananmen Square, Beijing, and at those trying to block the military's advance. Estimates of the death toll vary from several hundred to 10 000, with thousands more wounded.

September 2020

"There are some hideous plague-like diseases that can race through the small towns due to no running water, poor sewage, poor hygiene and general filth." Mr Wo's words resonate strongly.

Changsha, Hunan is 360kms from Wuhan. Wuhan is the global epicentre of the COVID-19 viral pandemic that is advancing around the world. A pandemic was declared by the World Health Organisation on 12th March 2020. International borders are largely closed and world economies crushed. At the time of writing, over 30 million people have been infected, and almost one million have died worldwide.

SINO-SITIS

ANDREW DRUMMOND
CHINA THE NINETIES

Television news and documentaries present us all with a pretty good idea of what modern China looks like – in the cities anyway. Writing as I am at the time of the coronavirus lockdown, readers will be familiar with scenes of apparent great speed at building hospitals, of well-equipped wards, fleets of new ambulances, appropriately garbed medical personnel and so on. We have seen footage of bullet trains, cities full of modern skyscrapers, fantastic bridges and other infrastructure and probably read of China's plans to connect Beijing to Europe via a high-speed rail line.

This modernity and progress has not always been so and this story will endeavour to describe how China looked and acted in the 90s: a time when most people struggled to just survive; when the Communist system so stifled financial, social and mental development; and when China was on its economic knees with little of anything in abundance except people.

My first visit to China was as a tourist with my wife in late 1991. We spent a week or so around the Pearl River Delta and Guangzhou, previously Canton, which is 150 kms or so inland from both Hong Kong and the old Portuguese foothold of Macau. We arrived at the latter via a high-speed ferry from Hong Kong and spent a few hours sightseeing the old city with its colonial architecture. We then entered China proper on foot passing through turnstiles and having our papers checked by a fairly sullen and visibly armed collection of immigration officials, who somehow forgot to say "have a nice day!" but, instead, officiously stamped our passports and let our group walk to our tourist bus.

The first hotel we stayed at in Guangzhou, called the White

Swan, was genuinely excellent. Multi-storied, spotlessly clean, staff in abundance and no local non-employees allowed to enter. We walked downtown that evening and were struck by several observations. We are of average Australian height but, walking along the footpath in the midst of a throng of locals, perhaps thousands, we could easily see over the heads of almost all of them and hence a long way down the street. Generations, or maybe centuries, of undernourishment had resulted in a relatively stunted population. It was almost embarrassing while we were walking when we realised that many younger adults were scurrying past us and then turning their heads back while still walking in front of us so that they could look at our faces and pretty much getting a view of our nostrils. When pedestrian traffic was lighter, we noticed some on the other side of the road, who were walking at a pace to keep level with us so they could continue to stare and take in the strange sight. Staring is not considered rude in China we soon realised.

We saw very few cars, not even taxis, but thousands of bicycles pretty much filled both sides of the road. Motorised traffic was confined to the occasional sky-blue truck and a rare motor scooter or light bike. Readers of my story concerning Siberia in the 90s will note that I remarked on the sky-blue window frames on all of the old timber buildings. Whether it is a coincidence that these two neighbouring countries both favour that same colour, or whether it is some Communist thing, or maybe it is just supposed to cheer people up, I don't know - but all Chinese trucks were sky-blue, as opposed to Henry Ford's mandatory black.

This segues into another development which we remarked upon at the time. From our hotel window we had a view of a newly constructed multilane freeway going to goodness knows where. Every couple of minutes, a blue truck would trundle along but there was no other traffic. The conclusion had to be that government central was planning and investing for the future, and had realised that large-scale developments could only be attained with an adequate transport infrastructure base. China was building infrastructure in anticipation of demand, rather than as a perpetual and ever-failing catch-up as retards Australia. On that trip, and in my work-related ones a few years later, I saw plenty of these freeways being built. Almost all of them at those times by hand labour:

the locals spread the gravelly base, mixed the concrete and levelled it on the surface. It often seemed that villagers were required to do a certain section as part of a government-ordained work program, which involved men, women and children - apparently irrespective of age. A sort of work for the no-dole scheme, I guess.

Our tour involved obligatory references to the wonders of the wisdom and achievements of Chairman Mao, whose pictures beamed at us from all sorts of places, and to the benevolent progress being engineered by the Communist Party for the benefit of all, who should be thoroughly grateful. We visited the village where one of the Chinese Communist founders, Sun Yat Sen, had been raised or educated, I forget which. In case we forgot in which country we were, most men in the countryside still wore those familiar Mao suits: the vast majority of them were very worn, patched and threadbare: evidence of insufficient funds for many to purchase newer clothes.

We toured a couple of villages, and the inhabitants were obviously poor. Apart from their shabby clothing, there were few consumer durables evident. Perhaps a couple of bicycles, a communal two-wheeled motorised tractor, no electricity or reticulated water. Peasants tilled the village paddy fields, perhaps with a buffalo if not possessing one of those tractors, with planting and harvesting done by hand. Livestock was limited to the village water buffalo and pig and ducks, which were herded around the rice stubble in dry paddies by one of the older village people ushering them along with a cane. So much work was done by manual labour. At one stage, beside a river, we saw a barge being loaded with river gravels to use as construction aggregate. All of it was shovelled into wheelbarrows by hand before being pushed to the barge, and it would have all been unloaded by shovel at its destination, too.

An interesting lesson was learned when we boarded the bus after we all checked out one morning. The bus did not move for about 20 minutes, and no explanation was given. Then a group of hotel staff appeared and this resulted in some of our fellow travellers being shamed. Every item in every hotel room we had used was checked while we were waiting and before we were allowed to leave. A red-faced couple had souvenired something like a towel for which they were made to pay. Another couple of ladies had mopped up some spilt coffee with a towel: they refused the demand for payment for it, returned to their room and demonstrated that the stains could simply be rinsed out, incurring another delay for the rest of us. Someone else had opened a bag of local crisps, decided they did not like the flavour, pushed the seam back together to make it look unopened to the casual eye, and returned it to the shelf: it was checked and found out - they were also made to pay and to look embarrassed.

There was some evidence of Mao's wonderful Cultural Revolution which, along with his earlier Great Leap Forward, resulted in millions of deaths by starvation and in destruction of much Chinese cultural heritage. We were taken to a site where there was once a large Buddhist temple which backed up against a cliff made of fairly soft rock. Over the centuries, thousands of images of the face of Buddha, each about playing card size, had been carved by the monks. The weathered nature of the carvings indicated that they were quite old. But one could clearly see, from lesser areas of fresh rock, that they had been defaced - literally: every one as far as we could see smashed by something like hammers. Somewhat mischievously, I asked our guide, who spoke excellent English, what had happened to the carvings. He wouldn't answer me eye to eye, nor admit the truth, and mumbled something about some unfortunate accident long ago.

Fast forward a few years to 1995 to 1997 when I returned to China as MD of Zephyr Minerals NL, for work rather than a holiday. In the interim, I had completed a significant amount of work in Siberia for that company, which is the basis of another story in this book. Generalising of course, but most Russians seemed a pretty racist mob especially considering that the newly emergent Russia had until recently been part of the Soviet Empire. The Former Soviet Union included many non-Slavic countries or regions which had recently gained independence and whose names often ended in 'stan, as well as others such as Chechnya and Georgia. They derogatorily called Chechens "blacks", and didn't have much time for the indigenous Siberian people, the Buryats, either. The general opinion was that the Chinese occupied a much lower social and development strata than did the Russians. They were good for trading with, but not for much else, and were criticised for their relative lack of development, education, achievements, and sophistication. The Russians saw themselves as Europeans whereas the Chinese were more lowly Asiatics.

After Zephyr's Russian adventures, the company received an injection of fresh capital for exploration and it came from Sinophile investors. The money was of course welcome but the catch was that the funds were expected to be used predominantly on acquisition and evaluation of worthwhile projects in China. We were advised that these quality opportunities did exist, China would generally welcome our direct investment, and it needed our mining and treatment technology. Having heard negative Russian refrain fairly constantly over a couple of years, and with my previous holiday experience in a small corner of this huge country, when Zephyr determined it would try its luck in China, I was not expecting too much when I first visited aiming to identify and acquire worthwhile mineral projects.

However, one of the first and best things that the new investors did was to introduce Zephyr to a Malaysian-Chinese mining engineer, Whye Kwong Lee, who was semi-retired and based in Melbourne. He spoke and read Chinese fluently and was able to quickly develop excellent contacts with various regional mining departments within China. He became the key member of our team. Once we decided that mainly gold, or gold as a co-product with base metals, were our target mineral deposits, he worked his contacts to come up with projects which might interest us and on which I might be able to negotiate a satisfactory deal. In summary, Why Kwong did a great job for us and I enjoyed travelling with him, learning a lot about the country and its systems.

Over several visits, we covered a lot of country from Yunnan, "the flower capital of the world", in the tropical south and just north of the Thai border, to somewhere west of Hohhot, the capital of Inner Mongolia, and many places in between.

I found rural China still very backward, with Hohhot a good example. It was a centre for steel making and pollution was terrible. Probably not so much from industry, but from domestic sources. Most cooking and home warming used a coal briquette of about 15 cm diameter and 10 cm thick, shaped like a wagon wheel with spokes which were also that 10 cm depth. Burned in braziers in the open or in kitchens, briquettes all smoked strongly. Starting from the first meal of the day, probably more of than one million fires were burning at once - one million separate coal fires polluting the local and regional air. When there was no wind, which apparently is often the case for several months of the year, the yellow to black smog hung over the city and spread laterally. As we drove through the city one morning, someone pointed beyond the windscreen and asked me if I could see the steel smelters. I couldn't for a while, but then just discerned an outline through the smog – and it was a huge industrial complex.

We travelled by road to an operating gold mine which was available for dealing. Production was from milling of quartz reefs and also by recovering gold nuggets and grains from bouldery riverbeds nearby. Mining of the latter was incredibly primitive, with the only protective gear being gumboots for some of the workers. They had to move the boulders with crowbars to access the gravels and sands underneath. For the really big boulders, they levered them up and held them in place with a wooden pole or two while somebody crawled underneath to retrieve those gravels with a spade or bucket: incredibly dangerous.

We moved to a nearby fairly high hill which overlooked the mine and the countryside and, through Whye Kwong's interpreting, I rather diplomatically remarked about the amount of haze, as opposed to pollution, that was obscuring the view. The local replied that on a clear day one could see the Yellow River, which is one of the biggest rivers on earth. So, I told him that I surmised it must then be 20 or 30 km away and he replied it was only three or four! When assessing the potential of a project, one has to take into account that we might need to send an expatriate Australian to manage or co-manage it. I decided that, if we were to go ahead there, we might as well try to hire some single bloke of fairly advanced years who had smoked all of his adulthood. It simply would not have worked trying to send a younger non-smoker and we certainly could not employ one who might want to take his family to site where he worked, or to the nearest large town or city - which was the almost invisible Hohhot.

I would like to think that by now the city has cleansed itself by providing an electric power supply to all of the homes so as to get rid of those burning briquettes. On the other hand, it may now be chockfull of polluting cars instead, in which case progress will be arguable.

On one hand there was a city with people having to live with dangerously polluted air. On the other hand, over the years I have read many stories which paint Chinese culture in a rather rosy light, or which portray the vaguely romantic and mysterious Orient in which the people live in Confucian harmony. I have read numerous articles in the West about feng shui and how Chinese consult on that and take it into account before they design their homes and work places and then build them so as to enhance the vibes which result. May I take the rest of this paragraph to disavow the readers of all this. I saw no evidence whatsoever in my travels, albeit before the more recent spectacular high-rise development in the big cities, and think that notion can be safely confined to the scrapheap. Mr and Mrs Rural Average Chinese struggle to obtain a subsistence income and perhaps to produce a little something which they could sell or barter for a different type of food or a cheap consumer item, let alone daydream about feng shui principles of home design and orientation. Proof of this is that, in town after town through which we travelled, most homes were built abutting each other in terrace style and were oriented simply to face the adjacent road at right angles, irrespective of its direction. As for design, after consulting the feng shui oracles, all chose exactly the same two storeyed houses! Each was exactly a roller door wide, all the same width; plus a share of the adjoining walls with the neighbours on each side, and made with locally or home-made bricks and mortar. Business was carried out on the ground floor in front of or behind the roller door and may have been anything from machinery repairs, carving or woodwork, metalwork, or an eating place with meals for sale: the family cooking and sleeping was on the first floor above.

I had occasion to visit an old goldmining area. There the weathered gold-bearing rocks were covered by barren soil to a depth of a couple of metres. This waste overburden was mined by a team of women from the nearest village. They apparently accepted a contract, or were required, to remove this dirt over a defined area which was evident on the ground by its boundaries being the vertical walls being made as the pit was excavated. As the earth was dug up it was loaded into woven baskets, which were picked up a pair at a time, attached to either end of a bamboo yoke, and carried on the women's shoulders to the edge of a ravine. There they tipped the earth over the edge and then returned to where the material was being dug. They dropped the empty baskets on the ground to be refilled by those with shovels, attached full ones which awaited them, to each end of the yoke and repeated the trip without rest – over and over again for each of the team, perhaps 50 of them.

As we traversed the country, the most notable feature was the abundance of people in all but the most remote and arid parts. An integral part of dealmaking with the Chinese hosts was banqueting, with the management of a possible target project really extending themselves to make us welcome and feeding us to a quantity and quality way beyond that which ordinary citizens were able to experience. One can't discuss China without mentioning the food which, to generalise, always tastes superb. Return to our first holiday trip to China, our group of tourists was always adequately fed, but only just sufficiently from our point of view. Shay and I are not particularly big eaters but several of our party were rather typically very overweight North Americans. The tour group generally used to sit at tables of eight and food was brought to the middle of the table in bowls so that we could all help ourselves. Shay and I learned to hop in early to get our fair share, as there were no seconds and no opportunity to purchase snacks along the way. I am sure we were given more to eat than the average Chinese could afford, and I suspect our North American cousins lost some weight on the trip.

When being fed as we toured potential projects, I often wondered what I was eating, especially the lumpy bits in the sauce between the vegetables, but I never had a dish which was not delicious. The Chinese are masters at making food mouth-wateringly tasty, irrespective of the original basic ingredients. Those readers who have walked in the bush while working or camping will be familiar with the instinctive reaction to leap sideways if, at the lateral periphery of vision, one sees something move and especially if it is accompanied by some sort of noise. Most of us hate snakes. I had a similar reaction once as I walked through the door into a restaurant. A movement and noise to my left made me instinctively hop to the right: when I looked back towards the floor there was a stainless steel wok which, upwards, became quite steep-sided. In it were thousands of live scorpions (not that I actually counted them) and every so often one would try to escape from the top of the heap by climbing up the sides of the bowl: when it became too steep for it to grip, it slipped back down into the morass of its mates. Darn, I forgot to order some, but did watch several patrons eat a scoop of deep-fried scorpions - rather like the mesh scoops for fish and chips we have at home. This high temperature cooking apparently neutralises the venom. On another occasion, the menu was also in English which was rather handy as it meant that I did not inadvertently order the bull's penis soup.

In the southern parts of China, the Szechuan area, a lot of the foods have chillies. A favourite form of banquet dining afforded to us was steamboat. Here a group of us would sit around a table in the middle of which was a deep wok with a middle divider to provide two halves. In each half was boiling water with one half having lots of chillies already placed in. Meats and veggies were then progressively added by either the mine managers or the waitstaff. Once these were cooked, each of us simply fished out the target morsel with chopsticks, over and over. I guess this was relatively hygienic when the water was still boiling but, as the lamps below eventually ran out of heating spirit, the temperature in the steamboat progressively dropped and the diners' licked chopsticks dipped in and out repeatedly. At that time in China there was no access to the toothbrush and toothpastes that we have and most of our hosts had fairly impressively black teeth. Anyway, soldier on and remember that is what I was being paid to do.

We eventually signed a deal on a project in Changsha Province, and it was about a five-hour road trip from the airport. We used to travel in a Kombi van equivalent and, on one trip, we stopped to fuel it up. Near the pumps were some chickens in cages, and the proprietor removed one. Once the tank was filled, I was motioned inside to what could loosely be termed a roadhouse and we sat down and awaited a meal. Within 20 minutes from when we arrived, that chicken was beheaded, plucked, chopped up and cooked with a collection of vegetables and spices and was absolutely delicious. I have no idea how it was done so quickly.

When the meal was over I asked for directions to the toilet for a quick bladder break before we departed. I was told to go upstairs to the first floor where I found the lady of the house and her daughter were cleaning up in the kitchen. They indicated a rather grubby cotton curtain behind which I stood and aimed towards a channel in the floor. This led to a drain hole in the low parapet wall beyond which was the backyard. Having done my thing, I walked to the wall, looked over and saw quite a few chickens in the small yard below. Hanging around the bottom of the drain hole they were hoping that some food scraps would be flushed down. Had I attended to a more serious business, that would have flushed out into the yard aided by a pail of water, the chickens would have eaten it, and one of them was probably tomorrow's lunch for somebody. Our lunch today was....? Sometimes it does not pay to be too inquisitive!

The sole reason I was in China was to find a good project for the company. In Australia, if you want one there are essentially three ways of doing it. The first is to identify prospective ground, which does not belong to anybody else, apply for it and, when it is granted and native title clearances et cetera have been gained, one then begins exploration. Statistically, most of the last is unsuccessful and, especially for a junior with limited funds, management needs to know when to cut losses and walk away, or when to decide to try to bring in a partner, or when to vend it off. The second is to deal in on a prospective property from an existing owner where the latter has lost enthusiasm for one or more of any number of reasons, or simply does not have the money to proceed any further. The third is to buy a property outright in one way or another, which could be by cash, shares, production royalty, et cetera, from an owner who no longer considers it a key asset or wishes to spend their funds elsewhere.

In China, really only the second option was available. Every known decent occurrence of mineralisation was owned in some sort of capacity, whether the property was actively mined or not. The first job for Whye Kwong was to identify a potentially available property which may meet our corporate criteria, and the owners with whom we would have to negotiate: the latter occurring after I inspected the property and made an assessment of its worth to Zephyr. Under the broad heading of inspection, I needed to assess a host of matters including availability of water, electricity, labour and transport infrastructure; likely difficulty or ease of mining and then ore processing; local environmental and social factors; the likely taxation regime relevant to the operation; assessment of the enthusiasm or otherwise for a partnership with a foreign company; whether the processing of ore to obtain a saleable product was relatively straightforward or would need advanced technology and a higher level of capital investment; how was product to be marketed and how would Zephyr earn its rightful share of gross income and then repatriate it, after taxes, back to Australia; whether we could expect to have our hands pretty firmly on the control levers for the operation; did the people with whom I was negotiating have rightful and sole ownership of the property; which other Chinese government instrumentalities were involved in the negotiation process and in signing off on a deal: and so on.

Anyway, many projects were brought to my attention over an 18-month period. Most of them had ceased production: sometimes due to depletion of the easily recovered or easily treatable ore; sometimes because the metallurgy required a completely different treatment plant and the owner did not have the funds to finance it; and sometimes because the whole operation simply needed to be scaled up to become profitable. My overall aim was to identify that which I considered to be a good project, and then negotiate to secure a deal which would be workable through the exploration, development and life of mine phases.

The Chinese owners' viewpoint generally came down to whether they viewed that the capital requirement they needed could be provided by Zephyr and how much of the actual game, in its broad sense, were they prepared to part with to attain that aim.

In some cases, my conclusion was that, had the project been in Australia, it would be relatively straightforward to get it into profitable production, and that a relatively small number of stakeholders would be involved in any decision-making process. In China there was always an additional dimension which, in summary, was that the deal would need to incorporate the ability to employ, or to provide demonstrable benefit to, just about everybody in the entire district - and that involved a lot of people.

As an example, we held good discussions with a company's management who were at pains to point out that they were a great mining company and hence a great partner for us. Apart from the mine and outdated processing plant, which were not operating anyway, proof of greatness was that they had 750 employees, and owned two villages, a school and an aid station. However, with no saleable production in sight and hence no income, they had no cash to pay for or do anything. I could not find out what they owed to employees and for accumulated unpaid taxes. So, if we were an incoming party, we would pick up the tab for those accumulated debts and effectively replace the State in providing the income source for the district.

Meetings we had when we tried to negotiate a mutually acceptable deal were often prolonged and stop-start but did have their funny sides too, even allowing for all of the discussions having to go through Whye Kwong as interpreter. We two would commonly sit together on one side of a long table, with perhaps 20 Chinese managers and officials sitting along the other three sides. They were always men, and always wore suits on which the label of the tailor was visibly displayed on one cuff. The more important ones carried a briefcase and the rest usually had smaller black zipped up vinyl bags to hold some paper and pens. Why so many? Well, in addition to the managers of the entity owning the actual mineral deposits there were representatives of a swag of different governing entities: they might include the equivalents of our local council, a foreign investment review committee, regional government, mines department, committee for reviewing the exploitation of state resources, and more. If there were not representatives of the Communist Party actually included around the table, one could only suppose that all appointees had Party approval anyway. I genuinely got the impression that the aim was to provide a decision by committee consensus: then if the deal later collapsed for some reason, no single individual could be held responsible or to blame.

In fact, in all my travels, I did not meet anybody who was introduced as an official from the Party. I never had any security issues, was never subject to any overt pressures and, in general, didn't experience anything untoward when I was trying to negotiate.

A possible, but not definite, exception was one night when I was talked into having a Chinese massage, which was always being recommended. During this first ever massage in my life, I wore a pair of canvas shorts which were supplied and I was alone in a room, except for the masseuse who was a young Chinese lady - maybe in her early 20s. The room had a cotton curtain along one side and a table on which I lay while she worked on me. It was sort of enjoyable, but not my cup of tea. Anyway, towards the end she pointed to the front of my shorts and, by sign language, indicated that she would be prepared to give me some sort of happy finish. I declined and she then immediately said something in Chinese, the first words of the evening. Four guys who were standing, soundless, all the time behind the curtain and unbeknown to me, then appeared and exited the room. Whether they were there to protect her from the foreign devil, or in the hope of catching me in a honey trap and photographing me involved in something embarrassing, I simply do not know. If the latter, they were disappointed.

Working as we were in generally remote parts of China, there was usually no mobile phone coverage and not much chance to ring back to Australia on a landline. It was also in the early days of personal computers, so I was not carrying one. This meant that, even if the hotel rooms were bugged, which was a pretty remote possibility away from the big cities anyway, little information could be gained from us.

Mobile phone coverage was just starting within China, but was still restricted to the big cities and only the wealthy at that time could afford a phone. It was quite hilarious to watch the nouveau riche Chinese businessmen in action in the lobbies of foreign-class hotels. They would turn up in foreign luxury cars driven by a chauffeur, accompanied by their wives who were beautifully dressed and wearing plenty of genuine gold and sparkling jewellery, and perhaps accompanied by a child. They then sat in comfortable leather chairs around tables in the lobby and simply talked on their mobiles, totally ignoring their families while big-noting themselves loudly and in semi–public. The families looked bored to sobs.

The problem I initially encountered in my negotiation meetings was that Chinese collectively could not understand that I had the power and legal right to be the sole negotiator on behalf of my company, and to sign off on a deal. Having said that, the initial sentence or paragraph of any heads of agreement I composed did always state "subject to the agreement of the Board of Zephyr Minerals", so I had an escape clause if I, or the Board, had second thoughts. I was forever learning on the job. In China, the chairman is the absolute boss and I was only a managing director so how could I sign off for the company? To get around that, after one trip I went to Ted Ellyard, our Chairman, and said "sign this, Ted". It essentially said that: "I, Ted the chairman, delegate all my responsibilities and empower Andrew to sign anything he sees fit". Presentation of that letter solved the problem for me when I next returned to the negotiating table.

Australia dispensed with the need for stamping of documents sometime previously but, it seems to me, the more backward the country, the more the need for stamps. We reached the end of one round of negotiations, but they were not able to be signed off as I could not stamp the document, having not bought one to China. When I returned to Australia I found our then redundant company seal in a drawer and, with a scalpel, removed the words "Company Seal", and left the rest of the name and border trimmings intact. That was then my fit for purpose stamp for subsequent trips.

I inadvertently transgressed some social and meeting protocols at times. On one occasion at one of those big meetings, progress was made, but agonisingly slowly. We haggled then agreed on equity percentages, how each party could take and sell its share of product, a management structure and so on. One reason we were advancing so slowly, was that one of their blokes just argued every little point, never taking no for an answer or trying to adopt a middle ground. He was always trying to add a few more percentage points to the Chinese side. An annoying ploy he used was to return the next morning to what had been agreed upon yesterday and then try to rework the percentages or management responsibilities, or whatever, and always in the Chinese favour. A relatively young guy, he probably had ambitious eyes on getting a top job at some time in the future.

At the dinnertime banquet, a multitude of toasts always covered such items as "to China, to Australia, to China - Australian brotherhood and relations, to profitable and happy dealings" - you get the picture. Usually each member of management would propose a toast at some stage during the evening and we all downed a shot of some Chinese herbal liqueur. But this young bloke jumped up for half a dozen of them. Whether to big-note himself, or just trying to get me drunk so I would be sub-par on the next day, I don't know. In short, he was a pain.

Because of his antics, next day I nearly reached the stage of deciding that we should just walk away from the negotiating table and go home but, instead, suddenly slapped the table loudly and yelled "Enough!" in an endeavour to shut him up. It worked on everyone at the table. In fact, there was stunned silence all round. I received a very disapproving look from Whye Kwong, who quietly advised that I should never do such a thing. I caused the young twerp a huge loss of face, and it would not bode well if we entered into an agreement and he was somehow involved from the other side in the future. Whoops!

Having to always work through a translator causes obvious difficulties. As far as I knew, only one of the Chinese I ever met had travelled beyond the boundaries of China, although perhaps a few had gone to Moscow or Mongolia at some time. This chap had had a spell in Tennant Creek, of all places, and there developed a taste for Schweppes Bitter Lemon: we brought some back for him on our next trip.

Just as I knew so little about them and their country, they knew so little of me and mine. Both sides had to encounter and then overlook different social habits, and it required some effort and goodwill from each side.

For instance, at one stage on one trip I suffered a heavy head cold and streaming nose. Fortunately, at each meal, a small handy pack of tissues is placed on the table for each diner. This provided a packet for me, and I also grabbed the half empty packs left by the others at the end of the meal. When I blew my nose, I would do so in one of the tissues and, as there was no visible wastebasket nearby, I simply put the used tissue in my pocket. This was apparently considered quite vulgar by the

Chinese and they had to put up with my poor form. They get over the problem of a runny nose by simply turning to one side in their chair, placing a finger against one nostril and blowing hard to get rid of the material onto the restaurant floor: then the other nostril. This rather complements the apparently compulsory and repetitious throat clearing and hawking on the floor. Young staff with a wet rag-mop attended to the mess every so often by simply swishing it all from side to side, not to remove it, but to evenly coat the floor with a damp sheen.

In similar vein, negotiations with another group were not completed by the end of the day and we broke up for dinner. The next morning, half a dozen came to my hotel accommodation, which consisted of a bedroom with a double bed, and an adjoining bathroom/toilet. We all sat around the edges of my bed twisting our necks towards the middle so we could look at each other as we spoke. Every so often, one of my prospective partners turned his head and cleared his nose or throat onto my bedroom floor. The more discreet ones went into the bathroom and did it on that floor – why not at least the toilet bowl! For me, this meant that after they left, I always had to wear shoes in my room and be very careful negotiating my path.

Generally, the possible potential mining partners went to extremes to make me feel welcome, often with some difficulty owing to our language and cultural differences. To meet one group, we undertook quite a long drive in a sort of Kombi van and stopped to refuel. I was beckoned inside the accompanying shop and shown a range of soft drinks on display. I pointed to an orange one, which wasn't the real thing but let's just call it Fanta. I assumed everybody was going to get a drink, but it turned out that my Fanta was the only one bought and was given to me when I boarded the van. It seemed that the bosses had not given the troops sufficient money for them to have a treat also. In all conscience, I could not drink the whole bottle myself without offering to share it around, but I thought that latter course might end up in a gastric disaster so couldn't. I nursed the unopened bottle all the way to the destination.

We arrived there in the afternoon and met a very enthusiastic management. There was a bit more money around than was usually the case and the female secretary for the boss was well dressed in a blouse, skirt and high heels. After the first round of meetings and the obligatory banquet dinner, we rolled up our sleeves the next morning to check out the prospects. Word had passed around that "the foreigner drinks Fanta" and four litre-sized bottles had been purchased, and split between two disposable plastic bags. The poor secretary, again in her high heels, which became much the worse for wear, walked behind us all morning as we navigated our way on foot along narrow earth walls between rice paddies and up-and-down stony tracks in intervening fairly hilly terrain. At least this time at lunchtime I was able to distribute three bottles for everyone else to share, while I retained one.

The meetings were promising the previous day and, before we set out on our field inspections that morning, I asked the manager if someone could write up for me in Chinese characters, in our absence, a simple A4-sized statement that I could frame when I returned to Perth office. To say something along the lines of "Chinese Mining Company XXX and Australia's Zephyr Minerals NL YYY Project Joint Venture". At least that is what I thought I asked. The mining company's offices were a couple of storeys high and built around a paved square When we returned I found my requested signage was painted on red paper which had been tacked along the rails of the first floor landing on three sides of the square: it was probably 30 m long by 60 cm high, so a bit too big to frame in my office. Not a big deal in itself, but a flag raised that there might be tremendous difficulties in the future occasioned by each side really not understanding what the other side wanted or considered should be done.

I began this story by stating that I was expecting China to be even less advanced, and more difficult to work in, than Siberia. In fact, it was often the opposite. In Russia, I found that the aftermath of Communist state control still hungover Siberia. Every town has the same concrete oblong architecture, whether it was for the administrative offices or the residents' flats. The centre of town was always a square with a larger than life statue of Lenin with one of two approved poses. In the first, he looked sternly forwards and held on to the lapels of his overcoat with each hand; in the second, his right arm is extended rather majestically palm upwards as though exhorting the obvious benefits of Communism to all. In China, at least to someone who cannot read Chinese characters, the only obvious manifestation of state control were the portraits of Chairman Mao. However, lots of posters and signs in their script may have exerted more control over the citizens, for all I knew, and goodness knows what they were exhorted to do over television and radio.

Things just happened in China. Everyone always seems to be working, doing something, even if incredibly menial and having little to work with but their bare hands or some very basic tools. Trying to travel by air in Russia was a drama, but never was in China. Planes were Western, even if second hand (funny to be in China and see instructions on the back of the seat in front written in French and English, then realising that it used to be an Air Canada plane), and they generally took off on schedule. Working monitors were found at all but the most remote airports that I used, and advice was in both Chinese script and English.

A country can't go from extreme backwardness to state-of-the-art modernity without a few hiccups. After my Zephyr days, we at Minemakers flew into Shanghai in 2012 for a fertiliser conference. We were picked up at the airport by a bus which had been organised for the occasion. I rather suspect that the driver was a uni student and his female navigator, who sat on the steps beside him working her mobile phone, was of about the same age. An excellent multi-lane freeway runs from the airport to the high-rise downtown sector of Shanghai and we set off merrily along it with the skyscrapers in the distance in front of the windscreen. After a while, the freeway started to veer to the left and we witnessed the skyscrapers through the side windows and then eventually behind us. The driver realised he was lost and asked the navigator to find some instructions over the phone, and which she was duly able to get for him. Somehow, he had missed an exit but the solution for the driver was obvious. On this very busy five-lane freeway, he simply did a U-turn, staying on the same, and now wrong, side of the freeway and started travelling back the way we came against all the traffic. Might is right and most other traffic fortunately were cars which had to get out of our way as we headed back towards downtown. Luckily, we were seated close to the rear of the bus but quite a few squeals came from the passengers in the front.

The annual road toll in China must be enormous. On several occasions, I asked various Chinese hosts if they knew what the national one was: no one had a clue and so I guess it is not the sort of information that the People's Republic announces. Roads were not only for motorised vehicles, but also buffaloes, pedestrians and wheelbarrows, and were generally densely packed. They were often used for the drying of rice, which was spread on cotton sheets and pulled off each time a vehicle came along.

We did eventually enter into an acceptable deal over what was apparently a gold system which covered several square kilometres, and situated about five hours drive away from Changsha city, the capital of Hunan Province. It was signed off in the latter half of 1996 and Zephyr explored it for some time into 1997, and after I left the company in March of that year.

Going from the Changsha airport to our project, we always saw road accidents and counted ourselves rather lucky we were not involved. For something to pass the time in a rather macabre way, we competed to see who saw the most accidents on a trip, with an accident only being counted if we heard the bang or if people were still milling around the damaged vehicle or vehicles. We were usually good for four or five and I think the record might have reached seven.

The project had several drill-ready targets, and our partner owned some drilling rigs we could use, but difficulties were encountered. For example, we employed a Chinese-speaking expat project manager and bought him a Hi-Lux vehicle so he could do his work. That inadvertently caused affront, because it was a joint venture and we failed to buy one for the manager from the Chinese side – even though he retained his original old company vehicle. And why should Australians get paid more than their Chinese counterparts? Things were not going to be easy. Some drilling was completed, without obtaining spectacular results, and then the project folded. From nowhere, another State-mandated party turned up and claimed a higher authority to own the mineral rights to the same land on which Zephyr had entered into its joint venture in good faith, and apparently with all appropriate government entity approvals.

This story provides a good illustration of the corporate hazards of working in countries that do not have a very formalised legal system. My accumulated experience was enough for me not to venture back to China ever afterwards chasing another deal.

BURMESE DAZE

SUSAN BELFORD
THE BAWDWIN MINE
MYANMAR (BURMA) 1997

Susan Belford graduated with a BSc Hons from UWA in the late seventies and went on to work for Esso Minerals, Newmont/Newcrest and Aztec Mining (exploring for orogenic gold and volcanic-associated base metal deposits), plus a short stint mapping with the GSWA. In 1990 she resigned her job with Aztec and went to Italy to study Paper Conservation in Florence. Returning to Australia she set up in private practice working as a paper conservator and a consultant geologist. Since then she has completed two further degrees, an MSc at UWA and a PhD at CODES, UTas and has worked in Australia, Indonesia, Burma/Myanmar, The Solomons, Fiji, Sardegna and Macedonia.

Her consultancy specializes in gaining an understanding of volcanic rock suites, their settings and chemistry to direct exploration efforts for deposits hosted by them.

In mid-1997 I was engaged by Mandalay Mining Company (an Australian Company with interests in Myanmar) to review all available geological and related reports on the Bawdwin Mine and surrounds in the Shan State of Northern Burma, and to undertake where necessary (and possible) field checking of the data. As the only source of much of the unpublished data was records held at the Bawdwin Mine itself, all reading was completed on site. At this time the mine was run by Mining Enterprise No 1, wholly owned by the Myanmar government which was seeking an injection of capital via a joint venture.

At this time Myanmar was under the rule of a particularly authoritarian junta, The State Law and Order Reconstruction Council, aka the SLORC, under the command of General Than Shwe. The acronym seemed comical, a throwback to the old TV spy spoof series *Get Smart* and *The*

Man From UNCLE but in truth there was nothing funny about the brutal way they controlled the country.

I arrived in Yangon in early September and passed through Immigration unscathed, and was surprised that, contrary to my expectation, I was not compelled to change any money to local currency. However, Customs did make me fill out a new declaration form stating that I would take my computer out of the country when I left.

I was met at Yangon airport and taken to the company's office where I was introduced to the local staff, given a large stash of low denomination chat currency, had lunch and was finally taken on a whirlwind tour round Yangon. Plenty of art-deco buildings, most now dilapidated and in disrepair. Dominating many intersections were prominent red billboards, often in English with the following slogan.

People's Desire

Oppose those relying on external elements, acting as stooges, holding negative views.

Oppose those trying to jeopardise stability of the State and progress of the nation.

Oppose foreign nations interfering in internal affairs of state.

Crush all internal and external destructive elements as the common enemy.

This was something to think about during my stay in-country.

The next morning I was woken at 5am, given bananas and toast and driven back to the airport to catch a flight to Mandalay. Once at the airport, having had to show my ticket to get access to the terminal, my baggage was seized by eager handlers and I was shepherded through to a counter. I then paid each handler 200 chats. New helper then seized my ticket, took me to baggage Xray, then through to luggage search, then to metal detector doorway, then body whisk, then to sit in the departure area. More money was dispensed at each station. All of this is pretty standard security procedure these days, but back then I thought it was quite exceptional.

The plane, with about 25% occupancy, left on time and flew first to Pagan - I only saw the extensive temple sites from the air. We flew low over an amazing plain of pagodas; the upkept ones, white and glinting gold; and the old and ruined, tumbling and crumbling red brick edifices.

Outside Mandalay airport I was met by a man holding a sign with my name, having assured yet another person that I did not work for Ivanhoe Gold, and was not the person they were seeking. We got straight into an old Land Rover and Saw James (my new minder, who was a Karen), and our driver (an ethnic Burman who wore his hair a in traditional knot) set off at a great speed for the next stage of the journey. Along the way the red slogan billboards reappeared at irregular intervals.

After driving a while, we began to ascend to the plateau in the east, continually passing other drivers, the road surface consistently appallingly bad. The other vehicles were a sight to behold. The trucks were all long past their use-by-date and tray-backs, converted to people carriers, were filled with multitudes hanging off the back and yet more stuff loaded high on top.

After two and a half hours we stopped for lunch at a wayside Chinese eating place - then another two and a half hours before a toilet stop and a stretching of legs. We passed through Pyin U Lwin, a quaint township with lots of painted stagecoaches - during the time of the British Raj it was a hill town/station where families retreated during the monsoon season.

Eventually we reached a military checkpoint, I'm guessing maybe at the Shan State border. We stopped here while the military checked my passport details to see if it matched the lists they had for permitted travellers - and all was in order so we were waved onwards. It felt strange that they were 'expecting me' and I was their work for the day. Another two hours on to Hsipaw, and only 168 miles back to Mandalay. I already thought that it had been a long trip for such a short distance but the next stage was even slower. We left the main road at Hsipaw and collected our (my) military escort for the next (and last) 44 miles - which took us three hours to traverse. This road was even worse than before; poor surface, narrow, winding with much up and down plus lots of decrepit trucks hauling massive teak logs all travelling out, so in the opposite direction to us.

By this stage I was starting to feel decidedly unwell; the trip seemed never ending and it was just not possible to sleep, with the three of us in the front of the Land Rover, and the minutes just crawled by. I revived a little when we passed through a town that made paper, and saw the deckles drying in the sun.

Finally we arrived at Namtu (or NamTu), the sun had set and it was about 10 hours since we left Mandalay. We were met by a gaggle of geologists - I made my greetings, excused myself from dinner, went to my room, finally lost the contents of my stomach, showered, and went to bed.

The next morning I was woken by the first workers' siren at 5:30 am, followed by the second at 6:00. An early morning walk revealed a smelter partially concealed in mist across the valley. Namtu is the old smelter town for the ore from Bawdwin, built by the British after they wrested control of the region specifically to get hold of the Bawdwin mine. After breakfast at 7:00 we went down to the railway station to board a tiny VIP train coach for the 11 mile, one and a half hour journey up the narrow 2' gauge railway to BawDwin village. The diesel engine was a converted Hino truck that ran on rails, our coach was a box car with wooden chairs and tables and little floral curtains over the paneless windows. Into the other coach went our helpers and military escort, they didn't even get a roof.

This was a relaxing little trip compared with the horrendous excursion of the day before. The train line winds up the course of the Nam Pangyun river valley gaining height through a little spiral, and through one reverse switch back section near the Tiger tunnel. Halfway up we stopped for water for the little diesel. The track was very steep and always upwards.

Finally we arrived at Bawdwin and met the Mine Superintendent, where I was introduced as a foreign expert (always somewhat embarrassing). Next I was introduced to the chief geologist (U Myint Lwin) and all the other geologists (including U Tin Maung Oo and U Chandra Kumar). I was also introduced to a man who fought off a leopard attack in the jungle near Bawdwin in his youth. And those scars looked pretty real to me.

For give some general background to the Bawdwin Mine I have extracted the following edited paragraphs from the introduction to my Bawdwin report.

The Bawdwin Mine is located at 23° 07' N and 97° 18' E in the Northern

Shan State of the Union of Myanmar. It is one of the oldest continuously worked Pb-Ag-Zn massive sulphide bodies in the world. A recalculated estimation of the original in situ high grade section of the ore lode is 11,000,000 tons at 19.1 oz/ton Ag, 22.7 % Pb, 13.8 % Zn, 1.05 % Cu. This tonnage does not include the 'low grade' stringer and disseminated ore nor the associated Cu mineralization.

Mining commenced in the Bawdwin area around 1412 by Chinese miners from Yunnan, who mined the argentiferous lead ore and then smelted it on site to recover the silver which was carted by mule back to Yunnan. They were skilled miners and metallurgists and it is estimated that they mined some one million tonnes of rock and left behind 125,000 tons of Pb-rich slag. The Chinese continued to work the mines until 1868. At this time Chinese control of the mine and trade routes had been severely weakened during the long Panthay Rebellion (1856 - 1873) in the Yunnan Province, and this lead ultimately to their withdrawal. Following the Chinese retreat, the Burmese Kings sent armies to work the mines, but were not successful due to ongoing territorial instability in the region involving constant warring with the Shan and Kachin people.

By the late 1890's Upper Burma had been annexed by Great Britain, and a British company was granted the concession to the Bawdwin mine area. In 1906 testing commenced, followed by reprocessing of the Chinese slags commenced and, by 1908, this had progressed to underground mining. In 1914 the company was reorganized and renamed the Burma Corporation Limited. The peak production was reached in 1927-1930 when annual production was about 500,000 tons mined, and 78,000 tons of refined lead and 7,000,000 ozs of silver were produced per annum. During the Second World War the mine was occupied and worked by the Japanese. After the war Burma became an independent nation and in 1951 the Burma Corporation formed a Joint Venture with the Burmese Government to work the mine. Production figures declined steadily post WW2 to an annual production of less than 100,000 tonnes per annum, and in 1997 production was about 60,000 tonnes per annum.

The topography is mountainous and extremely rugged with steep incised ravines. Mean relief in the immediate vicinity of the mine is 300 - 400 metres. Slope angles are commonly 45° and often as steep as 60°. Local landslips during the monsoon season are not uncommon. The RL of the Marmion Shaft is 995 metres above sea level and the highest local point is Mt Herschel at 1433 metres asl.

The vegetation immediately around Bawdwin is mostly grasses. All of the natural vegetation, originally a dense tangled jungle of tropical deciduous trees and bamboo clusters was destroyed or logged during Chinese mining operations, either for underground timber or as fuel for their smelting hearths. The ongoing gathering of firewood and annual burning of the grasses prevents tree regrowth.

The climate is tropical monsoon with three seasons, winter, summer and monsoon. Winter temperatures vary from 5°C to 27°C and it extends from November until February. Summer follows with temperatures reaching into the 30's. In April rain showers start to occur and by June the monsoon is established with rain and intermittent hot humid days. Recorded rainfall figures from Lashio, which would receive less rain than Bawdwin, are 1571 mm per year.

After meeting the locals we were taken to our guest house to drop off our luggage accompanied on the walk by nice steady monsoon rain.

Then back down to the mine office to start collecting reports - gradually they appeared one by one, most maps etc were missing and it turned out that the leopard survivor was the "keeper of the records" brought out of retirement for my visit.

I put the technology level at mid 50's - early 60's. Nothing could be taken for granted. Don't even think about simple computers, there was no phone contact to the outside world, no photocopier (there was however a copyist whose job was solely to type out copies of what was required), and no plan printer - in theory they could make 'white prints' using sun exposure, but had run out of chemicals so they made 'blue prints'. The only other time I had seen a blueprint in my adult life was when I did some paper conservation work on one. Many of the maps and plans were still on pale blue linen paper, meticulously updated by hand (or not updated as the case may be).

Next was a visit to the open pit, to look at the tuffs and volcanics - a strange experience for several reasons. The first being our security; as it was still almost in the town we just got two men with very large guns, one in front of me and one behind. The second reason was that everybody did what I did - if I stopped they all stopped. If I stepped back from a face to look up so did everybody else. It was like a silent game of "Simon says".

Our guest house was made of teak and painted white on the inside, the ceilings were about six metres up. The bathrooms were historic, I showered by standing in the bath and pouring hot water over myself with a ladle - not so bad really as it was quite warm and humid. Amazingly the water was so soft that it could hardly remove the soap, I guess it was pure rainwater.

On the first night there, right after dinner two local geologists (U Tin Maung Oo and U Chandra Kumar) arrived with their Scrabble set and proceeded to wipe the floor with the native English speakers.

My work proceeded, reading and collecting data, and I finally chose to do this work on a table in the guest house because my presence in the office seemed to unsettle the inhabitants.

The Burmese are very formal, everyone is always addressed with an honorific, I am 'Miss Susan' and the various other staff are either U, Daw, Ma, Saw plus others. The following day, my companions on the train trip here, (Mr Brian, U Aung Bo, Saw James and U Tin Lin) went back to Namtu to begin sampling the slag dumps and Miss Susan was left with U Than Tun (my new minder/butler who serves my food, announces arrivals and departures, and liaises with security), my cook, my personal maid (who washes my clothes and changes my sheets), my driver, and my two security guards sitting in the gate house. I found this retinue quite unsettling. The geologist assigned to me, U Chandra Kumar, came up each morning as well. He was a great help in this work and I'm told by the current owners of the mine that he is now the on-site, senior geologist in charge.

I moved 'my office' onto the eastern veranda - the light was a lot better, and I could at least look up and out and try to focus in the distance across the ravine. Late one afternoon I heard the sound of a procession down in the valley beneath, drums and bells and music - it was Buddhist Lent and this was the weekly procession for alms for the monastery during this period. I could just see them passing by on the valley floor at least 100 metres below. While U Aung Bo was here I noticed that he was observing Lent and not eating meat for the duration. U Than Tun told me that soon the Fire Festival would be starting - at the end of Lent, fire is offered up to the Nats. The Nats are spirits worshipped in Myanmar and neighbouring countries in conjunction with Buddhism, and there are 37 Great Nats, plus all the rest. He also told me that about 3000 people live in BawDwin, which was more than I thought.

As mentioned earlier, there are three seasons in Myanmar, and I was there at the end of monsoon, slowly sliding into winter. I was very intrigued by the Buddhist calendar. I was told a week has eight days which fits in the same number of earthly revolutions as our seven days. Then the year has 12 x 4-week lunar months. As you would detect this gets the calendar a bit out of sync with the seasons so, every couple of years, they are supposed to add in another month (by my reasoning that was still short 29 days a year so that didn't seem quite enough - but it's their year after all).

I decided I needed a field trip to look at the rocks in the pit to sort out some of the volcanic relationships so this had to be arranged ahead of time with my security; and then there was great excitement later in the afternoon when U Tin Lin arrived from Namtu with a photocopier for me; it had been sent up from Yangon, to speed up the process of copying the data, especially the sketches, maps and sections.

The next morning, immediately after breakfast, I was picked up and we set off for my little field trip. There was mist about the tops of the hills that morning, the first in Bawdwin for the season, and U Chandra Kumar told me that that meant it was now winter. We were accompanied on this trip by my two security guards. Only two because apparently the chance of insurgents in the open pit was considered small. We visited the backs of the open pit (which had you asked me - I would have said did not look geomechanically sound) and then up to the top of the world thereabouts (1400 m asl) at the top of the pit. From there I could see a bit more of BawDwin including the old Chinese bridges further up the valley. The rugged topography is impressive; I have previously mentioned that the collar of the shaft in the valley below was at 995m asl, and the slopes pitched at inclines of 45-60°. I had never worked anywhere so incised before. It reminded me of a 33 hour train journey in China from Kunming to Chendu that I had made ten years before in 1987, and my thoughts at the time when looking at the topography were: "thank God I don't have to work here" - well now I was.

After looking at the mine outcrops we travelled up the Nam Pangyun valley through the closely packed in houses and the many, many children, because, being the Buddhist sabbath, it was not a school day. I was a little surprised when I noticed the toilets, which were simple privies built out over the Nam Pangyun with basic board-style footrests and gaps in the floor. I thought about the E. Coli levels and made a note NOT to fall in the water. It was a fast flowing stream, but I wondered what happened in the dry season, as the head of the Nam Pangyun wasn't that far away, so the water flow must eventually cease in the dry season - or did they release water from their reservoirs, if there were any, as a flush to the system? We drank bottled water but, after that, I stopped cleaning my teeth in the tap water.

Observing the people as we moved about I noted that the Shan women carried loads on their heads with magnificent balance and that they carried really large and bulky loads on their backs tied with a strap across their foreheads. I don't know what this did to their necks but I did note that on the whole the women (and men) had excellent posture, even into their old age. The hump of osteoarthritis didn't appear to exist - what the doctors tell us about bone strength and load - bearing work must be true. Whenever I was about everybody, and particularly the children, stared at me, a stranger in their midst (perhaps it was not a good idea to arrive with both white and purple streaks in my hair - but I didn't really think it would matter).

Another field trip to sort out the geology had us driving down the Nam Pangyun valley to look at more contact relationships and then we walked down through the section to the Tiger camp. It is a small settlement of mine workers a quarter-way between BawDwin and Namtu and also the location of the portal to the Tiger Tunnel. The tunnel, built around 1908 and now abandoned, was an adit and along it ran an underground electric railway to part of the mine. When we passed the football field, built on mine fill, the girls' football team was practising and half of them were in long skirts. We had three security that day, one in front and two behind: I asked if this was a more dangerous area and when translated for the sergeant he just laughed. I think they got pretty bored hanging about the town and they enjoyed the excursions and the walk.

There was a full Army Company based locally as security for the Bawdwin mine, but I don't think they were kept very busy. In other more remote parts of the country, I gathered they were busy shaking down the local villages.

The next day's trip was accompanied by four security; it seemed the further away from the township I went, the more security I got. That excursion I wanted to visit several barite-quartz outcrops, unfortunately located right on the tops of steep hills that had to be climbed in a zig-zag fashion. The morning was warm, with the mist about and the ground was saturated and slimy on the top - just right for slip-sliding away. I ended up breathing very heavily (and sweating down my back) which seemed to amuse my sergeant greatly, he couldn't keep a grin off his face. Needless to say the security were all extremely fit and hardy types. On our way out we had to drive by the headquarters of the Army Company located here to guard the mine and we stopped so they could introduce me to the Major, who spoke some English and took me to see a Russian- built swimming pool - circular, about 15m diameter and four metres deep. He invited me for a swim which I declined and he seemed very concerned to know if I could swim long distances— why, I wondered? He also asked would I come to a party if they caught some animals in the jungle - I said maybe when Mr Brian comes back from Namtu - (the idea of a party where I was the guest of honour and only about two other people spoke the same language as me, wasn't my idea of fun).

Then there was a period when it rained solidly for three days, a legacy of a cyclone out in the Bay of Bengal and it started to get pretty cold. I needed to wear my jumper and coat most of the time, plus I was given an eiderdown for nights. The rain also saturated the house and we had some pretty constant drips in some parts. One I discovered when I determined to track down a large drip in my room that I'd been hearing for two nights but not seen. It turned out to be just on the end of my bed coming down between the mosquito net and the frame and landing at the foot. The mattress at the end was quite sodden. The configuration of the mattresses was unusual (to me): they are stuffed with kapok and are in two halves, and each half is made up of two parts themselves, stitched together along three hinge lines so they fold up compactly. They are all hand finished with tight edge quilting to maintain their box- like construction. All of this explained why it was so uncomfortable because, just where my hip reaches in the bed, is the join of the two halves, which are two reinforced ridges.

For about three weeks in Bawdwin my reading and compilations had continued and I now had a good understanding of the historical work and had developed a thesis for the formation of the mineralisation and was starting to write it up, identify targets and plan an exploration program.

The continuing wet weather almost washed out the road to Namtu; when they drove the Land Rover down, they had to temporarily repair two bad wash-aways before they could proceed. I was hoping the train line was okay as that was now the only way out. I was also thinking about the condition of the road from Namtu to Hispaw - specifically I was thinking about the scary wooden bridge at the bottom of the plus 300 metre deep gorge. I was told that the roads into China (which was very close, about 60 km away), aka the Burma Road, had been closed to all truck traffic as a precaution.

As the rain continued to fall, I began to consider that I (and of course my security) could always walk down the railway line to Namtu, it was after all only 11 miles away and all downhill. I decided it would probably take up to five hours but there was an old dirt airstrip at Namtu where they could land a helicopter (one of those big Russian jobs that seemed to be used for internal transport) if we were stuck.

Eventually there was a break in the rain and the MMC Country Manager finally arrived from Yangon after travelling up the little railway. Another week passed with more field visits and technical discussions and, as we prepared to leave, a landslip occurred on the railway line. Some poor unfortunate workers were sent to dig it out overnight. This was not just for our departure because the rail line was also used for the concentrate trains going to the mill. The road was no longer an escape option, it having slid away several days previously.

The digging continued through the night until the track was clear and the 'trolley' was brought through in the early hours. However there was another small slip after the trolley arrived. It was decided to send another train up from Namtu to the other side of the slip and we would go to our side and walk around the slump to the other train and then proceed to Namtu. This apparently meant that our security was compromised (horrors) as we couldn't get all the men with guns in our second trolley. All went as planned and the slip was nearly all dug out when we got there. The ride back to Namtu after that was uneventful.

The next morning the plan was to drive straight to Lashio 44 miles away. First we were told the road was not passable and we would have to go via Hsipaw which caused some consternation because of the extra time needed. Then we learnt the bridge between Hsipaw and Lashio was closed to trucks and the road was 'slow' - and here we were talking about the only road into landlocked Yunnan - the Burma Road - and there were currently in excess of 300 trucks stopped in each of Lashio and Hsipaw waiting to move. Then news came that the more direct route was passable by 'small car' read 'light, high clearance'. So we left in mild drizzle at 8:30am and that 44 miles took more than 3.5 hours and not because of traffic - just the road condition. In that distance we passed four bogged trucks, (all coming the other way and taking nerves of steel to drive around, considering the sides of the road), one truck on its side, lots of bullock carts, people on bicycles (they seemed to wheel them a lot) and one Corolla. Eventually we made it to Lashio and met our "local contact". After tea and bananas he took us to the railway station to negotiate the hire of a train to take 500 tonnes of slag to Yangon for on-shipping to a smelter somewhere (South Africa I think was the destination) to test the metal recoverability.

We had a little time to spare before our flight, so we spent half an hour at the local gold market. Lashio is a boom town from, firstly, smuggling opium (called Number 4 and, no, I don't know what Numbers 1-3 are); secondly, being the only road into landlocked southern China and, thirdly, some legitimate farming business - tea etc. The people have money to spend and banks are out of the question. The jewellery IS gold, and mostly 24 carat (a brassy yellow), and is sold on weight and the gold price with a few percent markup for workmanship. Also rubies. I didn't buy any of either.

Eventually we went to the airport for check-in at 2pm. As the plane left at 4pm this seemed a little early to me. First we got a permit for the Land Rover to go into the airport. Next we (well 'the contact') took our baggage and passports and collected our tickets and got labels for the bags - they were then put on a trolley. While we waited around in the sun for a bit a policeman spotted my camera, and I was ordered to remove the battery and place it in my hold luggage!!! Then we were moved on to a desk where they counted our baggage and laboriously copied out all our passport details, plus subjected us to a little interrogation. Next we went to a bench where they searched all hold and hand baggage. We moved slowly to another bench where they inspected the labels and our passports and stamped each label three times, one on top of the other, plus an extra label for hand baggage - and then opened it again just to confirm that nothing had changed. At last we exited that building and were finally separated from our hold baggage.

Next was a 500m walk to the terminal but, because we got a vehicle permit, our Land Rover picked us up and we drove this distance. The sun was now out and, as it felt like a typical Pilbara day, I had no objections.

Then we checked in: they looked at our tickets and passports and removed the flight voucher - but no boarding pass, as it was free seating. Next move was in to a little booth where we got frisked and patted down: the two women who patted me giggled and one held my hand (not to stop me doing anything, but to comfort me through this violation of my person). Next the metal detector and then the final hand baggage search and another interrogation and we were inside the terminal. The plane arrived on schedule and we flew via Mandalay and Bagan before reaching Yangon. From my window seat it was pretty obvious that central Burma was flooded and the rice crop lost.

A week later, when packing my bags to leave, I thought about my rock samples and realised that it might just be difficult getting them out of the country. I thought about the fact that education was revered in Myanmar, and proceeded to get some nice clean white calico sample bags and wrote 'Teaching Specimens' in large black letters on each bag, then wrapped my rocks up. On my way out everything was searched several times over, but the possibility that the rocks were to be used to teach people about the geology of Burma was all that was necessary for me to be allowed to leave with them. Six weeks after departing Australia I arrived home with my rocks, papers and a taste for chili noodles at breakfast, although it was probably at least a year until I felt like eating another banana.

A SCATOLOGICAL INCIDENT

ROB DUNCAN
CHINA ca 1998

Over three years in the late '90s, I worked around central and south-western China visiting and assessing gold mines and prospects where Western capital and modern technical expertise might find a synergistic partnership with the Chinese owners of known deposits. Based at Brigade 711 of the Chinese Gold Bureau in the Qinling Mountains of Xi'an Province, I was supervising step-out diamond drilling of the Jian Cha Ling gold mine on behalf of an Australian gold exploration company. This entailed a daily 4WD trip followed by a scaling of the 'Green Wall', a scree slope of slippery serpentinite to a saddle overlooking the beautiful rural CCG valley. A rooster at a village far down below would crow an alarm as we ascended steep mountain slopes to drill rigs manually assembled on precarious hand-excavated ledges. With the help of an interpreter, I would assess overnight progress and make rough logs of the drill-core before it was packed in wooden boxes for transport to Brigade 711 headquarters in Mienxian.

This particular day, I was inspecting and logging the core in the backyard of 711 before going down to lunch at a local restaurant.

Although food does not rate highly in my daily priorities, it is still a necessity of life, and while travelling and working in China, regular stops at the trough were an essential part of social and business life. Usually I dined with Chinese colleagues who picked carefully through the menu, questioned the waiter at length, and often demand inspection of a plastic bag of freshly flipping prawns or a sinuously uncoiling eel. At such times, I took refuge in my ignorance and relied on someone at the table with experience of Western squeamishness to apply an appropriate degree of culinary censorship to the exotic fare.

But this afternoon, being alone at table in the somewhat pretentious restaurant, I idly reviewed an English translation of a few pages of the lengthy and otherwise thoroughly incomprehensible menu book. When dining alone in China, I usually scanned the pages for the vaguely recognizable characters for chicken and rice and avoided dishes described by long lines of characters which I suspected tried to oversell something derived from reptiles or unskinned puppies.

But back to the menu. The English translation is very literal, and the list of dishes deserves serious attention!

Appetizers;

> Stewed pigs tripe
>
> Poached dogs meat
>
> Duck chin with sweet sauce

Braised food and soup;

> Stewed duck with caterpillar fungus (or donkey hide gelatine)
>
> Wax gourd and fish maw sauce.
>
> Spicy goldfish soup

Crock pots;

> Pigs tripe with ginko and dried bean curd roll
>
> Frog and chicken pot

Miscellaneous;

> Fried pigs tripe
>
> Pigs intestines with ginger
>
> Steamed kidney balls and jellyfish
>
> Crab ovum and frog oil flour

I suspect translation errors for some of the above – a duck must be the original chinless wonder, but the number of variations on goose intestines further into the menu made the pigs bowels a certainty, so I eventually chose the innocuous sounding Spicy goldfish soup. And excellent it was, highly spiced and full fish-flavoured, and I ordered a second bowl. The kitchen must have been delighted at the repeat order, for no less than a two litre crock pot of the same soup came forth, decorated with swirls of cream and herbal garnish and I made a near thing of finishing the lot.

It must have been an hour later, or maybe a little more, for my sense of time was becoming vague, as were most of my other senses, and leaving the trays of diamond drill core I was logging, I started on an increasingly erratic journey towards my room at the commune guesthouse down the hill. Still some distance from the guesthouse and just about floating off the planet, I spotted a door with a Chinese character for toilet, female as it turned out, with its cross-legged character, but vacant, and suddenly critical to my immediate future. Bursting into the first large-tiled cubicle with shoulder height walls and a squat hole in the centre, I frantically stripped off everything but my boots. And the world exploded. Spinning this way and that, I lost the battle with orientation, lost all control of everything, and blasted forth from both ends simultaneously.

When the spasms died away, that little room was spray-painted pink from wall to wall with aromatic spicy goldfish soup.

Too exhausted to be concerned with modesty, I staggered from the tiled pig-pen, uncoiled the fire hose from the wall, and water-blasted my body, my boots and every part of that white-tiled hell-hole and, via the back door, quietly slipped into the guesthouse and bed for the rest of the day and night.

I have rarely been ill in my life, but that incident must have shocked my immune system into overdrive, for in the succeeding years of foreign work and travel, from the "Stans" of Central Asia to the jungles of the Congo River, I've not had so much as a cold, and when I see my companions being medevac'd out of Kashgar, or crying out for Lomotil in Lomié, I wonder if Spicy Goldfish Soup might someday be revealed as the universal panacea.

"YES, COMMANDER. WHATEVER YOU RECKON"

JOHN NETHERY
THE PHILIPPINES 2001

January 2001: I was approached by Jim Wall on behalf of WRF Securities to do a short reconnaissance assessment exercise for gold potential in the Philippines in a joint venture with a local tribal group: "Well maybe. Where in the Philippines? Mindanao, eh? Where in Mindanao? East coast, eh? Well okay, it's not West Mindanao after all." As it turned out it was in the peaceful precinct of Boston, Davao, but inland and adjacent to it is the notorious Diwalwal Goldfield, where anything goes.

I visited Diwalwal to get a feel for the type of gold deposits being mined in the district. It is one of those strange anomalies that occur in some underdeveloped countries. One assumes it is a multi-million ounce gold vein deposit, discovered, so the story goes, by someone bulldozing a new road. It is not really known how large it is because it is totally run by illegal miners, so there are no official records. There was never a formal exploration drilling programme. Anarchy ruled. There were many separate operations, ranging from small one-person gouging shows up to well established underground operations run by individual warlords. Private armed militias fought for a piece of the action. Murder was commonplace, particularly in the early days. Men armed with sub-machine guns escorted us everywhere at the mine I visited. The place was bristling with guns and there was a distinct air of tension.

A series of adits (horizontal mining tunnels), one above the other, were driven into the hillside on this steeply dipping high-grade quartz carbonate vein system by these various militias, who commonly squabbled among themselves. It is rumoured that at one time a multilevel collapse killed two thousand miners in a matter of minutes.

After discovery, illegal miners had descended on the place and, before the authorities knew it, a shanty city of forty thousand people had replaced the jungle in a rambling, shambling way across the steep hillsides. It is amazing that this city actually functions. There is no water supply, sewerage, electricity, curbing and gutters, road sealing or even basic maintenance. The main thoroughfare through the centre of town was a steep gradient track, one truck-width and eroded down to uneven, sharp-fractured, rock basement. It was even more amazing that the trucks actually still negotiated it. I watched one truck loaded with bagged ore take just on ten minutes to negotiate a steep two hundred metre section of the "main street". A standard sedan motorcar would be out of the question and, even for a large high-clearance standard 4WD, such as ours, it was a serious effort. This main street was lined with shanty town buildings occupied by tavernas, bars, gold buyers, brothels, fuel outlets and weapons salesmen. I saw no council buildings, post offices, schools, hospitals, medical facilities, pharmacies and definitely no police stations. The place seemed chaotic and lawless but somehow seemed to operate. I was told the government drew an imaginary virtual boundary around the place and left them to it.

Boston, on the coast, and the small inland village of Caatijan where I stayed, were quiet, tidy and orderly in contrast to their neighbour. The primary school at Caatijan had a neat cut grass playground, bordering gardens and the national flag flying proudly. It was a pleasure to see the kids in tidy clean uniforms line up in rows and sing their anthem before class. A large neat sign painted on the roof announced to the world that this was the Caatijan Elementary School. The kids were happy and smiling and loved the opportunity to gather for a group photograph.

The area of the tenement included that village and covered 160 square kilometres of very hilly rainforest country, but with a reasonable network of forestry tracks. One of the very large international gold mining companies had done some exploratory work but obviously decided there was not an opportunity for a deposit of their target size parameters. I assume they were looking for a Diwalwal repeat. Several of the local men had worked on that programme and it was obvious from the outset that they already knew the locations of all significant target alteration zones within the tenement. I was pleasantly surprised at their basic modern exploration know-how. I welcomed their contribution to the field programme.

The arrangement with the local people was that I would stay in the village in the head-man's family home. They were so hospitable and waited on me like I was royalty. It was verging on embarrassing. It was one of those family situations of a middle-aged couple whose offspring had flown the coop, married and had family themselves, and all lived nearby. That sort of extended family in such cultures where the grandchildren are part of the household is a wonderful arrangement, and in my mind sure beats our nuclear family culture. Members of the younger generation were in the exploration team.

The exploration programme involved traversing all the substantial creeks, and examining the areas known to the local crew. We also visited several small shallow pitting gold mining operations outside of the tenement. Every morning a convoy of motorbikes would leave the village and meet at a prearranged spot. I was intrigued that on our way we would occasionally come to a halt and a bag of foodstuffs would be left beside the track under a tree or near a hut. The location varied. I thought that odd and so asked for an explanation. The response was that it was for the Western Boys. Nothing more was said so I was still in the dark, but thought no more about it. I assumed some local arrangement with a group of forestry workers.

One evening towards the end of my three-week stint the headman asked that we vary our schedule the next day. The Western Boys wanted to meet me. Great! It will be nice to meet these mysterious bush dwellers. Tomorrow will solve the mystery. Next morning only three of us on bikes left the village; the headman, a young woman who I'd noticed before because she seemed to flit in and out of the local social scene, and myself. Interesting. So we floated off out into the scrub then down a dead-end track to a small clearing in the jungle. A small open shelter in the middle of the clearing was occupied by one man. Four athletic-looking young men carrying M16 automatic rifles were strategically placed around the clearing and closed in as we dismounted. The man in the shelter walked towards me and it was obvious that he and the armed guards were in some sort of military apparel. "Hello John. Welcome. Nice to meet you. I am Commander Jomar from the New Peoples' Army. Come. We talk."

We made ourselves comfortable on benches on either side of a bush table, and Commander Jomar presented a bottle of ginseng tea. "No, I had not experienced ginseng before, but thank you, I would love to try it." Commander Jomar had a good command of English and said he wanted to talk about my situation. The four lads had positioned themselves at each corner of the shelter. I noticed they were still holding their M 16s and had not slung them over their shoulders. I should be careful about my choice of words. The village headman and the young woman, who I assume was a facilitator, did not join us, but sat under a tree on the edge of the clearing.

So we rattled on for an hour or more. Jomar was pleased that my employers, the Australian company, had initiated a joint venture with the local people. There was minimal involvement of the "corrupt capitalist pigs of a government in Manila". What did I, a tool of the capitalist system, think of that arrangement? Well of course I thought it was a marvelous idea that the local people should benefit from any discovery locally, and that the Australian company was here to contribute funding and expertise to try to get something up and running. Of course there were no guarantees that an economic deposit of gold would be found, but we will try our best. I am thinking: there is Diwalwal over the ridge where forty thousand illegal miners from all over the country have descended on this rich patch of ground and anarchy ruled, not communism. But best not say that. He did not bring up the subject of Diwalwal so I suspect there may have been some sort of cosy funding arrangement involved – a little mafia-type deal?

"Why is there no revolution in Australia?" "Well over two centuries since European settlement we have sorted out a system where there is very little class distinction. The workers participate in the political system and have Trade Unions to protect their interests. Everyone has the opportunity to vote for whatever governments they wish. Everyone had the right to an education to enable them to do what they wished in life. It is one of the most democratic countries on the face of the planet. There was a political attempt to ban communism in the 1950s but it failed, and it is perfectly legal for communists to stand for election to parliaments." He nodded wisely throughout my diatribe. I suspect I chose my words fairly well.

Eventually he called a halt. He had to go. Would I promise to procure him some mobile phones and a pair of binoculars? Whew is that all? "Of course your wish is my command, Commander" was my thought, though not my words."I will certainly try to get them." Away we headed back to the village. No more was said about the visit except that supplying mobile phones was obviously not feasible. The young woman had slipped away as fleetingly as she had come.

On my last day, a gathering of prominent people, including the mayor, wished me farewell and thanked me for my expertise at a formal ceremony on the village stage where there was also a gathering of villagers including the school-pupils. I, along with others, gave a little speech outlining our hopes for the future. The contrasts of culture that I had seen in my short visit were at the forefront of my mind: a joint venture between capitalism and local scale democracy at the tribal level, communist revolution out in the scrub, and "might is right" anarchy just over the hill.

INDONESIA TIMOR
AND
NEW GUINEA

THE EXPEDITION INTO IRIAN JAYA

NEIL STUART
INDONESIA 1972

Neil Stuart originally gained qualifications in agriculture, before turning to geology. He has a BSc from the University of Melbourne and an MSc from James Cook University in Townsville. He began exploration in Nickel Boom times, initially working for Utah Development Company and searching for copper, uranium and coal in Australia, Africa and Indonesia.

He then turned successfully to coal exploration in Queensland, managing the coal section at Marathon Petroleum Australia, and finding the Darling Downs deposit, which supplies coal to a large power station in Southeast Queensland.

In 1980, he established a consultancy and also identified and acquired projects which led to the establishment of a major gold mine and a large lithium brine operation, both in Argentina. He has been a director of several ASX-listed companies over many years.

1972, Indonesia. It had only been a few years since Sukarno, the first President of the new Republic of Indonesia, had been replaced by the army strongman Suharto. Suharto starts to open up the country to foreign investments – and particularly mining ventures. At this time I was working as a young geologist on a copper exploration program for Utah Development Company in South Australia. Utah's senior geologists and mining engineers were travelling the world looking for new projects. It was looking like Whitlam would be elected as Prime Minister of Australia and his proposed Mines/Resources Minister, Rex Connor, was talking about nationalizing the whole mining industry – so Utah, among other companies, was looking overseas for new projects. Anyhow one of the engineers happened to be in Jakarta and bumped into people who said they had a great copper prospect in Irian Jaya (Western Papua), just on the northern side of the range that hosted the newly discovered, massive copper and gold Ertzberg/Grassberg deposit, found and owned by Freeport Copper. So Utah needed a couple of geologists to check it out. Enter Senior geologist Gerry and Neil (the junior).

It was my first time in Jakarta, the capital of Indonesia - in fact it was the first time I had been to Asia! Because of the recent change in government there was a large military presence in Jakarta – soldiers everywhere; troop trucks, armoured vehicles, et cetera, were driving around town all the time. We ended up spending three or four weeks in Jakarta as we had to ascertain the legitimacy of the people who were offering the project; there were lots of "hangers-on" and people wanting slices of the action. Also, because of the military and associated sensitivities, it took much time (and money) to acquire the necessary permits for travel to Irian Jaya.

Going around town with our Indonesian partners/minders there was a lot of "flick this guy xx Rupiahs", or words to that effect, to be able to make any progress. We knew we had a tough jungle trip ahead of us and this wandering around meeting people was making us soft, so we initiated long walks around town – and sometimes ended up in interesting places. In the evenings we amused ourselves by hiring bejaks (the pedal-bike, rickshaw-like transport) and paying a bit extra so that we pedalled the bikes (one each) and the owners would be the passengers. Then we would have races. This terrified the bejak owners – I don't know why!

Eventually we received all the permits and paperwork et cetera to enable us to get to Irian Jaya. However we had little success in acquiring any maps, aerial photos or whatever as these things were restricted because of "security risks". We ended up acquiring (from the American embassy, I think) some maps at useful scales, that were derived from USAF aerial photos taken circa 1944. Unfortunately the maps had large blanks where there must have been clouds when the original photos were taken. The maps were of the kind where you might expect to see an arrowed comment saying "here be dragons" or some such. At least they did have the coastline, major rivers and some village names (phonetic) – which later proved useful. There were no roads or any marked tracks in the areas that we were going to visit.

The main Indonesian/Irian Jayan man (let's call him "the Prospector"), who was the instigator of the project and knew where the "mineral" was located, had attached himself to some Jakarta-based Indonesian entrepreneurs who had then alerted companies to the possibilities – and they would be part of the deal. We asked the "Prospector" if he could outline the general area of his prospect on our (1944) maps. He proceeded to wave his thumb around a huge area of our 1:500,000 scale maps, which I thought was not all that encouraging or useful, so much so that we later called Utah's head office in Melbourne to convey our concerns about the whole thing. However, it was a boom time for the search for Porphyry Copper deposits at the time and we were instructed to "carry on". And so we did.

The plan was that Gerry would fly to Biak (an island just to the north of Irian Jaya) with the Prospector and obtain the permits and make arrangements for the expedition, while I stayed behind in Jakarta for a few days in order to collect a large sum of money that was to be "wired" to me from Australia; this was the days before credit cards had really been established. In due course I collected, from a bank, millions of rupiah (truly a "King's Ransom") in a number of brown paper bags and wandered back through the streets of Jakarta to the hotel. Safely in the hotel I arranged for my flight to Biak.

It was leaving the next morning at about dawn. It was a "milk run" stopping at many places before arriving in Biak about 8 hours later. The plane was a four-engined, propeller-driven Vickers Viscount. I duly arrived at the airport (which was a military one in those days) pre-dawn and fronted the ticket counter only to be told that the flight was cancelled – "come back tomorrow". So next day, same time, same counter only to be told that the flight would not be going that day either. Further enquiries revealed that on the previous return flight to Jakarta there was an attempted hi-jack; these were the days when hi-jackers wanted to be taken to Cuba or whatever. Anyhow at that time pilots in Indonesia were mostly military or ex-military people and commonly carried side-arms while flying – as you do. So, when the hi-jacker burst into the cockpit waving a knife/sword and shouting "take me to Cuba", or whatever, the pilot just up and shot him, but a few stray bullets penetrated the fuselage and consequently the holes/damage needed to be repaired before it was allowed to be flown again.

So the third day I turned up at the airport (pre-dawn) again, but still the plane was not going: "come back tomorrow". When I arrived back at the hotel prepared to re-book for another night the reception staff said that they hadn't bothered to make up my room or book me out as they thought (knew?) that I would probably be back. The next morning, at the airport – yes! the plane is flying! But it was "chocka"– it seemed that four flights' worth of passengers and all their gear were going to be on that plane. When I got to my (window) seat there was already someone else's gear under the seat in front of me such that my knees were under my chin and my carry-on bag was on my lap – sort of. The central aisle was full of people's gear of incredible variety. Anyhow, off we go, getting up speed. I'm looking out the window and after a while the parallel white lines at the end of the airstrip go past, and we are still on the ground! Not to worry, I thought, "not so bad it's only paddy fields, so should be a soft landing". Somehow we were in the air, flying very low and it took a long time before we had enough altitude to allow the aircraft to execute a turn. Landing at Biak, which featured in World War II and had an American-built airstrip (which was now the island's civilian airstrip), there were a lot of old (and some not so old!) aircraft wrecks that had obviously crashed on landing and had been bulldozed off the strip into the edge of the surrounding jungle. A disturbing sight.

We had another two weeks or so in Biak obtaining permits et cetera for travel to mainland Irian Jaya. Disturbingly we had to surrender our passports to the authorities – hopefully being able to collect them again when we arrived back in Biak. Finally, flying in a Twin Otter, we were off to Nabire, a small village/town on the north coast of Irian Jaya and from where we would launch the expedition. Once there we rented a cabin owned by the Mission Aviation Fellowship, a Christian ministry organization that, among other things, attended to the health of local people and especially to those in remote villages in the mountains. As part of their equipment they had a single engined Cessna to assist in these endeavours. Interestingly we learned that only a few years previously they lost one of their missionaries who was *killed and eaten* in an area not that far from our area of interest.! Anyhow we hired their Cessna and did a reconnaissance flight over the prospect areas that we might be visiting on the ground. This led us to hire the plane again to drop some food and supplies in the area as our calculations showed that, if we were to be doing a three-week trek we would need a certain amount of food, porters, et cetera; and, if a longer time, would need even more food, therefore more porters and so on, and it becomes unwieldy. We acquired rice and dried local fish from the local markets (the fish were just laid out to dry along the road verges). We wrapped all this up in hessian bags and had the pilot drop them out of his plane as he flew low and slow over some river flats that we had spotted on our reconnaissance flight. Interestingly when we later arrived by foot in that area the people from the nearest village had already collected the bags and stored them for whoever it was that were going to be coming to their place.

It was at this stage that our Prospector racked off with a dozen or so porters and most of the food and supplies that we had bought as he had additional agendas (mostly political – there were elections coming up). We were to meet at a certain village later. So we raced around and employed a couple of locals to accompany us and to act as interpreters. We also had to organize some food, for who knows how many days, before we met with the Prospector again. He had even taken all the cooking gear that we had organized. However we did have our own packs and minimal survival gear and bought some dry biscuits and a few tins of tuna or whatever; the empty tins became our billys for making tea. The local guys turned out to be excellent. They stayed with us throughout the whole trip whereas the porters with the main party would only go to the first village and then wanted to go home. So the main party had to employ new porters, who also did not want to stray far from home, and had to be replaced at the next village, and so on.

Three days later we were able to hire a dug-out outrigger canoe with an outboard motor to take us along the coast, about a six hour trip, to where we stayed the night in a coastal village. Next day it was walking for miles along a vaguely visible track through the mangroves – mud, mangrove roots, mosquitoes – not all that pleasant. It was essential to traverse the mangroves before dark otherwise we would be forced to sleep up a mangrove tree as the place was infested with crocodiles. Luckily we completed the traverse in about five hours and were able to make a camp on dry, crocodile-free land – oh joy!

Gerry and I had army-style hammocks (with mosquito nets), which we hung up between trees high enough to be above any marauding wild pigs that might be wandering around at night. Our local (coastal) guides had "sleeping sheets"; the porters from the inland villages did not bother with these, but relied on only getting a good fire going. A fire at a camp was a high priority for the locals – and they were very good at it. They would always carry some dry grass or kindling with them, so that even when it was raining (which was pretty much every evening) they would be able to get one started. At a camp they would construct a "lean-to" shelter made of bushes and large leaves which kept the rain off and, together with the fire, was quite comfy. In general you would start to see a fantastic build-up of cumulus clouds about 2.00 pm, hear a bit of thunder around 3.00pm and be drenched with rain by 4.00pm. We were always walking through creeks, wading or swimming through rivers and generally were always wet. I would try and dry my boots near the fire at night and it was with great pleasure that I could pull on a pair of dry boots in the morning. Dry boots lasted until the next creek crossing – usually about 10 minutes from leaving camp.

From the first camp at the edge of the mangroves the trek to the first village took two days walking along and across rivers and then up steep, muddy ridges (crawling, not walking) and then down (sliding, not walking). In this terrain having a geologist's pick was quite handy and helped to drag oneself up slopes or slow one's descent downwards. Eventually we arrived at the first village and met up with our "advance" party (and all our supplies) and the local villagers, who were quite friendly. They were expecting us and were wondering why we took so long! This was the village that had collected our supplies and also where we were expected to attend to their medical problems – I don't know where they got the idea that we were doctors, health workers or whatever. We did in fact have one backpack dedicated to medical supplies, but that was supposed to be for us and the porters and guides. However Gerry and I had attended a tropical medicine course before travelling to Indonesia. So next day there is a line of a dozen or so people for us to see. In fact we could treat quite a few of them who had cuts, malaria etc. One guy presented with a terrible looking tropical ulcer, which we treated with antibiotics and told him to get down to the coast where there was a clinic. When we returned to this village two weeks later he was still there. We strongly suggested he still should leave.

The people of the inland/mountainous villages are small in stature – perhaps adults at 145-150cm tall. Gerry and I, at a very average Australian 175cm, towered over them. Clothing was very minimal – the women wore "lap laps", the men were naked except for their "Kotikas" (penis gourds pulled up and held with a piece of string around the waist). The children wore nothing. Each village generally had a "guest" hut for travellers. These were similar to all the other buildings and were on pole stilts about 2m high with a thatch roof and a floor of thin sticks laid parallel across some main beams. This floor, of course, had many cracks and food scraps and other things would fall through these and feed the chooks, pigs, dogs, whatever, living underneath. It was very hard to get a good night's sleep in these villages – there were plenty of whining mosquitoes, grunting pigs, crowing cocks, crying babies, old men hacking and coughing, couples arguing. Sleeping in the jungle was much preferred.

We spent a couple of days in that village organizing expeditions to the "mineral" areas with the Prospector and his mates/guides. Eventually we were off to find the "bonanzas". This took us many days of serious hill climbing and river crossing – I was reminded of the Aussie Diggers in WWII fighting the Japanese on the Kokoda Trail and elsewhere. They must have struggled up similar terrain, but at least no one was shooting at us too.

We eventually did come across two occurrences of "mineral". Both were highly pyritic shear zones in metasediments, but with no signs of copper, zinc, lead or anything exciting. The shiny, metallic looking, fresh pyrite was what excited the Prospector and his mates. Fool's gold strikes again! However, we took some samples for later assay (they did not return any significant copper). We also made geological observations.

Particularly, in the rivers, there was an abundance of pebbles and cobbles of very interesting rocks. They presumably were derived from the high central range we could occasionally see in the distance if the clouds lifted long enough – there are mountains of plus 4000m up there and they are commonly snow covered. At one camp we must have been quite high (we didn't have altimeters) as we were in a stunted, moss-covered cloud forest. The night was cold and our local guides, who were pretty much naked, wrapped themselves with leaves and huddled around the fire. Our guys from the coast had shorts and T shirts, but everyone was keen to get down to lower altitudes as soon as possible.

Later, in another village, we wanted to "do a loop" – about a five-day walk, which took in a different village that we would pass through in about three days. The name of this village was written (phonetically) on our old USAF maps and when we mentioned the name everyone recognized it and said "yes, yes, we know that one" or words to that effect. We decided that we would only take a small, fast party and travel light. So Gerry, myself, our two coastal guys and one local guy set off. We had arranged (or we thought we had arranged) for a few of the locals to go directly to the village in question and drop off some food there for us to collect on our way through – which would allow us to travel light. Fine, fine, fine we were told. Here I must point out that we would speak in English to our coastal guys, who then spoke in Indonesian to someone in the village that could understand Indonesian and then he would translate to the village guys who had their own dialect of the native language – this would vary from village to village. Chinese Whispers!

Anyhow we did our geological traverses and I remember thinking on the night before we were to reach the next village that we may as well have a big feed as we would be picking up the food drop the next day. Tomorrow comes and we arrive at the village. Lo! It is an abandoned village – with the slash and burn agriculture system the village gardens lose their fertility after some years and commonly the village is progressively abandoned and people establish a new one in a different area. The village retains its name and perhaps many years later is re-settled. Of course, there was no food drop waiting for us in this village and we had a few hungry days before arriving back at our starting point. When I questioned the guys who were meant to make the food drop, they replied that there would have been no point as the bush rats and other things would have eaten it all. "But why didn't you tell us?" fell on deaf ears; I suspect they didn't want to tell us as it would not have been good news: best to be happy (initially) then hungry, rather than be disappointed.

We completed our surveys (such as they were) and prepared for our three to four-day trek back to the coast. The Prospector and his mates did not want to accompany us as they were politicking with the locals. So Gerry and I and the two coastal guides headed off and eventually arrived at the coast village and sent out a "bush-telegraph" for a dug-out canoe to take us back to Nabire. It arrived the next morning.

Health-wise our group was quite lucky. We had numerous cut feet, but a bit of antiseptic and a band-aid usually solved that. Everyone got leech bites (every day). If they were on your leg you could scrape them off with a jungle knife, but they sometimes got into your boots and when you took them off at night you would pour out water-diluted blood from them. We had one guy who had a leech inside his mouth. We removed it with salt. Among the team we had three centipede bites, one scorpion sting and one snake bite. The snake bite can't have been too venomous — he wasn't actually with Gerry and me at the time, but we organized for him to be sent back to the coast via the quickest route with a couple of the guys as escorts/helpers and heard no more. One guy came down with malaria, but we had some spare quinine tablets, which we were able to give him. Probably the majority of the population had malaria in their system and occasionally it would break out and give them the fever.

Back in Nabire. Over the last few weeks I had this vision of getting back somewhere and socking down a beer — I had previously noticed that the local general store in Nabire stocked Heineken beer so I was very much looking forward to it. However among the first people we met was the Mission Aviation Fellowship leader who invited us back to his place for a "nice cup of tea". He was a genuine good bloke. The Heineken would have to wait.

A few days in Nabire and then we were off to Biak via the Twin

Otter – but when everyone was on board they could only get one engine started, so we all had to come back the next day when a replacement aircraft arrived. At Biak there were some hassles getting back our passports – another day or so and then, glory be – a DC-9 jet direct to Jakarta.

From Jakarta we decided to return to Oz via Bali – we reckoned we deserved a bit of R & R. Surfing, drinking, riding motor bikes (no helmets of course). Bali 1972 – the local girls still went around bare-breasted. Nice.

THE SALT OF THE EARTH

GARY MADDOCKS
SULAWESI 1975

Gary Maddocks was born in 1942 and graduated from Melbourne University as a geologist in 1966. After a few years of postgraduate geological and geochemical research, he began a career in mineral exploration working first in NZ with Kennecott during 1968-69 before returning to exploration- focused research in the Northern Territory. In 1975 he returned to direct mineral exploration, management and consultancy for base and precious metals with major mining houses and junior explorers, continuing this career until retirement in 2015. While mostly involved with exploration throughout Australia, his overseas experiences include assignments in New Zealand, Indonesia, Bolivia, India and Mozambique.

An Indonesian Novice

My first exposure to exploration in Indonesia came about by default during 1975 after I had re-joined Kennecott Australia to become involved with exploration based out of Kennecott's Ok Tedi porphyry copper discovery in Papua New Guinea. At the time I was told the project was expected to go ahead after negotiations with the PNG Government were finalised. As it eventuated, I never did get to Ok Tedi. The PNG government never granted Kennecott a mining lease application over its discovery or compensated the company for that discovery. Sovereign risk, it is called. After several years of international legal argument, the property was tendered out by the PNG government with it retaining minority equity in the mining phase of the project.

So, after working for a month or more in Kennecott's Sydney office as a "FIFO" employee living in Melbourne, I was assigned to a fill-in job on the company's Tapadaa project in Northern Sulawesi doing six weeks on/two weeks off trips while the PNG negotiations continued. I went on my first trip to Indonesia as a novice traveller arriving alone at Jakarta's then main and somewhat chaotic airport to be lost in a mob of taxi drivers all wanting to take me to wherever I wanted to go. Wet behind the ears regarding Asia, I chose a driver and got to the Jakarta office only to be told I had paid about 4 times the usual fare for my ride!

I left on my first flight from Jakarta to Gorontalo, Northern Sulawesi, a few days later, via the "milk run" through Banjarmasin, Balikpapan, and Pula. Very soon I had completed my first six weeks stint in Indonesia and, during this first trip, I was informed that PNG negotiations had hit a few snags so I would be working in Indonesia rather than PNG for the foreseeable future. This was at a time when our only field communication with the outside world was by shortwave radio to Jakarta office, which had a telex, and the only telephone service was a very limited one extending to a few government buildings and business within Gorontalo. So I decided to bring my wife and 20 month old daughter to Gorontalo rather than continue "6n2" week stints and forego regular contact with my family.

The Tapadaa porphyry copper exploration project was located several kilometres northeast from Gorontalo along a poorly constructed road crossing a few significant rivers. Access to site was by way of the tray of an old Dodge truck, or by motorbike. When I arrived the site had been the subject of several years of exploration during the early Seventies, first by Endeavour Resources, followed by a Kennecott/ Endeavour joint venture team comprised of a few expat geologists, surveyors and field assistants, and a lot of local labourers. During this time, stream sediment and soil geochemistry samples had been collected, contour trails along the undulating terrain had been constructed, and hundreds of regularly spaced exploration pits had been hand-dug to get some geological information concerning the covered basement rocks. Initial exploratory drilling of three main geochemical and geological targets had commenced using a single diamond rig and a smaller, more portable, Winke rig. My initial involvement included compiling a mass of data collected over the exploration period onto metric scaled maps as much of the earlier work was on imperial scale and Australia was then in a period of changing to metric measurement. These were the days of hand drafting, light tables for hand copying and sepia film print machines rather than the computer-assisted drafting now available. However, I was soon seconded to direct field work mainly relating to mapping and sampling of recently dug pits. I was generally staying a few days overnight in the purpose-built camp before returning to Gorontalo until my wife, then pregnant with our second child, didn't like being left alone overnight in Gorontalo.

As a consequence, I started taking a motorbike to site so that I could return each night. This was generally fine but, on one particular trip home after a long day, my return started after sunset and soon I was in total darkness under an overcast sky. I knew I had to cross the main flowing river using a submerged ford about half way back to Gorontalo, which was generally not a problem in daylight, but this time I was only part way across when the bike headlight extinguished. Not being able to stop in the darkness without potentially losing the bike due to the steady water flow, I hoped for the best and held a steady course to the opposite riverbank without riding off the side of ford. On dry land again, I managed to get the lights working again and continued on my way along the bumpy road towards Gorontalo. However, my ordeal was not yet over as a few minutes later the motorbike engine cut out, presumably due to river water having seeped into the electricals.

So here I was, in total darkness, with no torch or matches and a motorbike that I could not restart, on a road that was not a regular thoroughfare during the day, let alone at night, and still a long way from Gorontalo. Remembering that there was a small village set off the road a few kilometres closer to Gorontalo from which I may be able to get some form of transport, I started walking along the road into the darkness, leaving the bike in the middle of the road. After several kilometres of walking the road, and without seeing any sign of a turnoff to, or lights from, habitation in the forest, I decided to return to the abandoned motorbike. My next issue, however, was to find the bike in the darkness without a moon. After walking back to the area and several times walking back and forwards over a few hundred metres of the road, I finally found the bike by nearly falling over it. To my amazement it restarted within a few kicks, presumably because the electrics had dried out while I was walking, so off I went feeling very relieved. By the time I got back to Gorontalo the time was close to 10pm and my wife was highly concerned about my absence. Needless to say, I returned to travelling to and from site by daylight rather than to have another instance like that night again.

Like most exploration projects, Tapadaa never became a mine but the influence the exploration activity over many years had a lasting effect on the inhabitants of the relatively small township of Gorontalo. The period concerned was not long after some of the internal troubles stemming from the end of the Sukarno era and, in that region, the bloody start to the anti-communist Suharto regime. The presence of Europeans coming and going from the airport and the few children of expats, including my toddler first child, living amongst the locals was a source of wonder to children and adults alike. Also, the opportunities for a regularly paid income caused most of the town's teachers to join our local staff and made an affordable reality of such things as motorbikes, transistor radios, other European goods, and the wealth to afford more than one wife. I have never had the chance to return to that part of Indonesia but I am sure that there are many elderly Gorontalo men who have great memories of those days.

Muro and Moro

My first Indonesian reconnaissance for the Newmont Mining Corporation came in 1983 when I was seconded to help out with investigations of overseas project offerings, while remaining domiciled in Australia, and it related to newly discovered gold in the Muro region of Kalimantan. At that time, I was a lone exploration geologist venturing into the country for Newmont to investigate epithermal gold opportunities for which Indonesia is well-known. A few years after my efforts, Newmont set up a Jakarta office and subsequently made significant gold/copper discoveries at Minahasa in North Sulawesi and at Batu Hijau in East Sumbawa.

I arrived at a more sophisticated airport terminal than the one I had encountered during 1975 and was met by a representative of a small geological consulting group, PT Mincon Abudi, who provided a local geologist as my assistant and general guide. That geologist was Sulawesi-born Omar Olii who turned out to be a very able geologist on this and subsequent sojourns into Kalimantan. After a few days of administrative arranging handled by the consultants' office, I had my permit to travel to the Moro region of Kalimantan. Prior to our departure by air from Jakarta to Banjarmasin, I was informed that, as a result of smoke haze from forest fires, the previously arranged helicopter transport would not be available and we would have to avail ourselves of other means of transport into the southern central Kalimantan region.

Undeterred, we proceeded by domestic aircraft to Banjarmasin where Omar and I boarded a small commuter aircraft of 10-12 person capacity to fly to Muara Teweh township on the Barito River. All appeared very straight forward except that much of South East Asia was in drought conditions, now recognised as being due to El Nino conditions in the Pacific Ocean. As a result, fires had started in the forests along the eastern coast of Kalimantan and a blanket cover of smoke and low cloud reigned over much of the island. There were also rumours circulating that other geologists had taken more than one attempt to fly into Muara Teweh only to be turned back when unable to land. Now we became fully aware of why the initial use of a helicopter had been dismissed. Undaunted, our young aircraft pilot took off and continued towards our destination but, given the cloud cover we were flying over and occasionally through, it wasn't long before I pulled out the one map of Kalimantan I carried to see if there were any high mountains in the vicinity of our presumed straight-line flight path. Thankfully, that map showed no immediate altitude issues.

We probably continued for about 1½ to 2 hrs before the pilot was satisfied he must be in the vicinity of the Muara Teweh township so he began circling for some long minutes looking for a break in the clouds. One was eventually found and he immediately dropped the plane into what was a steep descent, pulling up only when he got through the cloud break and within some 300-500 metres of the Earth's surface! Once under this cloud cover, and still within about 200-300m of the treetops, he proceeded to look for the main Barito River and, having found it, began following it at what appeared to be near tree top level. Within minutes there was a sudden shout from a few of the Indonesian passengers that must have told the pilot he was flying in the wrong direction. He immediately turned the aircraft into a sharp banking manoeuvre and flew back along the river until he spotted the township and airfield off to the right of the aircraft. However, instead of doing the conventional circle and straight line approach towards the strip, the pilot simply dipped the wing and appeared to fly sideways at the strip until the very last moment when he pulled the aircraft around to land on two of its wheels. Needless to say I emerged from the plane with a somewhat jelly-like feeling in the legs, thinking that I had just escaped death in a possible aircraft accident brought on by these "cowboy" flying tactics.

However, my journey had only just started as we then needed to arrange a speedboat to venture upstream along the generally fast flowing Barito River but, due to low water levels resulting from the drought conditions, there were a few more obstacles for fast moving boats than normal. These included hidden rocks at shallow depths, partly submerged single tree logs, and marooned expanses of timber log rafts on their way to the timber mills closer to the seacoast. Fortunately, we arrived at the small riverside village of Purukcahu without incidents of note and, the following day, gathered a few local field assistants to carry our gear before boarding a minivan for the next part of our sojourn into the depths of Kalimantan. Being the only Orang Putih, or white man, in the party, I was expected to sit up front with the driver.

Road conditions for the area were reasonable as all were constructed for the purpose of trucking the big timbers which, at this time, were being rapidly harvested from the tropical forests of Kalimantan. So off we raced, with a driver who seemed to have a similar attitude to life as our young pilot, towards the next stage of our journey to the Mt Moro region. However, it wasn't long before we were moving at a fast speed over one crest in the road only to see a truck, less one rear wheel, in our path. That missing wheel was on the middle of the relatively narrow road being repaired by a few people, so blocking our only path. Undaunted, our driver continued without hesitation at his normal breakneck speed towards the truck scattering the tyre repairers rapidly. It was only the quick thinking of one of the truck's attendants, who thought to run back just in time to pull the truck tyre out of our path, that saved us from a serious accident. As we sped by, the top half of the minivan [and my head] appeared to miss the tray of the truck by only inches. It is a visual memory that I will carry to my grave as to how close we all came to disaster. Omar announced only days after this incident that some locals had mentioned that our minivan driver had been involved in a fatal road accident in which he and others had been killed. One can only thank Allah that it wasn't our fate days earlier.

We finally reached the recently discovered Mt Moro goldfields after leaving the minivan and walking in along other forestry roads and bush tracks cleared by forestry workers and across huge logs over one significant river, probably the Mahakem, to find a scene reminiscent of what the early gold rush days of the Bendigo and Ballarat fields must have looked like during the mid-19th century. A make-shift township and marketplace had sprung up in a clearing of the forest. In this township, despite Indonesia being a Muslim country, the usual sources of debauchery - alcohol, gambling and obvious prostitution – were running rife.

The mining operations were simple with privateers having set up teams of locals to hand carry the gold-bearing broken rock from primitive shafts along the auriferous epithermal quartz reefs that had presumably been discovered after timber cutting had started in the region. In 1983, all was relatively unscarred without extensive waste rock dumps and much of the immediate area had been stripped of its tropical rain forest. The privateers had also brought in small Honda motors for use in driving small wooden stamp batteries, capped with steel, to crush the ore rock. These batteries were constructed alongside a small creek that provided the water which was much needed to wash the crushed rock from the battery plates and across towels and blankets as the means of collecting gravity gold. However, mercury was also being used on the stamp battery plates to assist with the collection of finer gold; that mercury being boiled off in open drums heated over wood fires to recover the gold while the nearby local attendants breathed in the fumes.

Oblivious to the dangers of that metal, the attendants were very likely destined to die from mercury poisoning in the not too distant future.

I inspected outcrops of rocks in the area, sampled the reefs in outcrop and underground, once again taking some risk entering drives off the bottoms of shallow shafts to see what the fresh rock exposures showed. At the time, most of these shafts had been dug by hand to about 10 to 20 metres depth. No explosives were available so any large hard rocks had to be left in place. Their only form of light for these underground operations was from candles and the means of raising the broken rock from the shafts varied from simple rope and bucket to deploying timber-built windlasses about 1m in diameter. Discussion between my Indonesian colleague, Omar, and the local miners soon located a few shafts from which cross drives had been dug. To give me a better perspective on reef width and attendant alteration, I focused my investigation on those shafts.

Omar and I continued on from Moro to visit a few other sites in Kalimantan reported to have gold potential. However, after much difficulty in accessing these areas through motorbike and timber truck rides, plain old walking, and without the hair-raising events of my initial trip to reach Mt Muro, none of the sites appealed as most of the samples collected and analysed back in Australia reported little geochemistry of interest.

In geological terms, the Mt Moro gold occurrence is classically epithermal veining in nature and associated with past magmatic and volcanic activity. At the time of inspection I did not see potential for a major multi-million ounce gold deposit forming from these occasional narrow singular veins, and didn't recommend it to Newmont as a quality exploration target. However, subsequent exploration by geologists from other companies over a period of several years from 1983 did locate about one million ounces of gold and silver scattered between a series of modest resources which were eventually mined during the 1990s as a series of open pits around a central processing plant.

Salt Mining

Another investigation undertaken by two other Newmont geologists, Dr David Tyrwhitt and John Dow, and myself during 1983 involved another speedboat ride into the Indonesian Never-Never, this time using the Kapuas River in Western Kalimantan and, once again, we managed to avoid hitting any floating debris that was likely to overturn our boat. Another time, another river, a Newmont geologist was not so lucky, although thankfully was not injured. We did our trip in the company of two German consultant geologists from Dr Otto Gold Consulting Engineers, who represented the Indonesian owners of the property, and Bob Shafer, a Newmont alluvial mining consultant from the USA.

We arrived at site in the dark to find a respectable camp built near the extensive alluvial gold workings which we were to investigate. The following morning we entered the workings to see several small groups of men using relatively simple sluicing arrangements and mechanical water pumping equipment, said to be financed by Chinese, to recover the alluvial gold. All our investigations went well for the first two days. Our panned samples of the alluvial materials, collected simply with bucket and geological pick from the river banks, were showing sufficient gold colours in the dulongs, the local wooden panning dishes carved from large trees, to allow Bob to believe we could be onto a great prospect.

By the third day, however, Bob had sneaked off without telling us what he was up to and came back with his own sampling and washing of several of the sites we had first investigated in the preceding two days. When he showed me his results, I immediately said each was not as good as the washed products from the local dulong operators that I had been supervising and systematically recording in those preceding days. They were all sufficiently disappointing to cause us to quietly excuse ourselves from our German hosts and we sat around the table to reconsider our position.

Generally, count results for Bob's samples were lower and he quietly suggested that they had been "salted" by some means during the taking and/or washing of the first of the samples collected. Bob then asked us to start taking and treating our own samples in each of the previously visited positions without any local or host assistance whatsoever. To do this, we had to reveal our comparative results to our German hosts and, even though we felt certain that they were most unlikely to be involved in any salting exercise, we had to ask them to also stay away from attending the diggings the next day. This understandably did not go down well with them, but they finally agreed that the purpose of our exercise was to be fully satisfied with our investigation. So we then made our arrangements for the next day.

That day, all Indonesians were asked to remain in camp while we ventured out on our own to test our suspicions, but we were asked to allow the police officer that had been present throughout the investigation to attend for our own protection. This was agreed so long as the police officer did not come within reach of our sampling and washing procedures. Each of the previous positions was resampled and the washings undertaken by our consultant confirmed the lower values, so our own due diligence on the project was considered complete.

It was decided to return to Jakarta without further investigation and respectfully decline the property rather than make any accusations about the salting. As mentioned previously, at the time of visiting, the Kapuas alluvial diggings were being worked by locals on a minor scale, so alluvial gold particles were available to anyone involved in those operations. We believed, but could not confirm, that the salting was probably being done by local Indonesians dropping gold dust into their dulongs from the "roll your own" cigarette smoking common amongst them, or dispensing such dust from under their finger nails during the washing process. This tended to be also supported by the fact that one particular dulong operator always had most of the best gold counts.

The Kapuas alluvial gold project was further investigated by other companies after Newmont declined the opportunity to be involved. These led to the project subsequently being mined and treated by an Australian listed company without it being a highly profitable operation.

I have not returned to Jakarta since my geological visits for Newmont during 1983, but my most impressive memories of that time, and in comparison to my 1975 experiences, were just how much the Sukarto regime had improved the appearance of the city and the wider communications through Indonesia, at least in the region of Kalimantan.

As a result of satellite coverage instigated by the Sukarto government, one was no longer totally isolated in the kampongs scattered through the multi-islanded country. All had television to watch government health messages and, despite some echoes and other odd sounds, I could find an occasional phone service to ring Australia direct and speak to my family, instead of writing letters as I had done on my first trip to Sulawesi in 1975.

POWERLESS

MICHAEL FELLOWS
IRIAN JAYA 1995

Michael Fellows obtained a BSc with Honours and a Master's degree in Geology from the University of Arizona. After 12 years in mineral exploration in the USA with ExxonMobil he and family migrated to Queensland in 1983. There he was involved in gold and base metal exploration for several foreign and domestic mining companies as a consultant. He finished his career working for the Queensland Department of Main Roads. Now retired, he lives in Castlemaine, Victoria.

In April 1995 I took a short contract to log core on a Normandy Anglo Asian PTE Ltd. project in the "Vogel Kop" (birds head) in the West Papua province of Irian Jaya. Normandy's West Delta prospect was 150 kilometres northeast of Sorong at the extreme northerly end of the island of New Guinea. I was then working for Elliott Exploration, sourcing contract geos and fieldies for exploration firms. I had no desire to take this particular job on myself, loathing the tropics as I do, but Normandy asked specifically for me and I went. Having never worked in such an environment before I had no idea what to expect. Overall, the experience was an exploration of what it means to be powerless.

Arrival

I spent a day at Normandy's office in Jakarta and then on to Sorong via Kalimantan and Ambon. The Sorong airport at the time was on an island 15 kilometres southwest of the town. At the Jefman terminal I was met by "Hans" who escorted me to the mole directly opposite the airfield. There we boarded a "pirogue," a boat about 7-8 metres long and about a metre wide. Once laden, the water level was about 15-25 centimetres below the gunwales. We set off for Sorong across the Sele Strait. Thankfully the sea was smooth. I couldn't imagine making the trip in anything but calm water. The boat was powered by two outboard motors both of which failed at least once on the journey across the strait. Just before landing in Sorong we passed Doom Island on our right. I hoped that wasn't an omen.

We landed directly on the beach. As I emerged from the boat, someone onshore cried out "white man!" It must have shocked me and I instinctively yelled something back. Normandy's local agent, Warwick, met me and installed me in the Grand Pacific Hotel. Sorong was a sleepy tropical town. Although it served as a base for servicing oil fields and logging operations in Irian Jaya, it was not then a booming place. In the evening Warwick took me to dinner with his local lady, Yeti, and Dave, the project's helicopter pilot. We ate in a casual open-air restaurant near the beach. The meal was of fresh seafood which I relished. Sorong was everyone's idea of a tropical paradise. Except mine; I hated the tropics.

The next day we went to a small airfield southeast of the town where I, along with all my baggage, was weighed before being loaded onto the Aerospatiale Lama helicopter. Dave said he needed to know the weight of the load to within 10 kilograms. After a safety briefing, Dave lifted off and headed for the village of Sansapour. We followed the coastline north-easterly; a shimmering sea of blue on the left, a solid sea of green on the right broken only by the occasional road or logging operation. At Sansapour we turned inland: the change from the sea air to the rich aroma of decaying vegetation from the jungle below was immediately noticeable.

We travelled about 60 kilometres due east to a small outlying fly-camp (belonging to Uni-Gulf) on the side of a ridge. As the helicopter idled, we off-loaded an Indonesian geologist and picked up an injured man and then proceeded northerly to Normandy's base camp at "West Delta." The camp was three kilometres from the coast, a mere 0° 23' south of the Equator. The site was at an elevation of about 150 metres. The ranges to the south, where small teams of geologists and labourers worked in fly-camps, reached above 1000 metres. The nearest village of any size at the time was Koor, 30 kilometres to the west. The north coast was sparsely inhabited. The local seafarers didn't like travelling the rough waters in their shallow draft boats.

Camp

The camp was much larger and "cushier" than I had anticipated. It was basically a small village in a cleared area at the junction of two small streams. It was base for about 80 people, as far as I was aware, all men. I heard hints that there was a woman in the kitchen, wife of one of the cooks, but I never saw her. At any one time, about half the people were in the several smaller fly-camps.

There were perhaps ten buildings. The largest was a long office with several individual rooms and a radio room. The office was well stocked with worktables and desks, three computers, and a large metal map cabinet. A veranda extended along the full length of the east side of the building. Adjacent were accommodation, dining, and toilet facilities for expats and the Indonesian geologists. The diamond drillers had separate accommodation and dining facilities a short distance away. The eating, dining, sleeping and toilet facilities for the native labourers was completely apart. Expats and Indonesians had individual rooms; local labourers slept in a communal "long house" on cots. Toilets were the porcelain "squatters" typical of South East Asia. The showers merely large plastic garbage buckets filled with water, and provided with a small scoop for pouring water over oneself. Behind the showers were "donkeys" for heating hot water. These at least were a familiar sight.

Other structures comprising the "village" were a large storeroom and core shed, a repair shop/work room, camp security office, generator shed, carpentry shop, and a crudely lighted badminton court!

The buildings were timber framed, set on wooden platforms and walled with plastic sheeting and fly screens. If you were an expat or Indonesian, your building would have a plywood ceiling with a sheet metal roof; if Irian Jayan, merely an orange tarp. The structures were simple, functional and, for the setting, comfortable. However they could not have withstood a strong wind. Wooden or gravel walkways connected the various buildings.

Other than the size of the camp, I was most impressed by the fact that the timber for the buildings and much of the furniture were manufactured from trees felled at the site. I was amazed with what they'd done with the wood and a few tools. To this day that is the most memorable feature of my brief experience there.

One day I watched a crew tackle a felled tree 1.5 metres in diameter. A native Papuan no taller than the chain saw he was wielding skilfully split the tree into successively smaller orthogonal beams and planks. Final sizing was in the wood-working shop with a circular saw and hand-held plane. Having attempted this task myself I knew how difficult it was to maintain uniform thickness and right angle cuts. I admired and envied them. The carpentry shop was run by "Jimmy Hendrix" the joiner. He was given that nickname on account of his "afro" and his finesse with the tools of his trade. He was taller than the average West Papuan. He looked regal even in orange sweatpants and a dirty white tee-shirt bearing an A&W Root Beer logo. After the Herculean task of fashioning the timber for the buildings he was kept perpetually busy building core trays and core racks.

My room was comfortable, carpeted, and outfitted with electric light and a power outlet. Besides the locally made bed, I had a handcrafted small table and a chair. A mozzie net for the bed worked most of the time. Mosquitoes were kept down by frequent fumigation of the compound. (A legacy of my time in Irian Jaya was a mild malaria-like fever which would afflict me from time to time. I never found out what it was, but it took twelve years to shake.)

I lacked little except contact with the "outside." It was an embarrassingly colonial set-up. I'd put my dirty clothes by the door to my room in the morning and in the evening they'd be folded on my bed. I'd brought a tee shirt with a West Indies motif (Pusser's Rum I believe), one of my favourites, a present from a nephew. It went missing the first time it was laundered; either misdirected or pinched. Without any firm reason, I suspected the latter. It re-appeared the day I was to depart, but I left it behind as a gift for the laundry boy.

The food was good; exceptional if you happened to like rice three times a day. (The camp consumed 1,500 kilograms of rice a month.

The cost of feeding the crew cost more than their wages.) A sign in the dining room indicated mealtimes in Indonesian and English. Meals were for the most part standard South East Asian fare; not a great variety but tolerable. The dinner table was provided with a half dozen bottles of various hot sauces appreciated by the Indonesians, as well as aerosol canisters of Baygon. Fresh vegies, fruit and fish were on the menu for several days after new supplies arrived. Prawns and lobster tails were a special treat. Occasionally the cook would attempt western dishes like fish-and-chips or steak. Tea was always available as was horrid instant coffee. Fortunately I'd brought along my percolator and some ground coffee. Every few days I'd boil up a pot of coffee on the "donkey" fire for the showers.

The cooks kept several small pigs in an enclosure but a dozen or more chickens roamed freely. To keep the rats at bay, the camp had several cats. Their tails were cropped in the Chinese custom. Believing them to be vain animals, the locals would chop off the ends of their tails to humble them. They looked horrible. The camp had been occupied for a considerable time before I arrived and there was maize growing near the helipad where the laundry was also put out to dry.

At night feral dogs came into camp howling and fighting. Occasionally in late evening someone would hide on the veranda and shoot them with an air rifle. It only provided temporary relief.

All heavy supplies came by boat from Sorong on a small coastal freighter, the *Ama Lohi*. It anchored just off the beach at the mouth of the stream on which the camp was situated. There was a trail from the coast five kilometres up to the camp but all supplies were transported to camp from the boat by helicopter.

Other than by the irregular visits by the *Ama Lohi*, the only way to get in or out was by helicopter or on foot along the coastline. It was about a six day walk to the nearest vestige of civilisation where one had an opportunity to re-connect with the wider world. The camp was in daily contact with Sorong by radio. The helicopter made regular trips to Sorong for supplies and to ferry out samples. The field programme cost about $10,000 per day. The major component of that was helicopter charges.

It rained nearly every afternoon or evening. Rainfall was over three metres per year. It was humid but not hot and I was always bathed in sweat during the day.

People

In charge of the camp was Patrick 'Paddy' Waters whom I'd known at Australian Consolidated Minerals (ACM). He said that to build the camp, blokes were lowered on a cable from a helicopter through the forest canopy to clear a landing spot. More woodcutters and gear would then be landed to enlarge the space. The other expats in the camp were Geoff the drill supervisor, Dave the American helicopter pilot and his offsider and, later on, geologist Luke Swift. The diamond drillers were from Papua New Guinea and, with the exception of the expat supervisor, kept to themselves.

There were three or four Indonesian geologists and about eight other Indonesian (Javanese) support people. Most of the expats and the Indonesian geologists were single. There was a definite colonial atmosphere with the standard hierarchy; expats at the top, followed by Indonesians geologists and officials, next Javanese workers and foremen, then city-based native Papuans and, at the bottom of the heap, village Papuans. In all there were about 60 local labourers. Wages for the labourers were $3-4 per day. Working for the five months the field programme lasted, they could expect to get $600-700. Most came from coastal villages but a few, hearing of the availability of work, trudged in from the interior. One healthy lad, a good worker from inland, went for break and never came back. We later learned he had died of TB. Paddy mentioned one worker, Peter, who fell ten metres when being lowered from the helicopter to clear a landing space; he was never the same afterwards. While out on a break he got drunk in Sansapour and missed the return boat. He got a lift to Koor and then walked the 35 kilometres along the coast back to camp.

Most of the natives were of short stature. Their work was physically demanding. They'd work day in and day out: climbing up and down the ridges carrying heavy loads to and from the drill; cutting tracks along the ridges for sampling; digging drill pads by hand. Here I was sitting on the veranda drinking tea, earning over 100 times what they did. I couldn't help wondering about the justice of it and what it means to have power, and to be powerless. (I also wondered if they all had perfect vision. I never saw any of them with spectacles!) The majority of the labourers were Christian; the Indonesians, predominantly Muslim, but there were a few Christians among them as well. I was quite surprised to see that on Sunday evenings the Christians held an evangelical service in the native dining room. It was a large gathering.

Keeping an eye on all the activities for security purposes were a policeman, a native Irian Jayan army officer, and an Indonesian special security forces major from Jakarta. Paddy nicknamed the latter Mr. Toad, no doubt for his appearance and disposition as well as for the odious reputation of the Indonesian special forces in Irian Jaya. The company was obliged to pay for these people. The army officer helped out around the camp but the other two just sat around all day on the veranda of the security office. Above that office hung a white sign board with the lettering - Pos Keamanan - indicating the building's function as if that were necessary. The red and white Indonesian flag flew on a pole out the front. I thought the military presence was more of a political statement than a serious concern about our security. I suspect that the authorities were fearful that, wherever large numbers of locals gathered together, there might be attempts to recruit members for the OPM (Free Papua Movement). This camp seemed peaceful enough but earlier that year the OPM attacked a Dominion Mining exploration camp killing several people. It is doubtful these three men could have protected the camp from any armed intruders. The policeman had a rifle, the other two perhaps a handgun each. The policeman did use his rifle to shoot a wallaby and a deer. Apparently deer were quite plentiful, introduced to the island by the Dutch. The deer meat was put on the sample-drying racks to make jerky. It was not served in the dining hall!

One day the lads killed a snake in the camp. It was a big python, four plus metres long without its head. The pythons usually dwell in the tree canopy, this one either fell out or came down to eat one of the chickens. The locals are of two dispositions; those who loathe them and dread them alive or dead, and those who don't mind snakes and will eat them.

The following day we wondered what happened to the python as we poked at our dinner.

Paddy told the story of one American geologist working for Ashton in Irian Jaya who fancied himself a herpetologist. He found a large python which grew to four metres and which he took to Jakarta as baggage on the airplane. It was kept in Ashton's mess and the local boys were kept busy scurrying around town to find rats and mice for it to eat. It was called Bruce. Bruce had a temper and would hiss at people. Another geologist found a slightly smaller python in Sumatra, only three metres long. It was brought back to the office in Jakarta as well. Bruce ended up eating the Sumatran snake!

Work

The project was interesting, but not wildly so. It was a porphyry system, no doubt, but weakly developed at the level Normandy were drilling. At the nearby Kali Sute prospect field crews collected a sample of outcrop which had advanced argillic alteration with bornite - chalcopyrite copper mineralization and free gold. It was exciting to think of what such a favourable sign might indicate but the single test hole drilled on this prospect had nothing like that.

My life revolved around logging core, eating, and sleeping. It took me a few days to re-acquaint myself with porphyry copper style mineralisation and alteration. More than once I thought about my university advisor, Professor Spencer R. Titley, who spent several seasons in the South West Pacific working for Sydney promoter Ike Shulman. He authored a number of papers on porphyry systems in Papua New Guinea. I also thought of Jack Thompson, ex-chief geologist at AMAX who spent many years in PNG and recognized Panguna in Bougainville for what it was.

It was a monotonous routine broken occasionally by reading technical papers. The drilling was slow and I was easily able to keep up with it. There was plenty of time to reflect on the location, and the situation. It was a civilised, albeit distinctly colonial, existence with breaks for morning and afternoon tea for which the cook would bake cakes or biscuits. However, I drew the line at greasy doughnuts with 100s and 1,000s stuck on with liberal dollops of margarine.

The core shed and logging area were built on an elevated platform on stumps, much higher than any of the other buildings. The floor was rough-cut wooden planking, the roof sheet-iron. Wooden railing extended around the perimeter. Wooden stairs descended to a lower platform area and to the ground. I liked the feel, sound, and smell of my work area. It gave me a sense of being on the deck of an old sailing ship. From the railing I could look out to a "sea" of fuel drums at the helipad or turn and watch the "crew" on deck preparing our cargo of core for stowing. Core was split on site and one half sent to Jakarta for assay. The programme was generating ½ tonne per day of core and rock chip samples. The freight bill for samples alone was $100,000.

I never ventured more than a kilometre from the camp to where the drill was operating. The pads were all on ridge crests. Rough timber ladders were placed at the steeper segments of the slopes of the ridge for easier access. After one particularly heavy rain, the drill started slipping off a pad. The lads were quickly despatched to secure the rig and dig a new site by hand.

Geologists and sampling teams would go out for several days at a time to several subsidiary fly-camps on the exploration tenement to conduct ridge and spur sampling and geologic reconnaissance. Rain or clouds would often force them to extend their planned sorties. There were times when they were stuck for several days until the Lama could reach them. It was invariably wet and cold at the higher elevations they worked. I appreciated the difficulties the Indonesian geologists faced.

I allotted myself one beer a day. I could easily have had more but didn't want to end up like the drilling supervisor. He seemed to be in a perpetual alcoholic haze. One afternoon he was strenuously directing the helicopter pilot to go to the drill immediately and retrieve his offsider (who was then standing but a few metres away). On shifting drilling gear from Uni-Gulf to base camp two tool boxes came off the helicopter sling and fell into the jungle. The next day the supervisor wanted to send out a search party to look for them! He was a pleasant enough fellow but, in the evening, drink would get the better of him and he would be either maudlin or voluble. In the morning he invariably apologised for his behaviour but the scene was repeated over and over. The word around the camp was that he did more drilling in the whorehouse in Sorong than on the project. It was sad to see such dissipation but I could see how easy it might be to succumb to the effects of such isolation without an interior sustaining power.

I saved each bottle cap of my Bir Bintang. They served as a reminder of how long I'd been away. They were my "chips" which I'd cash in when I hit Jakarta and billed Normandy - four hundred dollars a cap. In late April, the company accountant in Jakarta advised Paddy that he was over-budget and would have to cut the programme short. The drill rig moved to Kali Sute for one last hole and the *Ama Lohi* was summoned to begin taking workers back to their villages along the coast. Originally sentenced to eight weeks, I was released early for "good behaviour." At the end, Normandy's regional manager Mike Mackenzie (who I'd known at my time with Exxon), and volcanologist John Wright visited the project and I travelled with them back to Jakarta.

The Camp Camp

It was a surprise to discover that West Delta Camp was indeed 'camp!' Ten days after my arrival my eyebrows were raised slightly when I noticed Sami, the 'helicopter washer', and someone else go into the shower together after badminton. Mid-morning the next day I see Sami and an older person strolling arm in arm down by the core shed. They sat on a fence facing me with the older person having his hand on Sami's knee. There were others around me working on the deck of the warehouse but I wondered if the show was especially for my benefit; either just to let me know what was what, or to shock me. The next night Sami (who occupied the cubicle next to mine!) apparently had "visitors" early in the morning. Or so it seemed unless he both talked and walked in his sleep. For several days thereafter Sami unashamedly had his arm around the shoulders of the helicopter pilot's offsider.

Sami was also the camp medic. Paddy said he wouldn't be at all surprised, if visiting someone in sick bay, to find the medic either holding the patient's hand or playing with his member. I suppose what made it a bit disquieting was that Sami would also occasionally strum his guitar at the Sunday church services.

Stories

Paddy was great raconteur. He had loads of tales, many relating to the dissolute lives of fellow expat geologists in the flesh-pots and dives of the Orient and elsewhere. He also related some insightful stories about our employer, Normandy and its founder Robert 'Discrepancy' de Crespigny, Prior to my arrival, one of the expat geologists, while out on reconnaissance, sat down for a break. A tree branch came down and hit him. He was concussed and airlifted out. Fortunately his injuries were not life threatening and he eventually returned. An accident report was duly filed with Normandy. Safety officers assessing the incident forwarded instructions to Jakarta for avoiding similar incidents. Their advice; "Don't sit under a tree while taking a break." They obviously had no idea of the working environment in Irian Jaya.

Normandy also sent out a doctor to the project to conduct a medical audit. His advice was that besides monitoring physical health, people should undergo psychological assessment as well! I would second that recommendation!

Paddy noted cynically that Normandy's policy manual on dealing with the death of an employee had list after list of authorities and company personnel to be notified but was silent on procedures for notifying next of kin!

Paddy was with Robert 'Discrepancy' on Robert's first visit to Jakarta. He says Robert won a game of darts in a bar; the prize, two prostitutes. Discretion prevented him from relating the end of that story! Paddy also had occasion to socialise with 'Discrepancy' while he was working in Turkey. Apparently, Robert was just at ease talking one day with the President of Turkey, and the next drinking beer with Henry Vox and his drillers at Dikili! Unfortunately, I failed to record the full gist of Paddy's story of Dikili project manager Roger Craddock pulling a turkey onto his lap at a Christmas celebration in Izmir.

Atmosphere

One lesson I learned quickly was that in Irian Jaya, "nothing is for sure unless it happens!" It was no use worrying about things that didn't work out as anticipated. It was useless to rail against the Fates. Too many factors beyond one's control could influence the course of events, no matter how well one planned, and one was powerless to do much about them. The most obvious was the weather. Rain or cloud cover might ground the helicopter. The helicopter might have mechanical issues; which indeed happened on my way out.

Another factor was the isolation. It did one little good to think about the "outside", and that was hard to adjust to. I found it difficult depending on others to send letters or faxes back to Australia. It appeared that there was no one else in the camp other than myself with any desire or need to send personal messages. Had I stayed much longer I would have had to develop a much greater detachment from the "outside world." With a family in Oz, that would not be easy. I could see that this sort of life was more suited to those with few emotional attachments, whether near or far.

There was very little news that reached us from the "outside" that didn't pertain to work. Not that that mattered because there was little one could do about events elsewhere. Occasionally a newspaper would arrive but it would be weeks old. At least the crossword would be current if someone else hadn't already completed it.

This was my first experience of working in the jungle. It was both fascinating and intimidating. The dense vegetation and the terrain limited one's ability to move about. It restrained one's natural inclination to explore the surroundings. With the heavy forest canopy and thick undergrowth, it was a world of near perpetual darkness. On the edge of a clearing one could seldom see more than ten metres into the jungle. I was not used to this environment. I had hoped to see some of the tropical birds, but my attempts were mostly futile. I could often hear them, but rarely see them. Their Siren-like calls I thought were attempts to lure me into the darkness but I resisted the urge to seek them out.

The jungle did hold dark secrets. Paddy related that early on in exploring the tenement a sampling team came upon the wreckage of a WWII Japanese Zero on the side of a ridge with remains of the pilot still inside. They notified the Japanese embassy in Jakarta but they showed no interest.

Deus Est Machina

Unsurprisingly, virtually everything on the project depended on the helicopter. All food, fuel and other supplies were delivered by it. Movement of drill rigs and their daily operation was only possible by helicopter. It was the only way to transport geological crews into and out of the fly-camps. Long range movement on foot was impractical.

We were in a sense, willing participants in a "cargo cult." Albeit a sophisticated one, but we were utterly dependent on, and placed our faith in, those gods of the mountains and the jungle, the Lama, the Puma, the JetRanger. No less revered than the machines themselves were the pilot-priests who operated them. They were invariably accorded special status in the camps. Less exalted, but respected and envied for their subsidiary roles, were the technicians, these virtual vestal virgins with drums of Jet-A1, and jars of oil and hydraulic fluid, who attend the gods. Of lesser status were the goddesses of communication, aerial and antenna, who controlled our contact with the "other world." (Ariel was in fact the name of the radio man who maintained daily contact with Sorong and with the helicopter.)

It was a major event when, at the end of April, Dave and his offsider departed, taking the Lama to Goroka, PNG for a major service. The machine was replaced by a twin-engine MBB Bölkow 105 of Pelita Air. Their smartly dressed pilots in orange jump suits seemed very professional-looking in contrast with the laidback appearance and behaviour of Dave the American. The Pelita chief pilot's life goal was to go on Hajj to Mecca. Dave was perfectly happy to sit on the veranda drinking coffee and reading a Wilbur Smith or a Tom Clancy novel.

Thoughts

The colonial aspect of the enterprise was apparent. Here were Paddy, Dave and I sitting on the veranda to the dining room sipping tea, watching the second shift of "black boys", with mattocks and shovels in hand, setting out to construct a drill pad. A week later when the drill was moved, from our same vantage point, we watch them trudging into camp, each one carrying a six-metre length of galvanized pipe which they'd carted through the jungle for about a kilometre. They resembled a platoon of native warriors carrying long spears. The camp could just as easily have been a tropical trading station or plantation of the late 19th century. The irony was unavoidable. It was easy to be a part of the situation but, having recognized it for what it was, damned hard to know what to do about it if an individual would or could do anything. Powerless.

A month after returning home I attended an Association of Exploration Geochemists symposium in Townsville. The theme of the meeting was "Exploring the Tropics." I wasn't at all interested.

TRYING TO BE FIRST
CAB OFF THE RANK

TONY GATES EAST
TIMOR 1999

Mineral exploration and mine development is essentially a real estate matter in the first instance. There is no economic sense in spending time and money on exploration until you have mineral title to the land, or expect that you can get it by dealing on it, all before someone else does. And obviously a company does not try to establish a mine on ground over which it has no title.

A cashed-up company can buy its way into superior exploration ground, into an advanced mineral project, or a potential mine, but an opportunistic entrepreneur, exploration syndicate or a junior explorer can generally not afford to do so. For them, the greatest likelihood of a significant return on an early investment is to be the first to identify and secure prospective ground or a mineral deposit.

Enter our efforts in newly independent East Timor, but a bit of history first. A prolonged and often bloody struggle ensued after the previous colonial power, Portugal, walked away in the early 70s. In 1999 the region's claimant, Indonesia, bowed to a strong degree of international and economic pressure and allowed the peoples of East Timor a referendum on whether they wished to gain independence or remain part of the Republic of Indonesia. By an almost 80% majority, the people chose the former alternative. Pro-Indonesian elements of the local population and some of the Indonesian armed forces opposed the referendum and as a result there was considerable bloodshed and much damage to buildings and infrastructure during a period of unrest and fighting.

The United Nations acted and installed a peacekeeping force known as INTERFET. It was almost 10,000 strong, with the Australian contingent numbering over 40% of that and was headed by the Australian Major-General Peter Cosgrove. INTERFET arrived in late September 1999, and pretty much established martial law.

Meanwhile, back at the minerals acquisition effort. I headed up a syndicate which we called the East Timor Operations Group which had the aim of being the first to try to discern whether East Timor had genuine potential for significant future mines and, if so, to acquire that prospective ground before anyone else did. East Timor is part of the very extensive Indonesian archipelago and the latter hosts some world standard mineral deposits. There was apparently little to no past mining in East Timor and not much historical literature on mineralization in the new nation was available. However, some records had been written and they were obtained, translated and studied.

We were able to organise a visitor's visa for me through their embassy in Darwin. I also had a very productive and encouraging meeting with Mr. Xanana Guzmao in at a hotel in that city. He was keen on our entrepreneurial attitude and promised assistance in due course. He had been one of the leaders of the independence movement and in fact became the country's first president when it gained full recognition by the UN as an independent country. He understood that a newly emergent country would have to establish revenue bases and provide training and meaningful work opportunities for some of the population. Through him, we were able to make some applications for exploration permits over areas which seemed prospective.

Armed with the legal and entry paperwork, I was able to get on a flight to Dili, arriving a few weeks after INTERFET had arrived and was trying to settle the situation under a pretty much martial law mandate.

I arrived looking relatively presentable, if I may say so myself. Haircut, designer sunglasses, khaki shorts and khaki shirt with epaulets. At my age, I certainly wasn't quasi-military but I guess could be taken to be a potentially senior administrative type.

Anyway, as I was passing out of Immigration in Dili, I was approached by an Australian Army officer who queried if I was Mr

Tony Gates of the East Timor Operations Group. Having confirmed that, he asked me to accompany him to his military vehicle without any further explanation being given by him. I was then given an exclusive and one-on-one guided tour of Dili by him. There was much evidence of the peacekeeping force and he showed me many examples of the recent destruction to infrastructure and public and private buildings wrought after the independence referendum and prior to the arrival of INTERFET. There was obviously a big job for that organisation over the coming months and much had to be done to rebuild or repair the administrative, business and residential areas of Dili.

As the extensive tour was coming to an end, he rather sheepishly advised me that there seemed to have been some sort of administrative snafu and he had not been able to obtain information on the role of the East Timor Operations Group in the rebuilding of the ruined country. I explained that that was not my job at all and that I had come to inspect our mineral tenement applications and determine whether we should apply for more somewhere in the country. His reaction was a mix of incredulity and annoyance somewhat along the lines of:

"Shit! You have got to be joking! We are up here trying to get people to stop killing each other and to rebuild the country as quickly as possible, and you shouldn't bloody well be here getting in our way and distracting our efforts!" After he calmed down, he proposed a solution to me which was that I wouldn't be arrested by him, if I agreed to get on the first plane out and didn't broadcast the army stuff-up on my return to Australia. This seemed a pretty reasonable deal under the circumstances so I was duly dropped off at the airport the same day I had arrived.

As leaving East Timor wasn't going to advance our cause at all, and would mean a total wastage of our investment, I decided to hang around the airport for a while to see whether something might turn up. Eventually another plane landed and I was able to insinuate myself into the group of passengers working their way out the airport doors.

I got a taxi to the downtown hotel which had been recommended by their embassy back in Australia. From that base I was able to organise the hire of a vehicle and driver and was able to get out in the bush to try to get a handle on the geology of the areas we had selected. I inspected several quarries which had been used for road metal and building materials and took some samples for assay of relevant outcrops that I saw by the roadside as I travelled along. I did what I was able to do from a vehicle and left for Australia after about three weeks in the country having experienced no particular incidents of concern.

As I said at the start, mineral exploration success is all about holding the real estate and, for some unknown reason, the new government decided to grant the tenements to another company over the top of our original applications. That was the end of the efforts of the East Timor Operations Group. But you've got to be in it to win it!

THE AMERICAS

JUMPING OFF THE DEEP END

LINDA TOMPKINS
BRAZIL 1983 - 1990

Linda Tompkins is a geologist and former lawyer. She graduated with a BSc (Hons) and MSc in geology from the University of Massachusetts, USA, in 1980 and 1983, respectively, and with a PhD in geology and an LLB (Hons) in law from the University of Western Australia, in 1995 and 2006, respectively. Her entire working career has been in the minerals industry with extensive on-site experience in Australia, Brazil, China, and West and Southern Africa, working in exploration, development, and mining operations (open cut and underground) for diamonds, lead-zinc, nickel, and gold.

Rio de Janeiro

I was in the first-class cabin (those were the days!) of a Pan Am flight from New York to Rio de Janeiro listening to Peter Allen's song "I go to Rio" on my Walkman. A good friend of mine had sent me the cassette as a going away present. It was October 1983 and, as a recently graduated geologist, I was flying to a place I knew nothing about for my first interview for my first job with BP Mineral's Brazilian company BP MineraCão Ltda. I was petrified and very much alone. I had never heard a word of Portuguese in my life, never been south of the equator in my life, and had absolutely no idea what I was getting myself in for. Nevertheless, I was determined to make this work. Little did I know how much inner strength and determination it would ultimately take.

I was met at the airport by BP's "facilitator" as soon as I disembarked the plane. I was briefed on how to get through immigration and customs, which I had to do on my own. I was told to make sure I do not divulge that I would be working in Brazil as I was arriving as a tourist on a six-month tourist visa.

After making it through immigration and customs without incident, the facilitator handed me over to a driver to take me to my five-star accommodation on Copacabana beach, The Rio Palace. I would have a personal driver for the next few weeks and he would take me around Rio to the usual tourist spots. I could sign for all meals at the hotel. My room had a view of the beach. I would meet with the president of the company in the early afternoon.

The president of BP Minerals Brazil was an American. In fact, he was the father of a geology Professor at Amherst College, which had a cooperative research interest with the University of Massachusetts. I had done my master's degree on the Koidu kimberlite in Yengema, Sierra Leone. At that time, the Koidu kimberlite was owned and mined by the Sierra Leone National Diamond Mining Company (NDMC) which was jointly owned by the UK mining company Sierra Leone Selection Trust Mining, and the Sierra Leone government. When BP Oil, in its wisdom, decided to diversify into minerals through the creation of BP Minerals, the UK mining finance house Selection Trust Ltd was acquired by the latter. It is through all those connections that landed me the opportunity to work in Brazil with BP MineraCão Ltda to assist the team in their diamond exploration projects. To get a job doing diamond exploration overseas was my dream come true.

I recall my meeting with the president being brief, to the point, and friendly. He was a very busy person and was about to get on a plane to go to the UK and I think he just needed to see the whites of my eyes before handing me over to the exploration team. My next meeting was then with the vice president, a Japanese Brazilian geologist. The vice president was a difficult person to read. He never looked anyone in the eyes but always looked away or downwards and at times with his hand covering his face. I never understood what he was hiding from as, to me, he came across as generally a kind person but definitely an enigma.

It was after my meeting with the vice president that I was then thrown into the lion's den and what I also always refer to as the snake pit.

Political Background

It is important to put my seven years in Brazil in the context of the political situation of the time between 1983, when I arrived, and early 1990, when I left. This is because much of what I experienced in Brazil during this time was informed by the drastic political changes then occurring in the country.

The country was ruled by military dictators from 1964 to 1985. When I first arrived there, the last of the Brazilian military leaders, João Figueiredo, was president. Times were changing and many of the political opposition leaders who had been living in exile overseas were returning to Brazil around then. There were calls for a democracy and direct popular presidential elections by the opposition and in late 1984 their candidate Tancredo Neves was elected by the Brazilian congress. He was the first elected civilian president of the New Republic, which continues to the present. I remember that Tancredo was very popular among Brazilians. Unfortunately, he died before he could be sworn in as president. Following his death, José Sarney, who was Tancredo's running mate, became the first civilian president of Brazil.

From 1985 until 1989 Brazil went through a very difficult transition phase under the Sarney presidency. It rewrote its constitution and just about every law down to the dog act. The country became very xenophobic and there was a concerted effort by the federal legislature to push out foreign (*gringo*) companies and people. This was a time when the country and its people felt that they could and should do everything. As an example, the desk top computer revolution happened during this period, yet no foreign computers were allowed in Brazil. Companies were often raided to check if they had any illegal foreign computers. Only Brazilian made computers or other Brazilian made electronic equipment was permitted. Needless to say, that equipment was considerably more expensive and often did not work.

Monetary inflation was also rampant. The last few years that I lived there, inflation rose to 4% a day!

Despite the difficulties, Brazilians were resilient and strongly believed in a better future. Everyone was excited about the move to a democracy and just about every person I would encounter had a view on the new political situation and upcoming election and were more than happy to talk about it. I will never forget the tears in the eyes of a friend of mine when he went into a polling station for the first time in his life. A basic democratic right that I had always taken for granted.

In 1989 Fernando Collor de Mello became the first President of Brazil duly elected by a majority of the Brazilian population. It was an historic moment. The years leading up to this election were turbulent, but the judiciary responded quickly and justly to any misfeasance by any of the candidates during the presidential campaign. I left Brazil the day that he was sworn in.

Botafogo, Rio de Janeiro

I thought I had been hired by BP to assist in their diamond exploration programs. However, after my short discussion with the vice president, I was handed over to a Brazilian female geologist who was the company's petrologist. She was not welcoming to me at all and greatly resented my presence in her country and more so in the company. She was nasty and made the few weeks I had to work with her hell. I did not know any Portuguese and, although she knew English fairly well, refused to speak with me in that language most of the time. I soon realized that it was expected that since I came from overseas, I must be considered some form of expert. I was a recent University graduate and by no means an expert in anything. In her high-pitched voice I was ridiculed as being stupid, chastised for not being an expert and, therefore, having obviously lied about my qualifications. When I think of how young graduates are treated today going into a company the size of BP, I sometimes feel that they don't fully appreciate how good they have it. They now typically go through extensive graduate training courses, inductions, etc. I had none of that. It was sink or swim within a seemingly hostile environment with everyone determined to get me on the next plane back to the USA. Although I was very depressed at my situation, I was determined that I wasn't going to let the bastards get the better of me. I had made a commitment to the president of BP and I was going to keep it.

An unusual aspect of working in Brazil at the time is that any University graduate was referred to as Dr (male) or Dra (female). This led to some anxiety in my first few days at BP because everyone was referred to as doctor. I thought, "Oh my god everyone here has a PhD in something and I only have a master's degree. What am I doing here?". I later found out that only a few geologists had a PhD and that hardly anyone had a master's degree. Of course, I was never referred to as Dra, this seemed to be reserved for the Brazilians only. I was not concerned at all as I felt it inappropriate to be referred to as a doctor without a PhD. Even when in the field I insisted that everyone simply call me by my first name.

I spent the first three months working in BP's Botofogo office. The first few weeks were spent under the piercing eye of the female petrologist doing reflected light microscopy. Of course, at the end of each day I was told how incompetent I was. It was soul destroying to be put down so harshly and cruelly each day. Fortunately, after a few weeks, I was sent to the company's mineral separation laboratory, which was located away from the Botofogo office, to work with young Brazilian geologists on mineral identification.

I found mineral identification of concentrate very tedious and boring work when doing it all day every day, and even more so when every single sample I analysed was then redone by the Brazilian geologists. They were also tasked at identifying, and presumably correcting, all of my mistakes (which of course was every single sample) and reporting these errors to the petrologist in Botofogo. So, although I was working away from the head office, I still would get chastised every day by the petrologist in her high-pitched shrilling voice, either in person or by phone, for my continued and systemic incompetence. I learned not to trust the Brazilian geologists at the lab.

The situation got so bad for me that I eventually had no choice but to book a meeting with the president after about six weeks of the constant bullying. I explained to him that I wanted to be moved to the diamond exploration team. About ten days later I finally met with the head of the diamond exploration program, a Portuguese born and bred diamond geologist, who had worked many years in Angola on alluvial diamond projects. I was finally going to work with the diamond exploration team and I was truly excited.

Once in the diamond exploration team, I spent the next month in the Botofogo office preparing to go into the field. The company was preparing to start a new diamond exploration program in Iporá, Goiás, and I would be working on this project. I had to do a lot of reading. There were some reports in English, but many were in Portuguese. It was tedious work for me to read by looking up just about every word in a dictionary.

During my time in the Botafogo office BP did provide me with daily Portuguese lessons. They had an in-house language teacher who would teach both English and Portuguese. I quickly learned the days of the week and how to count and a few miscellaneous food words. The lessons worked on speaking skills only, but unfortunately not reading and writing skills.

A story I must tell is that in my first week I went to lunch on my own to a small sandwich bar around the corner from the company's office in Botafogo. Another geologist had showed me where it was. I recognized the word for pineapple (*abacaxi*) on the menu board and ordered what I thought was pineapple juice. It turns out that I pronounced the word abacaxi incorrectly and actually ordered an avocado (*abacate*) juice. When the green sludge arrived, I was quite shocked and thought it was made from some rotten fruit. I pushed it back saying "mal" (bad) and the person on the other side of the counter pushed it back to me saying "bon" (good). I tried a tiny bit and it definitely did not taste like pineapple juice or anything I recognized. I then thought that perhaps it wasn't a safe place to eat, so I paid the bill and left both the juice and my sandwich behind untouched. On return to the company on an empty stomach, I retold my experience to the Brazilian petrologist. It was perhaps the only time I ever saw her smile or even laugh as she explained to me my error. As I had never heard of avocado juice before, I also had to laugh. So, in my first week I learned how to pronounce these two Brazilian food words very carefully. It turns out that avocado juice is actually very tasty!

I had a few other food encounters during my first few weeks in Brazil. On a weekend I decided I would check out supermarkets so I could put some food in my room fridge as I was getting tired of the hotel food. I bought what I thought was a bottle of juice. After returning from a run very thirsty, I opened the bottle and drank about half the bottle. I immediately starting gagging and my throat went completely dry. I thought I was poisoned. I didn't know what was wrong but drank heaps of water and poured the rest of the juice down the drain. When I explained to someone at the hotel what had happened, they laughed and explained to me that it was juice concentrate after I showed them the bottle! I was supposed to heavily dilute it with water before drinking.

Overall, my only consolation during the difficult times working at BP's Botafogo office and mineral laboratory was when my driver picked me up at the end of the day and took me back to The Rio Palace. I could talk with English-speaking tourists who did not judge me harshly for being there, walk along the beach, and have excellent food overlooking the bay. Eventually, I was moved into an Apart-Hotel in Leblon. It had a kitchen, so I could cook if I wanted to. I found the beach in Leblon a lot cleaner than in Copacabana and the shopping and restaurants much better as well.

Iporá, Goiás

Finally, after more than three months in Rio, I was going to go into the field. I left Rio on my own and was met in Goiâna by a Brazilian technician, and five workers all from the state of Piauí in the northeast of Brazil. The only English words any of them knew were "beer" and "money", and the only Portuguese words I really knew were the days of the week, how to count to 20, and few miscellaneous others. I was in charge of this team and tasked with following up aeromagnetic anomalies in the region to find kimberlites.

I recall the drive from Goiâna to Iporá. I was now in the interior of Brazil in the heart of the agriculture district. On the road there were lots of large trucks top heavy with packed charcoal. These trucks were scary as they always seemed ready to tip over, and some would!

There were also many cattle grazing in large paddocks. The cows were very different from the black and white cows I was used to in the USA. Most of the Goiás cows had this large hump on the shoulder that looked to me like a cancerous growth. I thought what a poor state the animals were in and how this must reflect the health of not only the animals, but the people as well.

The first evening in Iporá we all went to a churrascaria (steakhouse) for dinner. Initially, the workers wanted to go separately to a different, cheaper, place to eat but I insisted that we all eat together. Churrascarias are famous for their rodizios (grill menus) which involves all-you-can-eat of a variety of rotisserie grilled meats served on your plate off the spit. It was the first time I had ever been to a restaurant like this. As I didn't know what was being served, I asked the technician as each meat was served "what is this?" in Portuguese. He patiently would try to explain to me what type and cut of meat was being offered. Then came this very fatty piece of meat. When I inquired what it was, it eventually was described to me as the hump of the cow or *cupim*. The same humps that I thought were cancerous growths. I almost threw up. I stopped eating and resorted to only eating rice. Everyone tried to tell me it was ok, but I didn't believe anyone. I only ate rice for the next three days not trusting any other food and waiting to get seriously ill from the food I ate at the churrascaria. Of course, it never happened.

I learned much later that Goiás was actually a wealthy state with some of the best cattle and agriculture farms in the country! The hump or zebu hump was typical of that breed of cattle and, although very fatty, was safe to eat. How different from my initial and quite naive first impressions.

After the churrascaria food incident, I worried that I was not going to be able to survive in Brazil. Apart from rice I would eat a few fruits that I recognized. But with time I eventually came to trust the food. In fact, by the time I left Brazil and went to Australia, I didn't know how I would survive in Australia without some of my favourite Brazilian dishes!

On my first trip I spent one month in the field based out of Iporá, was back in Rio for a week during Carnival, and then back out in the field for another five weeks before my visa would end and I would have to return to the USA. Every day, including Saturday and Sunday, we would get up early and drive to the area of a new aeromagnetic anomaly. The aeromagnetic survey was an older 1970's vintage survey that was widely spaced and not very accurate. When we could ground locate the likely source of the aeromagnetic anomaly, it was anywhere within a 500m to

1km radius from the actual centre of it.

Navigation equipment included very inaccurate topographic maps (1:250,000 and 1:100,000) that did not match the older 1960's vintage black and white aerial photographs that were at a different scale, and my hand-held compass. To make things even more difficult, everything was metric and, coming from the USA, I had been field trained in the imperial system. Because the topographic maps were often so inaccurate, we essentially had to remake the maps in real time as we walked upstream along principal rivers by keeping track of tributaries and any other recognizable land feature.

We did stream sediment sampling in all the drainages emanating from the anomalies. This meant that every morning the first thing I had to do was get wet. I would spend the rest of the day that way with soaking wet feet in my boots. I would often dream of being able to have dry feet for a day. The task of the five workers was to take the stream sediment samples, sieve the gravel, bag and number the concentrate, and then carry the samples for the rest of the day until we returned to the vehicles. It was back-breaking work and at times I don't know how they managed to do this work all day long, but they did.

Most days we were up at dawn and home after dark. We would walk somewhere between 20-50km a day up and down river valleys. Most of the river and streams were heavily vegetated, so it was necessary to often climb up hills to get a better view in order to orient ourselves better. At times the guys would have to cut a path through the bush with machetes. I loved the work and felt very lucky to be able to be outdoors in a magnificent part of the country with a good team.

We went out in most weather though, if there was heavy rain, we would stay in town and I would do desk work. I had to write monthly reports and it was expected that I would write these reports in Portuguese! Fortunately, I had a secretary in Rio who kindly translated my English written reports into Portuguese. With time, I did learn to prepare my own reports in basic Portuguese.

After six months in Brazil I had to return to the USA. The company agreed to keep me on and organized for me to get a temporary working visa. It would take about three months. During my time in the USA, I did analytical work at the University of Massachusetts on samples from Iporá, and from other previous diamond projects that the company had worked on before my arrival. This enabled me to keep involved in the company's diamond exploration efforts and, importantly, to continue to get paid.

I returned to Brazil with a temporary working visa as a "specialized prospector" because I was not able to work in the country as a geologist. The professional roles were (and still are) very unionized and over the seven years I lived and worked in Brazil, I tried to get professionally certified but was unable to do so. As mentioned earlier, the country was very xenophobic and was not open to certifying any gringo into a professional rank at that time. I doubt much has changed today.

Interestingly, when I returned with that temporary visa, I was able to bring a foreign computer with me as part of my personal goods. BP sent me money to buy a top-of-the-line desktop IBM computer to bring into the country. It was one of the few legal foreign computers that they had in the Botafogo office.

On my return to Brazil with a temporary working visa I was immediately sent back to Iporá as soon as I had all my Brazilian documentation in order (driver's licence, identification card, work book) and found an apartment to live in. I did this all in about 10 days. My apartment was a very small one-bedroom apartment in Ipanema near the Lagoa. It was a fantastic place to live, and to this day I can't believe that at 26 years of age I had an apartment in Ipanema, and was working for BP in kimberlite exploration in Brazil.

Whenever I was in Rio on a field break, I always met with my Brazilian friends at Posto 9 Ipanema beach at the stall (*barraca*) of Lula. Many years later when I returned to Ipanema, I found out that Lula was a real legend of the Rio beaches as he was the first person to have set up a barraca and sell beer and soft drinks on the beach in the 1980's when I was there. In 2012 when I was in Rio de Janeiro working as general counsel for an Australian mining company which had assets in Brazil, Lula was still at Posto 9. He remembered me after so many years and, although he claimed he was retired, he was still selling beer and soft drinks there!

On my return to Iporá I spent a little over three months straight doing kimberlite exploration. I now had about five teams to oversee and lead. Each team was led by a trained technician and had about four to five workers. I found the technicians very good value. Most of them had a formal technical degree as a geologist assistant and they were all very good in the field.

One technician who stood out, Sebastião, did not have a formal education and could barely read or write. However, he was the best person in the field. He had a natural ability to draw 3D maps in his head while walking in the bush. The first time I went out in the field with him I had my maps, aerial photos, compass, etc and kept asking him where we were. He would immediately locate our position on the map and when I questioned him about the location, he would proceed to show me all the errors on the map. He was always correct even though he never carried any navigation equipment or maps himself. It was always a pleasure to go in the field with him.

Sebastião lived in Coromandel, Minas Gerais and, after I left BP, I would always hire him to come into the field with me if I was in the Coromandel area. His elderly mother lived nearby in a simple house and when we would return from a day in the field his mother always had piping hot cheese bread (*pão de queijo*) waiting for us. She made the tapioca flour (*polvilho*) herself using a handmade stone mill, made the cheese herself, and baked it in a home-made brick oven heated with wood. Without a doubt her pão de queijo was to me by far the best in all of Brazil.

Needless to say, the one constant problem for me working in Brazil was the language. I was in a town where not a single person spoke or understood English. When I would talk to my boss by telephone, he refused to speak to me in English, although his English was very good. Portuguese grammar is complex and pronunciation is difficult for English speakers. I was totally immersed in the Portuguese language, and as I was in the field I initially learned the language from a team of uneducated men from the northeast of the country who spoke a mixture of Portuguese, native indigenous language, and an African language mix: and my boss who would only speak Portuguese as spoken in his native country of Portugal (which is very different from Brazilian Portuguese).

After three months in the field I was able to speak Portuguese with a person one-on-one in a sufficient manner to be understood. I was very excited about this and when I returned to Rio at the end of my three-month stint, I was proud to show off my language skills. Unfortunately, when I spoke with my secretary, she immediately shut all the doors and said to me that it was clear I didn't fully understand what I was saying. So, I had an emergency meeting with the language teacher and she initially spent about five minutes apologizing to me. She realized that she should have been teaching me survival skills rather than academic language skills. I then learned that the guys in the bush had taught me some very bad words and expressions. For the one week that I was in Rio on my break, I had a daily crash course on swear words, Brazilian slang, and other useful survival phrases. The guys in the bush had got me really good. I thought it very funny and when I returned to Iporá, while walking in the bush I confronted some of them about what they had taught me. They claimed that they thought I knew what I was saying, but of course they knew that I didn't. They were worried but then when I burst out laughing, they realized all was ok.

The language was the hardest aspect of living and working in Brazil for me to overcome. I had only intermediate Spanish before arriving in Brazil. I worked very hard at being able to read, speak, and write Portuguese. It ultimately took me close to two years though before I could fully understand conversational Portuguese and feel comfortable conversing in a group situation. As I never formally studied the grammar, my writing skills never developed as well as my ability to read, speak, and understand the language.

I spent over one year working out of Iporá. We identified numerous alkaline igneous rocks under the kimberlite exploration project, some already known, some not previously known. No kimberlites were ever found despite recovering some diamonds in the sampling. Some areas we worked had alluvial diamond workings or "garimpos". I always enjoyed visiting the active diamond garimpos and talking with the garimpeiros. In Goiás they were friendly and would always would show me their diamonds.

After some time, the main project changed from kimberlite exploration to full time alluvial diamond and gold exploration along the Rio Claro. The alluvial project was a large one with about 60 workers, two geologists (myself and a Brazilian), a full-time draftsman (everything was hand drafted in those days!), an administrator, a couple of full-time surveyors and numerous gringos from the UK head office always coming and going. At one time we also had a UK geologist on the project.

The alluvial project involved surface mapping of outcropping terrace gravels, overseeing drilling along transect lines across the river valley, mapping open cut pits, overseeing on-site processing of gravel bulk samples, diamond sorting of bulk sample concentrate, drawing up cross-sections and calculating diamond and gold grades.

The draftsman would draw up the surveyed surface topography of each of the transect lines, make a paper copy, and then give it to me to draw in the geology. He would then draft the geology onto the original. The draftsman was busy all day every day as he would also draft up any diagrams that we needed for our reports. There were no computers or even typewriters in the field. Everything was done by hand.

I enjoyed the project and found it interesting as I was learning new skills. I also had the honour to learn from Humphry Willis, a UK geologist who spent many years at Yengema, Sierra Leone, where I did my Master's thesis. Humphry spent a few months in Iporá. He accompanied me into the field every day and spent many hours with me in the field office going over all the data and reports. He was a fantastic mentor and taught me everything I know about alluvial exploration.

The downside of the alluvial project was that, due to its size, the project was more structured. I missed the freedom of walking in the bush and going to different places every day. I had a core team from the kimberlite exploration days that also missed the field work. Whenever I could come up with an excuse to do some more kimberlite exploration, I would call up these guys from the alluvial project and take them in the bush with me for a few days. They loved it as much as I did and always thanked me for getting them off the alluvial work.

Iporá was part of the wild west of Brazil and one of the memorable aspects of the town was the local bar. It was full of bullet holes, usually from fights between regional feuding families. There were times when we were sitting at a table at the bar and had to duck under the table and sneak away due to a pistol feud taking place. I learned to dodge many a shootout during my seven years in Brazil, mostly in Rio.

Every day, at the end of the day, we would go the bar and have a few very cold Antarctica pilsner beers and the popcorn man (*pipocara*) would always come by with his wheelie stall of freshly made popcorn. I had an account with the pipocara, which the company administrator paid for me at the end of every week! Apart from Coromandel in Minas Gerais, I considered Iporá had the best popcorn anywhere in the world.

Epilogue

Eventually BP decided to abandon diamond work all together. I was very sad about this, but understood. The project was literally closed in a day. I woke up one day with instructions that I had to pack all my stuff up and return to Rio for good. It was quite a shock for me as I hadn't seen it coming.

After Iporá I worked for a short time at Forteleza de Minas in Minas Gerais on a massive sulphide nickel project, and then was sent to Araputanga in Matto Grosso working on a greenstone-hosted gold deposit at the Cabacal-I project. I spent a little over one year in Araputanga.

In 1986 I was let go from BP. The tide was turning for the company and the Brazilian legislature was making it more difficult for foreign companies to stay in the country, particularly for BP. It eventually pulled out of Brazil a few years later. After leaving BP, I decided to stay in Brazil. I worked as a consultant for a few junior foreign companies but that didn't last very long due to the political climate in the country. Inflation was getting worse and it was just too difficult to do anything in Brazil. A few weeks before I left the country the new finance minister froze everyone's bank account and no one could draw out any cash from the bank. Fortunately, I had already paid for my move to Australia and had little money left in my account. I couldn't get out of the country quickly enough.

Before I left Brazil, I had started organizing the Fifth International Kimberlite Conference (IKC), on behalf of the organizing committee. The fifth IKC, which was held in Araxâ, Minas Gerais in 1991, was a huge success and had great support from the Brazilian Department of Mines in Brasília. I also had published a paper on my kimberlite exploration work in Iporá in 1987 and two papers with a Brazilian diamond colleague, Guilherme Gonzaga, who I actually only met after we both left BP. One was published in Economic Geology in 1989 and the other by the Brazilian Department of Mines in 1991. I still retain and treasure the friendship with Guilherme and with a few other very close friends I had in Rio. Although there were many difficult times for me during the years I lived in Brazil, I ultimately left with so many fond memories, a rich professional experience, some strong close friendships, and no regrets.

THE BRUNER RAT

JOHN HAMMOND
USA 1990

In late March 1990 ACM Gold sent mining engineer Mike Trumbull and myself to the United States to review some gold exploration opportunities. Our first stop was Denver where we house sat for a friend of Mike's while making contacts to set up field visits to properties we may have been interested in. We also did a day trip to American Gold Resources' (AGR) property at Ouray in South West Colorado and visited the USGS office in Lakewood to get advice and purchase relevant maps and publications. The house sitting was uneventful except it was winter with overnight snow and the basement boiler had been kept on to keep the place warm. We had been left instructions as to how to regulate the temperature but this involved going into the basement via a sealed door. Now the basement was occupied by a collection of rare tropical frogs and we had to be careful not to let them get out. Anyway we arrived home late one evening after a memorable meal and several margaritas at a restaurant famous for its "Rocky Mountain Oysters". The house was too warm, the frogs were forgotten and door was left open as we went to adjust the boiler. Needless to say many frogs escaped up into the main house and we spent most of the rest of the night rounding them up and getting them back to the basement where they belonged.

From Denver we flew out to Reno, Nevada. Before heading to the field we visited Dr Hal Bonham at The Nevada Bureau of Mines and Geology. Bonham was at the time the expert on Nevada gold deposits and their settings and he provided us newbies with some very useful pointers both technical and pragmatic. Next day we drove south and took in the sights of Lake Tahoe before completing our journey to picturesque Bridgeport, California where we stayed overnight. The following morning we drove to the Bodie Gold project (also in California) which at that time was held by Galactic Resources. The project was centered on the historic Bodie Mine and right on the boundary of a national park. (The old ghost town of Bodie was incorporated within the national park.) At this time the environmental movement in California generally was extreme and no less so with respect to activity adjacent to this particular national park. There was a drill rig working near the park boundary and it was completely surrounded by a structure made of bales of hay. This was a condition for the drilling permit to suppress noise and designed so as to not disturb critters across the boundary in the park. We spent a full day with the company geologists discussing, reviewing data and on field examination. That evening we drove on and stayed in Hawthorne in Nevada. Over dinner Mike and I reflected on what we had seen at Bodie and decided that, although the project had technical merit, its location plus the evident present and likely future environmental hurdles were just too great. We decided it was best to leave California to the Sierra Club!

The next day we started early and met up with geologists from a small company who had arranged a half day visit for us to the Round Mountain Gold Mine. (Round Mountain, owned by Homestake Mining, was at that time the largest heap leach gold operation in the world.) In the afternoon the Junior Explorer would show us around their Bruner project near Gabbs. Prior to this, Mike had contacted Dan Izdal of AGR and arranged to meet him with one of their geologists for dinner in Hawthorne that night. Mike did not mention our plans for the day trip to Round Mountain and Bruner with a competitor.

After a very interesting visit to Round Mountain, we arrived in Gabbs about lunchtime, grabbed a quick hamburger and were set to spend the afternoon in the field. Our contact had a big old Ford SUV with enough seats for us and another visitor from Canada who he was showing around. We left our hire car parked in the main (only) street in Gabbs, assured that it would be quite safe there, and headed out on a gravel road in the Ford to Bruner. It was a nice sunny afternoon and a pleasant temperature with no need for a jacket or jumper. We drove on the gravel road for about 20 Kms then turned off to go up the mountain on a track constructed as an extensive series of switchbacks. The switchbacks took us 8 or 10 Kms to the project area where we parked just down the hill from the shaft of an old silver mine. On the way up our driver had lost his cap, caught in a gust of wind and blown out of the window. "Never mind, we'll get it on the way back", he said. At the mine shaft we were shown project maps and sections and then examined the mine dump and some sparse outcrops. It was about 4pm and already getting cooler. We had seen what there was to see so all jumped back in the Ford for the journey back to Gabbs.

The adventure was about to begin. Our driver turned the key, the solenoid clicked but the motor did not turn over. Second and third tries to start it failed. We soon became aware that this was an automatic vehicle and a roll start in gear was not an option. Our host then rather sheepishly mentioned that he had been having a "bit of battery trouble". So there we were stuck on a mountain top in a remote part of Nevada with the sun starting to sink and the warm sunny afternoon quickly giving way to a distinct chill in the air.

Now the Ford was parked on a flat area near the remains of an old wooden mine building and, apart from the shaft on the hilltop, there was also an adit [a horizontal mine tunnel] into the hillside just behind the building. We had to come up with a plan! It was too late to walk for help and we had another immediate problem; Mike was diabetic and needed food. Between the four of us we scrounged around in our pockets for snacks and came up with some opened packets of peanuts, chocolate and a few muesli bars. There was just enough there to solve the immediate problem. Next it was obvious that if we were going to be up on the mountain for the night we would need some heat. There was plenty of scrap wood around and an empty 44 gallon drum would make a good fireplace that would also radiate heat. All that was needed then were matches to light a fire. No one had any in their pockets or packs and the glovebox in the SUV didn't have any either. Now on the way up the mountain we had seen an abandoned caravan about a kilometre down the hill next to the track. Mike and the Canadian walked down to see if they could find anything in the caravan. About half an hour later, when it was nearly dark, they arrived back to report that the caravan was empty except in the last drawer that they opened was a box of matches. What luck and what a relief!

We successfully got a good fire going in the drum and stood around it till about midnight when even with the fire it was simply too cold to be out in the open. Remember we were all dressed in light clothes for a nice sunny afternoon in the field. In desperation we all got back into the Ford, closed the windows and crammed up together trying to stay warm but to no avail. We had to figure out something better or we were going to freeze. I think Mike had the bright idea that it may be warmer in the adit. So about half an hour later we went and had a look by stoking the fire embers in the drum for light and levering it along to the mouth of the adit. Upon entering it we found it was a little warmer than in the open air or inside the vehicle. On further examination about 15 or 20 metres in, we located a winze [an upwards trending mine tunnel] above the old adit. The drum was levered along until it was under the winze which made a perfect chimney. Wood was hurriedly gathered and brought in to keep the fire going through the night. We lay on the adit floor with toes toward the drum then, every 20 minutes or so, we would have to turn with heads towards the drum to keep warm all over. The situation was not comfortable but it was adequate and infinitely better than outside. First light came around 5.30 am and we were all feeling pretty stiff from lying in the dirt, half cold. To add insult to injury there was a rat (we called it The Bruner Rat) that kept nibbling our boots when they were pointing away from the fire drum. Anyway by 5.45am we were all up from the floor and contemplating our next move.

As I have said previously, the track up the mountain was a switch back and it had quite a flat gradient on each leg. Our host decided he could roll the vehicle down the hill in neutral and use the hand brake if needed. I suggested that before we did this he try and start it one more time. The key was turned and the engine whirred into life. It was just before 6am. We all jumped in and headed off downhill. Our luck seemed to have turned until our driver spotted the cap that had blown off on the way up. The three passengers shouted in unison, "don't stop", but it was too late and the engine died. The cap was retrieved but of course the engine would not start again, so it was back to the plan of rolling the vehicle down the switchback with 5 or 6 Km still to go to the bottom. Remarkably that drive was achieved without further incident and we were back on the "main" gravel road and rolled a further kilometer along it before 6.30 am. Despite attempts to start the Ford again and complete the journey into Gabbs, we were out of luck and there was no traffic on this road in the early morning.

It was decided that our newfound Canadian friend and I would walk the remaining distance of just under 20 kilometres into Gabbs to get help. It was still very cold but the sun was starting to shine over the road. I guess we had been walking for about 20 minutes when a pickup drove slowly towards us from the opposite direction. We flagged it down. The driver was highly suspicious of two people dressed only in summer clothes walking down an outback road in the early morning. He wound his window down ever so slightly and we briefly explained our predicament. Our foreign accents caught his attention and he wanted to know where we were from. Once we had stated our nationalities the old fellow wound his window down further and we elaborated on our problem. It turned out that he was a local rancher; he had served in the Pacific in WWII, had good memories of visiting Australia, and was willing to help us. He took us back to his ranch and picked up some heavy duty tow chains. In the course of conversation he told us it had been below freezing overnight at the ranch and he estimated it would have been at least 5 degrees colder up on the mountain.

By 8am the rancher had driven us back to the Ford and towed it into Gabbs. We thanked him and tried to offer money for helping us out. When he wouldn't hear of it, I asked him what his "poison" was and he admitted he liked a bourbon or two sometimes. It seems strange to us but grocery stores often sell alcohol in the USA. I went over the street and bought two bottles of Jim Beam in brown paper bags. Our new friend was happy and, after a bit more chit chat, he headed back out to the ranch. While I was doing "the shopping" Mike had the sense to call Dan Izdal and lucky he did. When we had not shown up for dinner in Hawthorne as arranged and Dan had not heard from us he was concerned. Early next morning he phoned the motel in Hawthorne and learnt that we had not returned the night before. He figured we may be in trouble and had pressed the button on a full-scale search for us by relevant authorities, police et cetera. Mike's prompt call meant the search was quickly called off before it really got started. Needless to say we all enjoyed a hearty breakfast and coffee in Gabbs that morning. We had dodged a bullet more by some good luck than good management!

After breakfast Mike and I drove back to Hawthorne where we met up with Dan and his geologist. We headed to the field and our first stop was a small scale heap leach operation owned by AGR. They were not interested in another company becoming involved in this mining venture but might have considered an exploration farm-in over the surrounding landholding. We spent the next day and a half in the field and examining data. The tenement holding was small and turned out to be of little interest to us.

Retracing our journey we arrived back in Reno where we stayed the night before a very long day trip to the new, high grade, open pit, Sleeper Mine which is situated north of Winnemucca and was owned by AMAX Gold. We had a very comprehensive tour with Chief Geologist Bill Utterback but the security was the most extreme I have ever seen on a gold mine. There were armed guards at the two stage office doors and no geological sampling was allowed in the pit. Bill felt so bad we couldn't have samples from Sleeper that when we were leaving and outside the security area he gave me a small piece from a nearby prospect with visible gold in an adularia crystal. We headed back to Winnemucca where we stayed the night after dining on a long table at a traditional Basque restaurant.

Another old AMAX colleague had offered a prospect that he had pegged to the east of Carlin so from Winnemucca we made a field visit to an area the majors considered well "off trend". As an aside this chap had written a couple of years earlier a famous memo at AMAX titled "Down The Yellow Brick Road" (a reference to The Wizard of Oz) which described the newly developed BLEG geochemistry technique as having arrived in the USA from Australia "like a dose of the Spanish Flu". ACM did subsequently farm in to his prospect but, after limited shallow drilling, pulled out.

From Nevada Mike and I headed north to Salmon, Idaho and visited a few more properties including the Beartrack prospect, held by AGR in The Trans Challis Belt. The latter was later developed as a low grade heap leach operation. With our assessments finished in Idaho, we headed home to Australia.

ACM Gold subsequently set up an office in Denver later that year. As first order of business a team of ACM and local geoscientists carried out a BLEG geochemical orientation survey covering areas in Nevada, Oregon and Washington State. Our contact who had written the "Down The Yellow Brick Road" memo was one member of the local team. Near the Quartz Mountain prospect in Oregon the claim owner questioned why we were at the edge of his claim and what we were doing. Our local guide explained that he was showing round a bunch of crazy Aussies who believed they could detect gold to a limit of 1 part per billion in streams. The claim owner just laughed.

The 1990 gold trips to the Western USA were a great learning curve but the most important lesson was from Bruner. The main thing I learnt from that episode was that wherever I went in the world thereafter I would carry a pack that included not only my geological tools of trade but also some basic survival items. I would not rely totally on other people, especially when offshore.

AN UNEXPECTED TURN OF EVENTS

RAY CARY
USA 1993

Ray graduated from the University of Western Australia in 1969 with a Bachelor of Science majoring in Geology, and Physical and Inorganic Chemistry. He commenced has career as a field geologist in 1970, working for various Australian companies, and mainly in Australia.

In 1994 he formed a company to provide consulting and advisory services to the exploration, mining, resource financing, corporate advisory and service sectors of the industry. In that role, he has had extensive overseas experience, working in all continents except Antarctica.

The story is set in 1993 when I was working with Rothschild Australia Limited (RAL). RAL had a large shareholding in a small ASX-listed gold explorer, General Gold Resources (GGR). GGR was on the lookout for a gold development opportunity, but not necessarily in Australia.

In 1992 I visited the Lincoln Gold project, owned by Sutter Gold, near Sutter Creek in the historic Mother Lode district in California on behalf of Emperor Mines of Fiji. Lincoln lay between two major past producers, located along strike to the north and south, and it seemed a pretty logical thing to have a look at the quartz reef in between. Sutter Gold was developing an exploratory decline down the strike of the reef, which had come to a standstill due to permitting issues and a lack of cash. My host was an "engineer" who didn't seem to know very much about anything. Further, there was little in the way of data which might be of any real use.

Sutter Gold was then owned, or majority owned, by an American businessman and entrepreneur, Jack Larson. Jack lived in Riverton,

Wyoming, and had interests in all sorts of things including building, property investment, aviation and manufacturing, to name but a few. He had also corralled about 50% of the known uranium resources in the US at the time, under the umbrella of his company US Energy. Unfortunately, Jack wasn't there for the first visit, so my concerns about the lack of data couldn't be addressed.

The plan for the 1993 visit was straightforward enough. Fly to San Francisco with the MD of GGR, pick up a car at the airport, overnight with friends in Berkeley (the next suburb east of Oakland), drive to Sutter Creek to look at Lincoln, this time with Jack to be present, then return to San Francisco to continue alone to do a few other RAL-related things. These were to visit Butte, Montana, to look at a couple of Super Fund rehabilitation sites over then-closed open pit porphyry copper mines, stop by Denver to catch up with a group of ex-RAL employees who had set up Resource Capital Funds, on to Las Vegas to have a look at the Castle Hill heap leach gold project on the Californian side of the Nevada-California border, then home.

The first part of the trip proceeded according to plan with an overnight at my friends' home in Berkeley, before driving to Sutter Creek the next day. This time there was a "geologist" on site who hosted us on an underground tour of the same couple of hundred of metres of decline which had been developed on my last visit. The tour was accompanied by the hard sell on the virtues of the project, but produced very little in the way of useful information. The "geologist" didn't seem to know much more than the "engineer" who hosted me first time around.

We then visited the Sutter Gold field office, a beautifully restored two storey building in town, with the promise of seeing all the data of which, of course, there was still very little. Upon expressing my concern, I was advised that Jack was due to arrive the next day and would sort it all out. It later appeared that Jack was probably largely attracted by the prospect of meeting someone from a bank whom he might be able to convince of the worth of the project and hence attract funding. I say this because at every subsequent introduction I was proudly referred to as being a representative of Rothschilds, the London bank, who had come all the way from Australia just to see Jack.

Jack turned out to be a big guy, loud in that typical American West way, but very affable, very likeable and, as I was to about find out, very generous and very hospitable. He also had quite an air of authority and decisiveness about him. I liked Jack very much.

I don't think Jack really understood when I told him that the information I needed to make an evaluation was not in Sutter Creek. According to Jack, if this was the case, it was all in Riverton, and he insisted he was going to take me to the middle of Wyoming to look at it. No ifs or buts, no whys or wherefores. "If the data you need isn't here, then we're gonna take you to Riverton to get it".

As an aside, Sutter Gold Mining Inc. was, until May 2019, listed on the Toronto Stock Exchange - Venture Board, but was placed into receivership late in that month. It listed its principal asset as the Lincoln mine near Sutter Creek in Amador County, California. I'm not sure that the project had progressed very much further than when I saw it, apart from a number of development studies based on an Indicated Resource of 152,000 tons. Jack passed away in 2006, with glowing tributes for his love of family, business achievements, public service and generosity, many of which I was fortunate to experience first-hand.

Back to the main story:

Jack insists we are going to Riverton, no matter what. "How are we going to get there Jack?"

"The boys have the plane at Rancho Murietta and we are going to fly you to Riverton."

"But Jack, my bag is in Oakland at a friend's house, and I have to be in Butte in a couple of days."

"Well Ray, we're gonna get your bag, and we're gonna get you to Butte."

"How are we going to get my bag Jack?"

"The boys will fly us to Oakland and we'll drive up to your friend's house."

And so it came to pass.

Awaiting us At Rancho Murietta was a brand-new Cessna Citation executive jet, along with "the boys", both probably in their early thirties.

And off we went to Oakland airport, with barely enough time to get off the ground and gain some altitude before we had to land. At Oakland, a red carpet was run out across the tarmac from a light aircraft terminal, with which a young man dashed up to the plane and opened the door with much bowing and scraping and tugging at his forelock, saying something along the lines of "Yes Mr Lasson (the mis-spelling is intentional), no Mr Lasson, three bags full Mr Lasson." Jack explained that this was a fixed base air service which accords personalised service to private aircraft users. Jack operated one of these in Riverton.

Once inside the terminal, Jack headed for the rental car counter, threw a credit card and his driver's licence on the counter saying "I need a car for a few hours, a big car, and I don't want no insurance". Once we had found our car, Jack handed me the keys and said "you drive". Now I have driven plenty of left-hand drive cars both in the US and elsewhere, however, driving with Jack seemed to me to be an altogether different matter. However, Jack insisted, and away we went. No dramas.

Coffee with my friends, then back to Oakland airport. After becoming airborne and a couple of very generous bourbons, Jack suggested I join the boys in the cockpit to look at the Sierras as we flew over them. How beautiful, with the setting sun behind us bathing the snow-covered slopes of the mountains in a soft golden glow.

Jack had already told me that Riverton airport was closed due to runway repairs, and that we would have to fly to Lander, then drive to Riverton, about 40km away. Some way into the journey, we were advised that Lander was also closed due to low cloud. Not to worry, we will fly to Salt Lake City for the night. Jack was born in Salt Lake and visited there frequently, hence there was a car kept at the airport. After dropping off the boys, Jack took me on a guided tour of the city, before we checked into our motel.

The next morning, on the way out to the airport, the boys were ribbing each other about the night before. Apparently one of them was in the process of getting divorced, and the other accused him of eyeing off a pretty barmaid, threatening to tell his wife. The response? "Waal I might just tell her myself!"

Just before take-off we were notified that Lander was still closed due to fog, but that it was expected to lift by mid-morning. Plan B was to fly to Jackson Hole, adjacent to the Grand Teton mountains, wait for the fog to lift, then fly to Lander. The Tetons are truly spectacular, with the highest peak, Grand Teton, standing 1,990 metres above the surrounding landscape. As we resumed our flight back to Lander, Jack asked had I ever seen the Grand Teton? "Well no Jack, no I haven't."

"Boys, show Ray the Grand Teton", with which we banked into a wingtip climb with the Grand Teton seeming within reach on the left-hand side of the plane. In fact, so close, that we could see hikers mooning us from the slopes.

Once we arrived in Riverton, Jack took me to his offices to look at the available data. Not surprisingly there still wasn't much, so it didn't take long to review. Later in the afternoon, Jack came to tell me that he and his wife and family were heading to Laramie in the south-eastern corner of Wyoming to attend a ranch ropin', which is an open-range rodeo, that is, no fences. Would I like to come to Laramie, or stay in Riverton and visit the Buffalo Bill museum in Cody? "I think I'd like to come to Laramie Jack if that's OK".

Jack and his wife Lorrie (Lorraine) were going down to Laramie that afternoon, and it was arranged that I would come down with Hal and Lisa (son-in-law and daughter) the next morning. "In the mean-time you are going to need a car. Mine's in the parking garage under the building so you can use that. Hal and Lisa are going to take you out to dinner tonight and they will drive you to Laramie tomorrow morning." Jack's car? A brand new, top-of-the-wozza Lincoln Town Car for me to drive around. Which is what I mean by generosity, and more to come. Pity there wasn't much to see or do in Riverton with the Lincoln at my disposal.

Hal and Lisa duly arrived to take me out to dinner, with about a half a dozen friends in tow, all in a ginormous Chevy Suburban, and all with a huge paper cup full of beer. Hal explained that Wyoming was the only state in the US where you could drink and drive. "Ya can't drive drunk, but ya can have a drink while ya drive." So there ya go.

Dinner was to be a treat for me with Australian Lobster. In the middle of Wyoming? In Cody? Yep, there it was! Goodness knows what that cost. Then back to Riverton to see Sylvester Stallone in that epic movie,

Cliffhanger, more beer, then home.

When we arrived at the ranch ropin', I was given an effusive welcome by Jack and introduced to dozens of people, many of whom were either seriously wealthy and/or quite famous for one thing or another. The two that I can remember are Dr Fred Schoonmaker, who was then a leading heart surgeon and heart disease researcher from Fort Collins Colorado, and an Olympic athlete, Maury Coomers or something similar, if I recall correctly. Again, if I recall correctly, many were related to the Larsons, at least by marriage, and this was certainly the case with Fred Schoonmaker. I was introduced to most (with some reverence) as "the man from Rothschilds, the London Bank".

The ranch ropin' was a lot of fun, and was followed by a barbecue, lots of beer, lots of music, lots of fun and lots of noise. Boy, did they know how to party! Somehow they found room for me to sleep on the floor of the ranch house.

Next day, I was assigned to drive one of the boy's brand-new Cadillac Northstar back to Riverton. The Northstar was a recently released super high-performance engine (for the time). There was obviously a degree of nervousness about my driving it home as I had to make sure that I always had Hal and Lisa in the Lincoln in my rear-view mirror. The reason I was driving the Cadillac was that the boys had "made friends" with a couple of cow girls during the course of the previous night's festivities and had taken off with them in the Citation to see basketball games in Chicago and then Phoenix Arizona. I guess if you have flown a Citation to a ranch ropin' you don't have too much trouble when you ask a girl to go to the basketball.

The next day, back in Riverton, Jack took me to see a couple more of his interests, amongst which were the manufacture of Brunton instruments and Lakota knives. Waiting for me was my own personally inscribed Brunton compass of the kind that field geologists use. Brunton also produced very high-quality optical equipment, including high performance binoculars and telescopes. These had remarkable capabilities in low light conditions, and I still have the pair I purchased at the time. Regards the compass, when I tried it at home, instead of pointing north, the needle pointed doggedly downward. Apparently, I was given a northern hemisphere version; in the southern hemisphere the change in declination renders the compass useless. Still, it is the thought that counts.

I was also given several Lakota knives of the nasty hunting, skinning kind. These were big and incredibly sharp. I have no idea of how I got them into Australia, but I kept them for years before giving them to someone who might get some use from them, like a fisherman.

Later in the day, Jack took me to see several of his uranium interests held through US Energy. Wyoming is well known for roll front type deposits, and here Jack was in his element. As noted above, I believe that, at the time, US Energy held about 50% of US uranium resources when these were very much out of vogue. The projects I was shown ranged from low-confidence, in-ground resources to idled mines and treatment plants. Later that afternoon I was taken to Jack's home where I met his wife Lorrie again and many of his family and friends over a barbecue. His family was obviously very close-knit and I felt privileged to be welcomed into the gathering. I was humbled by Jack's generosity and hospitality.

First thing next day, I was off to Butte. Jack's nephew, Shane, was to fly me over in a Cessna 210 operated by Jack's light aircraft charter company. As it lies between Riverton and Butte, we had to fly over Yellowstone National Park, and before that, the jagged Beartooth mountains. As we flew toward a gap in the mountains, we could see low cloud and lightning. This didn't seem to faze Shane and through it all we went. On the other side, Shane asked if I had ever seen buffalo. "No, not really", and that was the signal to descend to tree-top height and chase them goddamn buffalo through the trees. By way of explanation for his actions, Shane exclaimed, "If they catch me, they gonna burn my ass, but heck it's fun", and indeed it was.

And so on to Butte. Butte was pretty routine and boring compared to the previous few days which had pretty much left me reeling in disbelief that it had all actually happened. It took quite some time to fully absorb my adventure.

Next stop Denver where I met my ex-RAL associates, then off to a bar for the obligatory reunion drinks. After a couple of beers, a bowl of bar snacks was offered with the invitation to try some. Next thing I knew I was on the floor with my head cradled in the ample bosom of a barmaid as she had me sip milk to soothe my badly burned throat to the laughter of my hosts. The bar snacks were fried jalapeño chillies.

Whilst writing of Denver, I would like to relate another tale from a previous visit.

It is about 1981 and a colleague and I were travelling around the US on the excuse of an "Educational Tour" to look at a variety of mining operations utilising particular mining methods. These were located in various places including South Dakota (Homestake), Missouri (lead belt), Colorado (Climax and Henderson molybdenum mines), California (Yosemite), Nevada (Grand Canyon) and Hawaii, to name but a few. Whilst in Denver, I thought it would be a good thing to catch up with an old work colleague from my first employer who we shall refer to by his nickname, Torgie, an abbreviation of his full surname.

Torgie lived in Arvada, then an outer suburb of Denver. We caught a cab from our hotel and were driven at seemingly breakneck speed toward our destination. Our driver was a big guy, if you get my drift, who seemed to know the ins, outs and back streets to get from downtown to Arvada. Suddenly, without warning, he mounted the kerb and aimed the headlights at a corner street sign. He reached into his shirt pocket and withdrew the right half of a pair of spectacles reminiscent of the bottom of a coke bottle. Said visual aid, badly scratched and covered in smeary fingerprints, was lifted to his eye in an effort to read the street sign through a tight squint before we resumed our high-speed journey. Very reassuring. I guess his knowledge of the way to get to Arvada enabled either him or the cab to drive on auto pilot.

On arrival, we were accorded a very warm welcome at Torgie's place. Torgie was on the wrong side of retirement age and was feeling somewhat melancholy on the night. Apparently, his wife, like himself, was a keen golfer, and also an enthusiastic member of a ladies sewing circle. It emerged that the reason Torgie was feeling sad and sorry for himself was that his wife had, only a few days before, broken the news to him that she was leaving him for another woman at the golf club. Out of the blue, just like that! Torgie was devastated and conveyed that devastation to his visitors over a rather simple dinner.

After dinner, we headed off to a few bars to swap fibs and provide a bit of solace to Torgie. Late in the evening, we crossed a series of railway tracks and pulled up outside a rather dubious looking establishment. Once inside it wasn't so bad, but by no means salubrious. After a few drinks, the music changed tempo, and to my astonishment, all the gals and guys stripped to the waist and began to dance cheek-to-cheek, if you know what I mean. Well, I never did! Never seen anything like it, before or after!

Many years later, I was introduced to the music of the late Leonard Cohen, a Canadian singer/songwriter/poet. Cohen is known for the beauty of his compositions, both musical and poetic, but as much for the obscure and often risqué nature of his words. Hallelujah is probably the best known of his works, but has pretty much defied all efforts to understand the meaning of the lyrics. One of my favourites is a song called "closing time", the first verse of which goes:

"Ah we're drinking and we're dancing

And the band is really happening

And the Johnny Walker wisdom running high

And my very sweet companion

She's the angel of compassion

She's rubbing half the world against her thigh

And every drinker every dancer

Lifts a happy face to thank her

The fiddler fiddles something so sublime

All the women tear their blouses off

And the men they dance on the polka-dots

And it's partner found, it's partner lost

And it's hell to pay when the fiddler stops

It's closing time

(Closing time)

(Closing time)

(Closing time)"

Well, I don't know what you think, but I reckon I know what he's talking about! The coincidence is just too great.

For anyone who is interested, look for the song on YouTube performed by Cohen during his "Live in London" concert.

Anyway, back to the main story.

After the episode with the fried jalapeños, it was on to Las Vegas where I met a representative of another member of the banking syndicate that funded the development of Viceroy Gold's Castle Mountain heap leach operation just over the border in California. Not much to relate here, except that on arrival at major airports in the US there were generally deals, specials on car rentals, usually involving top end luxury cars for ridiculously low rental rates. Despite the rules and regulations imposed by the bank on its employees (which were largely ignored anyway), how could I resist saving the bank some money by hiring my enormous snowy white Cadillac for less than the cost of a more pedestrian car. Remember, this is in the days when American cars were real American cars, more boat-like, not the wimpy little environmentally compliant cars of today.

Anyway, we had a very successful visit to Castle Mountain, after which we returned to Vegas and had the obligatory boys' night out on the strip. The next day I got caught up in a traffic jam on my way to the airport, then overshot the rental car return, and had to negotiate part of the traffic jam to get back to the airport. Of course, by this time, I had missed my flight to Los Angeles.

Now here is something I really like about the American airline system. Having missed my flight, I was immediately transferred at check-in to another operated by a different carrier, resulting in an eventual delay of only ten or fifteen minutes. Competitor airlines did then, and probably still do, honour each other's tickets, and I suppose, sort it all out later. What a great system!

And so home, wondering what on earth all that was about and how it came to happen, but with a raft of memories that took quite some time to fully digest. To quote again from Leonard Cohen's song, late in the second verse:

"And I swear it happened just like this..........

I can't say much has happened since

But closing time

(Closing time)

(Closing time)

(Closing time)"

DEFINITELY NOT AUSTRALIA

PHIL FILLIS
GUYANA 2010 - 2012

Phil Fillis graduated in geology from London University in 1972 and with a Masters in Mineral Exploration from Leicester University in 1974. He started his geological career initially on base metals in Canada, iron ore in Australia and uranium in South Africa before migrating to Australia in 1980. Based in Perth he has extensive experience mainly in gold exploration in Australasia; South East, Central and Far West Asia; West Africa, Europe and South America.

A great deal of my working life has been spent in the hot sweaty areas of the world and I've often thought that it would be nice instead to work in a cool mountain environment. But, as advised by Aesop, be careful what you wish for, lest it come true! Thus, my sentiments changed when, in November 2002, I found myself 4000m up pinned to a steep mountainside in a howling gale and sub-zero temperatures in far western China. Since then I don't whinge quite so much about the hot sweaty areas, so when I was given the opportunity to work on a mineral exploration project in Guyana for a couple of years, I jumped at the chance to experience life back in pristine rainforests. While it certainly proved to be an interesting experience, trying to conduct an exploration programme in remote, densely forested terrain certainly throws up, if not problems, then issues which have to be tackled in a sometimes unconventional way. Challenges included navigating the rules and regulations of Government departments, travelling around the country, adapting the ways we carried out mineral exploration, and personal health and safety issues.

Just getting to the country from Australia is daunting. Although Guyana is part of South America there is very little connection to the rest of the continent. The country looks to the north and is part of the Caribbean community – think cricket and Clive Lloyd. In fact, once the thrice-weekly Belem – Cayenne – Paramaribo – Georgetown flight was discontinued in 2009, there was no way to fly into any other South American country from there. I once had to make a trip south to Manaus in Brazil, which required a flight to Lethem in southern Guyana, then four taxis through bandit country to Boa Vista where I could connect to the South American network. Travelling from Australia is usually at least a two-day trip via the Caribbean ports or New York/Miami. Any which way, it's a long journey from Perth.

Approximately 80% of Guyana's landmass is covered by tropical rainforest. In 2009, Norway pledged $250 million if Guyana limited its annual deforestation rate to 0.056 percent between the years 2010 and 2015. Recent studies have shown that they succeeded in achieving that goal so much of the country is still in a pristine state – in marked contrast to Brazil to the south and Venezuela to the west. Not that forest destruction is not happening and, unfortunately, the main culprits are artisanal miners. Known locally as pork-knockers, the technique they use is to dig a deep pit at their gold prospect and using water pumped from the river, sluice the weathered rock containing gold into the pit. This slurry is then pumped over primitive collection tables and the wastewater run into very basic dams. These dams and wastewater are flooding huge tracts of forest and killing the trees. In addition, there is widespread use of mercury to extract gold from slurry and ground material even though this is banned by law in Guyana. Which is strange, given that I was told the only place to buy mercury in the country is from the Mines Department.

Having arrived in the capital, Georgetown, getting to site could also present problems – particularly in the rainy season. As we quickly found out, the dry and wet seasons dictated what exploration activities we could and could not do at any particular time and also how we moved people and equipment to site. There are two wet seasons, mid-November to mid-January and May to mid-July, but rainfall is highly variable and can be expected at any time. Because there are very few roads - and most of these are unsealed - access to the interior of the country is by river of which there are many. For example, the Essequibo is South America's third largest river and runs the entire length of the country from the wonderful and beautiful savannah highlands, on the Brazilian border in the south, to the Atlantic coast.

The project area I was working on is located on the Barama River, a tributary of the Wainu. In the dry season it was possible to get to site via Port Kaituma, a small town infamous for the 1978 Jonestown Massacre in which 909 US followers of evangelist Jim Jones died. Its airstrip was where US Congressman Leo Ryan and four others were murdered by followers of Jones. The flight to Port Kaituma from Georgetown takes about an hour and the subsequent drive to site can take anywhere from six to over 24 hours depending on the condition of the road. In the wet season the road is impassable, and we had to resort to using the rivers. The first time you do this it's a wonderful and very interesting trip but, after half a dozen times, it's just a very long trip on a hard seat and exposed to the hot sun – again taking up to 20 hours depending on river levels.

The trip starts with a drive across the Demerara River to Parika, thence a ferry across the 15 kilometre wide mouth of the Essequibo to Aurora, then another hour's drive to Charity on the Pomeroon River. Here you join the boat that takes you all the way to site. Initially it heads downstream along the broad Pomeroon before emerging into the Atlantic Ocean and then closely follows the coast for seven kilometres before turning abruptly into the forest and the hidden Maruka Creek. This is a spectacular section of the trip that winds upstream through a tunnel in the rainforest before emerging into open grasslands/wetlands with Amerindian villages on islands. The boat then crosses over the watershed in the wetlands before heading downstream back into the rainforest and into a major artery, the Waini River and thence via the Barama all the way to site. The Barama follows a particularly tortuous meandering course and, although only 70 kilometres as the crow flies, this section alone can take up to 12 hours depending on water levels. In the later stages of the project the operator of a neighbouring mine cleared an airstrip – in dry weather the plane trip from Georgetown to site was under an hour. Unfortunately, you could never guarantee the weather and we have had to wait up to a week before a plane could get in.

Most of the project area was covered by small mineral claims which could be held only by a Guyanese citizen. In theory, tenement title is relatively secure, but it did mean that any new tenements we pegged had to be in the name of one of our employees, presenting some risk. Small mineral claims are just that – small, up to 200m x 400m. They don't have to conform to a particular shape, don't have to be contiguous with adjacent claims and are pegged by marking each corner on a tree in the forest. The Mines Department periodically tries to record the claim locations with a GPS but there are many discrepancies, overlaps and gaps. At times it was very hard to know who you were supposed to be dealing with. In addition, once there is any activity in the area, particularly clearing of forest and creating access, you soon find you are not alone. For example, people start farms with the objective of claiming compensation, or to set up bars and brothels. This is supposed to be controlled by the Mines Department – we bring them in, they flush out the illegal operators, but once the officials return to town, the illegals flood back in. A mob of drunken locals in an isolated forest setting doesn't allow for much peace of mind to those of us sheltering in an obviously well stocked camp.

Much of our work relies on soil, trench and drill sampling. These samples were transported by plane or boat to Georgetown for processing. At the time there were only three or four laboratories in the country and these only carried out sample preparation, with analysis undertaken in Canada, Venezuela or Chile. Unfortunately, this means that turnaround - that is the time taken to receive results - was between one and two months: not very useful when you need those results to plan the next exploration step.

Drilling itself also presented a few challenges; firstly, in getting a very heavy drill rig to site and secondly, getting the rig to perform in the very difficult terrain. For the type of drilling we wanted to do we had to import a drill rig into the country. At the time, as we were coming up to the wet season, the plan was to transport the drill by boat all the way from Georgetown to site. However, a combination of delays getting the rig into the country and a poor wet season meant that by the time we were ready to mobilise, river levels were dangerously low. We got the rig to within 12 kilometres of site before the barge carrying it got stuck on a rock bar on the Barama River. We then spent two weeks cutting a track through virgin rainforest over very muddy ground and mosquito-infested swamps before eventually walking the track-mounted rig to site. We certainly hadn't planned for that and the unlooked-for track project allowed all team members in the pressure cooker of a small contained camp to voice their very strong and differing opinions as to the best route through the rainforest.

Personal health and safety at site were other aspects of life that required serious consideration, given the isolation. Malaria was rife and a number of crew, both local and expatriate, succumbed. Fortunately, serious accidents were rare as getting people out quickly for appropriate treatment could not be counted upon. As an example, on one occasion I was in a serious boat smash on my way to site, resulting in half a dozen broken ribs, a punctured lung and various gashes with some loss of blood – and this on one of the minor rivers in a very remote area. It took some time to get me by boat to an Indian settlement which had an airstrip and thence by air to intensive care in a hospital in Georgetown but, fortunately, all achieved within the same day. In this case I was lucky as the weather allowed a plane to land in the settlement – sometimes you can wait days for the weather to clear.

On another matter, I always find it surprising that locals have to my mind an irrational fear of potentially dangerous animals, particularly snakes – even though they must have grown up with them and be used to them. I was repeatedly warned about taking my early morning run down a particular track to the river as that was where the "tiger" lived. Turns out that a jaguar was spotted there about five years previously!

Despite all the difficulties and challenges of working in Guyana it was a very interesting and enjoyable experience and one that, given the opportunity, I would certainly do again. With destruction of rainforest on a large scale in many parts of the world, it is heartening to see that Guyana is doing a great job preserving pristine wilderness.

AFTERWORD

In 1985 I was in a small mining town called Meggen in West Germany. It is off the well-beaten tourist track, and a visit to it had been organised via TELEX from Australia, in those pre-FAX and pre- email days.

The mining company had organised accommodation for us in the local hotel and we had invited the chief mine geologist to join us for dinner on the evening of our arrival. We reached the dining room first and were the objects of quite some curiosity from the staff and clientele, none of whom apparently spoke English, although it was probably not hard for them to realise that we were doing so. Shortly afterwards the chief geologist arrived, very nicely turned out in polished shoes, a good haircut and a three-piece suit. To our total surprise, all the locals in the bar essentially stood to attention and bowed to him in welcome, with deferential greetings such as "Guten abend, Herr Doktor".

He apologised that he was not able to join us for dinner after all, but advised that he would ensure that we were well treated by the establishment. He explained to the staff and clientele who we were, that we were his honoured guests, and that we should be looked after appropriately. When he left, there was again a chorus of almost obsequious well-wishes and heel clicking to Herr Doktor. Someone was found who could translate the menu for us and the evening went very well indeed.

You will have read in this book a suite of stories about the trials that we, being mainly Aussie geologists, have had to content with as we play our part in finding the minerals the world needs, and on which the Australian economy so thoroughly relies. But I am not aware of any public statue of a geologist anywhere in Australia. I remember thinking in Meggen that if I were to be reincarnated and the new man again wanted to become a geologist, I would prefer to be a German one - then I might finally get some RESPECT!

Andrew Drummond, Perth 2020